PATHOLOGY OF LABORATORY MICE AND RATS

Other Books from the FEDERATION OF AMERICAN SOCIETIES FOR EXPERIMENTAL BIOLOGY:

Cell Biology
Environmental Biology
Biology Data Book, Second Edition (In three volumes)
Respiration and Circulation
Human Health and Disease
Inbred and Genetically Defined Strains of Laboratory Animals (In two volumes)

Available from Pergamon Press

PATHOLOGY OF LABORATORY MICE AND RATS

prepared by
THE BIOLOGY DATABOOK EDITORIAL BOARD

with the editorial assistance of
PHILIP L. ALTMAN
Federation of American Societies for Experimental Biology

and the technical assistance of
KATHLEEN McLANE and JOHN BROSSEAU
Berul Associates, Ltd.

The Biology Databook Series
is a joint project of

FEDERATION OF AMERICAN SOCIETIES FOR EXPERIMENTAL BIOLOGY
Bethesda, Maryland
and

PERGAMON INFOLINE INC.
McLean, Virginia

Pergamon Press Offices:

U.S.A.	Pergamon Press Inc., Maxwell House, Fairview Park, Elmsford, New York 10523, U.S.A.
U.K.	Pergamon Press Ltd., Headington Hill Hall, Oxford OX3 0BW, England
CANADA	Pergamon Press Canada Ltd., Suite 104, 150 Consumers Road, Willowdale, Ontario M2J 1P9, Canada
AUSTRALIA	Pergamon Press (Aust.) Pty. Ltd., P.O. Box 544, Potts Point, NSW 2011, Australia
FRANCE	Pergamon Press SARL, 24 rue des Ecoles, 75240 Paris, Cedex 05, France
FREDERAL REPUBLIC OF GERMANY	Pergamon Press GmbH, Hammerweg 6, D-6242 Kronberg-Taunus, Federal Republic of Germany

Library of Congress Cataloging in Publication Data
Main entry under title:

Pathology of laboratory mice and rats.

(Biology databook series)
"The Biology databook series is a joint project of the Federation of American Societies for Experimental Biology...and Pergamon Infoline Incorporated."
Includes index.
1. Laboratory animals--Diseases--Handbooks, manuals, etc. 2. Mice as laboratory animals--Handbooks, manuals, etc. 3. Mice--Diseases--Handbooks, manuals, etc. 4. Rats as laboratory animals--Handbooks, manuals, etc. 5. Rats--Diseases--Handbooks, manuals, etc. I. Altman, Philip L. II. Biology Databook Editorial Board (U.S.) III. Federation of American Societies for Experimental Biology. IV. Pergamon Infoline Incorporated. V. Series.
SF996.5.P38 1985 619'.93 85-6493
ISBN 0-08-030077-4

Printed in the United States of America

CONTENTS

ACKNOWLEDGMENT

The SNOVET codes were provided by Dr. James L. Palotay, Oregon Regional Primate Center, 505 NW 185th Avenue, Beaverton, Oregon 97006. The Biology Databook Editorial Board is indebted to Dr. Palotay for his assistance.

LABORATORY ANIMALS DATA SOURCES
COUNT OF MOUSE AND RAT PATHOLOGY RECORDS BY STRAIN
ACCORDING TO DATA SOURCE, SEX, AND AGE

STRAIN / DATA SOURCE	SEX: FEMALE, AGE IN MONTHS 0 < 12	12 < 18	18 < 24	24+	MALE 0 < 12	12 < 18	18 < 24	24+	TOTAL
BALB/C									
NAT. CENTER FOR TOXICOLOGICAL RESEARCH	212	369	193	71	162	300	142	65	1514
NATIONAL INSTITUTE ON AGING	0	0	0	0	9	14	5	29	57
*TOTAL BALB/C	212	369	193	71	171	314	147	94	1571
B6C3F1									
BIONETICS RESEARCH LABORATORIES	8	14	143	490	2	9	134	478	1278
DOW CHEMICAL COMPANY	1	2	13	82	2	0	19	78	197
GULF SOUTH RESEARCH INSTITUTE	2	7	82	122	3	5	89	123	433
HAZLETON LABORATORIES	9	31	402	129	51	123	241	115	1101
MASON RESEARCH INSTITUTE	5	13	350	100	5	5	315	97	890
SOUTHERN RESEARCH INSTITUTE	42	36	366	92	61	57	338	67	1059
*TOTAL B6C3F1	67	103	1356	1015	124	199	1136	958	4958
CB6F1									
NATIONAL INSTITUTE ON AGING	0	0	0	0	7	12	7	31	57
*TOTAL CB6F1	0	0	0	0	7	12	7	31	57
C57BL/6									
NAT. CENTER FOR TOXICOLOGICAL RESEARCH	176	367	192	75	121	286	154	74	1445
NATIONAL INSTITUTE ON AGING	0	0	0	0	10	14	7	28	59
*TOTAL C57BL/6	176	367	192	75	131	300	161	102	1504
FISCHER 344									
BIONETICS RESEARCH LABORATORIES	2	10	108	534	5	19	122	505	1305
DOW CHEMICAL COMPANY	0	0	15	85	0	3	9	88	200
GULF SOUTH RESEARCH INSTITUTE	1	3	44	102	0	7	32	110	299
HAZLETON LABORATORIES	1	1	10	88	2	2	13	81	198
MASON RESEARCH INSTITUTE	16	5	102	458	50	12	189	543	1375
NATIONAL INSTITUTE ON AGING	0	0	0	0	42	44	21	113	220
NORWICH-EATON PHARMACEUTICALS	3	2	12	81	0	1	21	77	197
SOUTHERN RESEARCH INSTITUTE	2	2	25	144	2	13	25	135	348
STANFORD RESEARCH INSTITUTE	1	1	11	83	0	0	5	91	192
*TOTAL FISCHER 344	26	24	327	1575	101	101	437	1743	4334

LABORATORY ANIMALS DATA SOURCES
COUNT OF MOUSE AND RAT PATHOLOGY RECORDS BY STRAIN
ACCORDING TO DATA SOURCE, SEX, AND AGE

STRAIN / DATA SOURCE	SEX FEMALE AGE IN MONTHS 0 < 12	12 < 18	18 < 24	24+	MALE 0 < 12	12 < 18	18 < 24	24+	TOTAL
ICR/HAM									
NORWICH-EATON PHARMACEUTICALS	2	11	17	20	9	16	15	10	100
*TOTAL ICR/HAM	2	11	17	20	9	16	15	10	100
LONG-EVANS									
NORWICH-EATON PHARMACEUTICALS	6	11	18	54	6	6	18	59	178
*TOTAL LONG-EVANS	6	11	18	54	6	6	18	59	178
OSBORNE MENDEL									
GULF SOUTH RESEARCH INSTITUTE	1	3	5	50	1	2	14	39	115
HAZLETON LABORATORIES	6	36	39	286	10	62	90	204	733
*TOTAL OSBORNE MENDEL	7	39	44	336	11	64	104	243	848
SPRAGUE DAWLEY									
CIBA-GEIGY	0	8	24	76	0	10	18	72	208
ENVIRONMENTAL PROTECTION AGENCY	22	14	28	35	14	28	27	7	175
G.D. SEARLE	15	0	0	0	15	0	0	0	30
NORWICH-EATON PHARMACEUTICALS	5	16	119	108	6	37	105	108	504
PROCTER & GAMBLE	63	0	0	0	54	0	0	0	117
SMITH KLINE AND FRENCH	67	7	0	0	77	7	0	0	158
SOUTHERN RESEARCH INSTITUTE	6	15	249	21	4	15	242	29	581
SQUIBB	3	0	0	0	0	0	0	0	3
*TOTAL SPRAGUE DAWLEY	181	60	420	240	170	97	392	216	1776
SWISS WEBSTER									
MCNEIL LABORATORIES	0	0	46	0	0	0	43	0	89
*TOTAL SWISS WEBSTER	0	0	46	0	0	0	43	0	89
WISTAR									
ICI AMERICA INC	0	5	12	0	1	6	14	0	38
WISTAR									
MCNEIL LABORATORIES	0	0	0	99	0	0	0	101	200
*TOTAL WISTAR	0	5	12	99	1	6	14	101	238
TOTAL	677	989	2625	3485	731	1115	2474	3557	15653

INTRODUCTION

The Laboratory Animal Data Bank (LADB) was initiated in the fall of 1974 as a computerized resource of comparative baseline data and reference information on species and strains of animals commonly used in biomedical research; it was managed by the National Library of Medicine (NLM). The stated purposes of the project were to provide a central repository for control animal data; to serve as a comparative base for possibly reducing the number of control animals necessary in experimental testing; to improve experimental design by predicting the expected effects of age, sex, and strain variability, as well as environmental and management factors, on test results; to determine incidence of spontaneous disease conditions and pathologic changes; and to monitor strain drift over time. However, LADB was a concept ahead of its time, and its potential was not realized because of a lack of funding.

Both the database and the concept languished until Pergamon International Information Corporation combined its knowledge of computerized databases with the ability of the Federation of American Societies for Experimental Biology (FASEB) to develop published databooks. Their combined efforts have resulted in this publication and the prospect of similar, equally valuable publications in the future. The readers' interest in subsequent publications will be assessed by their responses to requests for additional control data to expand and publish new databooks.

RETROSPECTIVE VIEW

The pathology data, selected for inclusion in this publication, were collected by Battelle Columbus Laboratories of Columbus, Ohio, for the NLM from a variety of governmental and private laboratory sources (*see* the table on Laboratory Animals Data Sources, p. xvii). The data constitute a numerical compilation of diagnoses rendered by many pathologists, and represent histopathological examinations of normal "negative" and vehicle control animals from six commonly used strains of mice and rats. Submission of the biological data to the LADB by each organization was accompanied by a detailed description of the care and environmental conditions experienced by the animals. Prior to entry into the data bank, the pathology data were reviewed by Battelle pathologists for consistency and completeness. The criteria for data acceptance and entry were examined by an independent user assessment panel under the direction of FASEB's Life Sciences Research Office (Altman, P.L. 1980. Review of data standards related to the Laboratory Animal Data Bank. Life Sciences Research Office, FASEB, Bethesda, MD. 114p.). The pathology data for those mouse and rat strains with sufficient numbers of animals in the LADB were retrieved and are presented here in tabular form.

Although the data were reviewed by the Battelle pathologists, the members of the Biology Databook Editorial Board did not have the opportunity to review the completeness of the gross examinations, the extent to which sections were taken and tissue examined, or the accuracy of the resultant histological examinations. Therefore, these limitations must be recognized by anyone who wishes to use these data for comparative purposes. In addition, consideration must be given to genetic "drift" and to the circumstances related to animal maintenance; it is well known that the several influences of atmosphere, humidity, diet, and bedding modify the status of the animals, and additionally the impact of a variety of other unidentified factors cannot be ignored. For these reasons, comparison of experimental results to these data are useful as a guide, but statistical comparison seems unwarranted and unwise.

STRAIN CATEGORIZATION AND VARIABILITY

For each strain represented in the databook, pathology data are cross-classified by sex and by age at death. The four categories for age at death are 0-<12 months, 12-<18 months, 18-<24 months, and 2 years or greater. In choosing age categories the goal was to select enough intervals to document shifts in the prevalence of lesions with age, subject to the constraint that intervals should be wide enough to guarantee a sufficient number of animals in each age category to form useful rates. Certain strains (e.g., BALB/C and C57BL/6 mice) are more heavily represented in the younger age categories, because the experiments from which those control data were obtained had planned, interim sacrifices. Other strains (e.g., the B6C3F1 mouse and the Fischer 344 rat) are more heavily represented in older age categories, because for the most part those control data came from experiments without planned, interim sacrifices. Data from five strains in the computer file are not included in this databook because the number of animals from these strains were insufficient to provide useful rates after cross-classification by sex and age at death. The five omitted strains are the CB6F1, ICR/HaM, and Swiss Webster mice, and the Long-Evans and Wistar rats. For those interested in these omitted strains, the data sources and numbers of animals, cross-classified by sex and age at death, are provided in the table on Laboratory Animals Data Sources.

Each data cell (i.e., entry for a specific pathologic diagnosis, cross-classified by strain, sex, and age at death) in the tables contains two numbers divided by a slant line. The larger number to the right is the number of animals of the strain and sex which died in the age category. The number to the left is the number of these animals dying in the age category for which the indicated pathologic diagnosis was made. The number of sources from which data were obtained varied from 2 to 9 for different strains. In order to assess the degree of variability among data sources associated with the rate of a particular diagnosis (where the rate is defined to be the number to the left in each data cell divided by the number to the right), the binomial variance test was computed for each data cell. Those data cells with variability significant at the 1% significance level are indicated by a footnote. The number to the right of each data cell may include some animals for which a given organ was not actually examined. Thus, rates calculated from the data presented in this book may underestimate the presence of a given lesion, particularly for smaller organs.

ORGAN SYSTEMS AND SUBDIVISIONS

A standardized grouping of organs within organ systems was used throughout the tables. Where sufficient data existed, organs were subdivided into component parts, such as heart into endocardium and pericardium. This subdivision allows the user of a table to examine subsets of diseases as well as the totals within organ systems and organs. The grouping of organ systems and organ subdivisions is as follows:

Body Cavities
- Abdominal Cavity
- Body Cavity (not specified)
- Thoracic Cavity

Cardiovascular System
- Blood Vessel
- Heart
  - Endocardium
  - Pericardium
- Lymphatic Vessel

Digestive System
- Accessory Organs
  - Gallbladder
  - Liver
    - Bile duct
  - Salivary gland
- Alimentary Tract
  - Esophagus
  - Intestine
  - Mouth
  - Stomach

Endocrine System
- Adrenal Gland
- Pancreas
- Parathyroid Gland
- Pineal Gland
- Pituitary Gland
- Thyroid Gland

Genital System: Females
- Cervix
- Clitoris
- Ovary
- Oviduct
- Uterus
- Vagina

Genital System: Males
- Ductus Deferens
- Epididymis
- Male Genital Accessory Glands
- Penis
- Prepuce
- Scrotum
- Testis

Hematopoietic and Lymphopoietic Systems
- Blood/Bone Marrow
- Lymph Node
- Spleen
- Thymus

Miscellaneous
- Adipose Tissue

Multiple Organs

Musculoskeletal System
- Bone
- Cartilage
- Joint
- Skeletal Muscle

Nervous System
- Central Nervous System
- Peripheral Nervous System
- Special Organs
  - Ear
    - Middle
  - Eye
    - Accessory glands
    - Cornea
    - Retina

Respiratory System
- Larynx
- Lung
  - Bronchus
  - Interstitium
- Nasal Passage/Sinus
- Trachea

Skin and Appendages
- Harderian Gland
- Mammary Gland
- Skin

Urinary System
- Kidney
- Ureter
- Urethra
- Urinary Bladder

Editor's Note:
All data for the Pancreas (Pages 131–145) should have been included under the Digestive System, except for the islet cell diseases which are correctly listed on pages 134, 139–142, 144 and 145.

DISEASE CLASSIFICATION AND PREFERRED TERMINOLOGY

Diagnostic terms have been classified according to categories of major change. Each specific pathologic change is a subset of one of the following disease classifications:

Mechanical Abnormalities and Cyst Formation

Degeneration, Necrosis and Deposition

Growth Alterations: Non-neoplastic
- Hypertrophy
- Hyperplasia
- Atrophy
- Metaplasia
- Dysplasia
- Hypoplasia

Developmental Anomalies

Inflammation and Repair

Growth Alterations: Neoplastic
- Benign
- Malignant
  - Primary
  - Metastatic
  - Primary site not determined
- Not Determined

Parasitism

The Editorial Board anticipated that among those using these data tables there will be considerable variation in knowledge of, and familiarity with, the diagnostic terminology utilized. In the original data, different terms were used to represent similar or identical lesions and syndromes. For these reasons, the Editorial Board considered it necessary to pool homologous diagnostic terms and summarize them under a single "preferred" diagnostic term where, in its judgment, there was a need for standardization and interpretation. The "preferred" terms have been classified also according to SNOVET (Palotay, J. L. and Rothwell, D. J., eds. 1984. Systematized nomenclature of veterinary medicine. American Veterinary Medical Association, Schaumberg, IL. 198p.), a standardized nomenclature derived from the Systematized Nomenclature of Medicine (SNOMED) and adapted for use in animals. These terms were considered by the Editorial Board to be most appropriate for a given type of tissue change as best determined from the original data submitted. The original diagnostic terms have been preserved and are listed in the appendix along with the corresponding "preferred" term. It is the hope of the Editorial Board and the publisher that this procedure will enhance the utility of this databook for pathologists and non-pathologists alike. However, the users of this book should be aware that though the members of the Board reviewed the terminology for consistency, they were unable to review the data per se.

PROSPECTIVE VIEW

Data collection is a continuous process. Additional new and significant data are becoming available as this volume is being completed. Such data will be evaluated and as needed new volumes may be considered. If you or your organization have or are collecting data on control animals and wish to have such data included, please submit a letter outlining your current program with details of how your data are being compiled and stored and send to: FASEB Biology Databook Office, 9650 Rockville Pike, Bethesda, MD 20814.

—The Biology Databook Editorial Board

PATHOLOGY OF LABORATORY MICE AND RATS

TEXT TABLES

ORGAN SYSTEM: BODY CAVITIES   ORGAN: ABDOMINAL CAVITY   STRAIN: BALB/C

SEX:	FEMALE				MALE			
AGE IN MONTHS: / DIAGNOSIS	0 < 12	12 < 18	18 < 24	24+	0 < 12	12 < 18	18 < 24	24+
*************** INFLAMMATION AND REPAIR ***************								
INFLAMMATION		1/369	3/193				2/147	

SEX:	FEMALE				MALE			
DIAGNOSIS / AGE IN MONTHS:	0 < 12	12 < 18	18 < 24	24+	0 < 12	12 < 18	18 < 24	24+
*************** MECHANICAL ABNORMALITIES AND CYST FORMATION ***************								
CYST				3/1015				1/958
HEMATOMA								1/958
HEMORRHAGE							1/1136	
INFARCT	1/67							
THROMBUS	1/67							
*************** INFLAMMATION AND REPAIR ***************								
FIBROSIS								1/958
INFLAMMATION	1/67		9/1356	2/1015	1/124		1/1136	2/958
INFLAMMATION, GRANULOMATOUS		2/103	3/1356	2/1015			2/1136	1/958
INFLAMMATION, SUPPURATIVE		1/103	1/1356		1/124		1/1136	
PERIVASCLLITIS/VASCULITIS			1/1356	1/1015				1/958
*************** DEGENERATION, NECROSIS, AND DEPOSITION ***************								
MINERALIZATION			1/1356					1/958
NECROSIS			2/1356					1/958
NECROSIS, FAT			7/1356	16/1015	1/124		5/1136	21/958
*************** GROWTH ALTERATIONS, NON-NEOPLASTIC, METAPLASIA ***************								
METAPLASIA, OSSEOUS					1/124			

SEX:	FEMALE				MALE			
DIAGNOSIS / AGE IN MONTHS:	0 < 12	12 < 18	18 < 24	24+	0 < 12	12 < 18	18 < 24	24+
*************** GROWTH ALTERATIONS, NEOPLASTIC, BENIGN ***************								
ANGIOMA				1/1015			1/1136	
LIPOMA			3/1356	4/1015			1/1136	5/958
*************** GROWTH ALTERATIONS, NEOPLASTIC, MALIGNANT, PRIMARY ***************								
ANGIOSARCOMA			2/1356	3/1015				
*************** GROWTH ALTERATIONS, NEOPLASTIC, MALIGNANT, METASTATIC ***************								
CARCINOMA								1/958
CARCINOMA, ALVEOLAR BRONCHIOLAR							1/1136	
CARCINOMA, CORTICAL							1/1136	
*************** GROWTH ALTERATIONS, NEOPLASTIC, MALIGNANT, PRIMARY SITE NOT DETERMINED ***************								
FIBROSARCOMA			1/1356					
LEIOMYOSARCOMA				2/1015			1/1136	
OSTEOSARCOMA			1/1356					
SARCOMA			1/1356				2/1136	

ORGAN SYSTEM: BODY CAVITIES   ORGAN: ABDOMINAL CAVITY   STRAIN: C57BL/6

SEX:	FEMALE				MALE			
DIAGNOSIS / AGE IN MONTHS:	0 < 12	12 < 18	18 < 24	24+	0 < 12	12 < 18	18 < 24	24+
*************** INFLAMMATION AND REPAIR ***************								
INFLAMMATION		1/367	8/192	2/75				

SEX:	FEMALE				MALE			
DIAGNOSIS / AGE IN MONTHS:	0 < 12	12 < 18	18 < 24	24+	0 < 12	12 < 18	18 < 24	24+
*************** MECHANICAL ABNORMALITIES AND CYST FORMATION ***************								
ASCITES				1/1575				
HEMATOMA				1/1575				
*************** INFLAMMATION AND REPAIR ***************								
ABSCESS							2/437	
FIBROSIS			1/327	1/1575			1/437	
INFLAMMATION			1/327	1/1575	1/101			
INFLAMMATION, GRANULOMATOUS				1/1575			1/437	3/1743
INFLAMMATION, SUPPURATIVE								2/1743
PERIVASCULITIS/VASCULITIS				2/1575				11/1743
*************** DEGENERATION, NECROSIS, AND DEPOSITICN ***************								
MINERALIZATION				2/1575				
NECROSIS, FAT			6/327	18/1575			3/437	26/1743
*************** GROWTH ALTERATIONS, NEOPLASTIC, BENIGN ***************								
LIPOMA			1/327					5/1743
*************** GROWTH ALTERATIONS, NEOPLASTIC, MALIGNANT, PRIMARY ***************								
LIPOSARCOMA	1/26							

SEX:	FEMALE				MALE			
AGE IN MONTHS: / DIAGNOSIS	0 < 12	12 < 18	18 < 24	24+	0 < 12	12 < 18	18 < 24	24+
MESOTHELIOMA				1/1575		1/101	6/437	5/1743
*************** GROWTH ALTERATIONS, NEOPLASTIC, MALIGNANT, METASTATIC ***************								
ADENOCARCINOMA			1/327	1/1575				
*************** GROWTH ALTERATIONS, NEOPLASTIC, MALIGNANT, PRIMARY SITE NOT DETERMINED ***************								
LEIOMYOSARCOMA				1/1575				
LEUKEMIA								2/1743
NEUROFIBROSARCOMA							1/437	
SARCOMA								1/1743

DIAGNOSIS	Female 0 < 12	Female 12 < 18	Female 18 < 24	Female 24+	Male 0 < 12	Male 12 < 18	Male 18 < 24	Male 24+
MECHANICAL ABNORMALITIES AND CYST FORMATION								
THROMBUS								1/243
INFLAMMATION AND REPAIR								
ABSCESS				1/336				
INFLAMMATION				1/336		1/64		1/243
PERIVASCULITIS/VASCULITIS			1/44	3/336			4/104	16/243
DEGENERATION, NECROSIS, AND DEPOSITION								
ARTERIOSCLEROSIS			2/44			2/64	6/104	3/243
NECROSIS, FAT				1/336			1/104	2/243
GROWTH ALTERATIONS, NEOPLASTIC, BENIGN								
ANGIOMA						1/64		
LIPOMA				1/336				1/243
GROWTH ALTERATIONS, NEOPLASTIC, MALIGNANT, PRIMARY								
ANGIOSARCOMA				1/336				
GROWTH ALTERATIONS, NEOPLASTIC, MALIGNANT, PRIMARY SITE NOT DETERMINED								
FIBROSARCOMA			1/44	1/336				

ORGAN SYSTEM: BODY CAVITIES ORGAN: ABDOMINAL CAVITY STRAIN: OSBORNE MENDEL

SEX:	FEMALE				MALE			
DIAGNOSIS / AGE IN MONTHS:	0 < 12	12 < 18	18 < 24	24+	0 < 12	12 < 18	18 < 24	24+
HISTIOCYTOMA, FIBROUS, MALIGNANT								1/243

DIAGNOSIS	FEMALE 0 < 12	FEMALE 12 < 18	FEMALE 18 < 24	FEMALE 24+	MALE 0 < 12	MALE 12 < 18	MALE 18 < 24	MALE 24+
*************** MECHANICAL ABNORMALITIES AND CYST FORMATION ***************								
HEMORRHAGE			1/420				3/392	
INFARCT				1/240				1/216
*************** INFLAMMATION AND REPAIR ***************								
INFLAMMATION	4/181	2/60	3/420			1/97	1/392	
INFLAMMATION, NECROTIZING							1/392	
INFLAMMATION, SUPPURATIVE			1/420			1/97	1/392	
PERIVASCULITIS/VASCULITIS			1/420	1/240		3/97	4/392	7/216
*************** DEGENERATION, NECROSIS, AND DEPOSITION ***************								
MINERALIZATION						1/97	4/392	2/216
MINERALIZATION, VASCULAR							2/392	2/216
NECROSIS, FAT				1/240			1/392	
*************** GROWTH ALTERATIONS, NEOPLASTIC, BENIGN ***************								
FIBROMA							1/392	
LIPOMA			1/420				2/392	
*************** GROWTH ALTERATIONS, NEOPLASTIC, MALIGNANT, METASTATIC ***************								
ADENOCARCINOMA		1/60						

SEX:	FEMALE				MALE			
DIAGNOSIS / AGE IN MONTHS:	0 < 12	12 < 18	18 < 24	24+	0 < 12	12 < 18	18 < 24	24+
RHABDOMYOSARCOMA								1/216
*************** GROWTH ALTERATIONS, NEOPLASTIC, MALIGNANT, PRIMARY SITE NOT DETERMINED ***************								
LYMPHOMA, RETICULUM CELL			2/420					
SARCOMA							1/392	

 ORGAN SYSTEM: BODY CAVITIES ORGAN: BODY CAVITY (NOT SPECIFIED) STRAIN: B6C3F1

SEX:	FEMALE				MALE			
AGE IN MONTHS:	0 < 12	12 < 18	18 < 24	24+	0 < 12	12 < 18	18 < 24	24+
DIAGNOSIS								
*************** GROWTH ALTERATIONS, NEOPLASTIC, MALIGNANT, PRIMARY ***************								
MESOTHELIOMA				1/1015				

SEX:	FEMALE				MALE			
AGE IN MONTHS: / DIAGNOSIS	0 < 12	12 < 18	18 < 24	24+	0 < 12	12 < 18	18 < 24	24+
*************** GROWTH ALTERATIONS, NEOPLASTIC, MALIGNANT, PRIMARY ***************								
MESOTHELIOMA				1/1575			5/437	13/1743

DIAGNOSIS	SEX:	FEMALE				MALE			
	AGE IN MONTHS:	0 < 12	12 < 18	18 < 24	24+	0 < 12	12 < 18	18 < 24	24+
*************** INFLAMMATION AND REPAIR ***************									
INFLAMMATION				1/193			1/314		

SEX:	FEMALE				MALE			
DIAGNOSIS / AGE IN MONTHS:	0 < 12	12 < 18	18 < 24	24+	0 < 12	12 < 18	18 < 24	24+
*************** MECHANICAL ABNORMALITIES AND CYST FORMATION ***************								
HYDROTHORAX				1/1015				
*************** INFLAMMATION AND REPAIR ***************								
ABSCESS				1/1015				
INFLAMMATION				1/1015				

 ORGAN SYSTEM: BODY CAVITIES ORGAN: THORACIC CAVITY STRAIN: C57BL/6

SEX:	FEMALE				MALE			
AGE IN MONTHS: / DIAGNOSIS	0 < 12	12 < 18	18 < 24	24+	0 < 12	12 < 18	18 < 24	24+
*************** INFLAMMATION AND REPAIR ***************								
INFLAMMATION			2/192					
INFLAMMATION, NONSUPPURATIVE								1/102

SEX:	FEMALE				MALE			
DIAGNOSIS / AGE IN MONTHS:	0 < 12	12 < 18	18 < 24	24+	0 < 12	12 < 18	18 < 24	24+
*************** MECHANICAL ABNORMALITIES AND CYST FORMATION ***************								
HYDROTHORAX							1/437	
*************** INFLAMMATION AND REPAIR ***************								
INFLAMMATION					1/101			
PERIVASCULITIS/VASCULITIS				1/1575				
*************** GROWTH ALTERATIONS, NEOPLASTIC, BENIGN ***************								
FIBROMA				1/1575				
MESOTHELIOMA, BENIGN							1/437	
*************** GROWTH ALTERATIONS, NEOPLASTIC, MALIGNANT, PRIMARY ***************								
ANGIOSARCOMA				1/1575				
MESOTHELIOMA				1/1575			1/437	
*************** GROWTH ALTERATIONS, NEOPLASTIC, MALIGNANT, METASTATIC ***************								
CARCINOMA, ALVEOLAR BRONCHIOLAR								1/1743
*************** GROWTH ALTERATIONS, NEOPLASTIC, MALIGNANT, PRIMARY SITE NOT DETERMINED ***************								
SARCOMA						1/101		

SEX:	FEMALE				MALE			
DIAGNOSIS / AGE IN MONTHS:	0 < 12	12 < 18	18 < 24	24+	0 < 12	12 < 18	18 < 24	24+
*************** INFLAMMATION AND REPAIR ***************								
PERIVASCULITIS/VASCULITIS								1/243
*************** GROWTH ALTERATIONS, NEOPLASTIC, MALIGNANT, PRIMARY ***************								
LIPOSARCOMA								1/243

SEX:	FEMALE				MALE			
DIAGNOSIS / AGE IN MONTHS:	0 < 12	12 < 18	18 < 24	24+	0 < 12	12 < 18	18 < 24	24+
*************** MECHANICAL ABNORMALITIES AND CYST FORMATION ***************								
HEMORRHAGE								1/216
*************** INFLAMMATION AND REPAIR ***************								
EMPYEMA	2/181				4/170			
INFLAMMATION, FIBRINOUS	6/181	3/60	6/420	3/240	3/170	4/97	5/392	
*************** GROWTH ALTERATIONS, NEOPLASTIC, MALIGNANT, PRIMARY SITE NOT DETERMINED ***************								
SARCOMA, HISTIOCYTIC							1/392	

ORGAN SYSTEM: CARDIOVASCULAR SYSTEM ORGAN: BLOOD VESSEL STRAIN: BALB/C

SEX:	FEMALE				MALE			
DIAGNOSIS / AGE IN MONTHS:	0 < 12	12 < 18	18 < 24	24+	0 < 12	12 < 18	18 < 24	24+
*************** MECHANICAL ABNORMALITIES AND CYST FORMATION ***************								
THROMBUS								2/94
*************** INFLAMMATION AND REPAIR ***************								
FIBROSIS						1/314		
INFLAMMATION				1/71		1/314		4/94
INFLAMMATION, GRANULOMATOUS		2/369						
*************** DEGENERATION, NECROSIS, AND DEPOSITION ***************								
MINERALIZATION						1/314	1/147	5/94
NECROSIS		1/369				1/314		
*************** GROWTH ALTERATIONS, NON-NEOPLASTIC, HYPERTROPHY ***************								
HYPERTROPHY	1/212							
*************** GROWTH ALTERATIONS, NON-NEOPLASTIC, METAPLASIA ***************								
METAPLASIA, CARTILAGINOUS				1/71				1/94
*************** GROWTH ALTERATIONS, NEOPLASTIC, MALIGNANT, METASTATIC ***************								
CARCINOSARCOMA, MAMMARY		1/369						

SEX:	FEMALE				MALE			
DIAGNOSIS / AGE IN MONTHS:	0 < 12	12 < 18	18 < 24	24+	0 < 12	12 < 18	18 < 24	24+
*************** GROWTH ALTERATIONS, NEOPLASTIC, MALIGNANT, PRIMARY SITE NOT DETERMINED ***************								
LYMPHOMA	1/212	2/369	1/193		1/171			1/94
LYMPHOMA, RETICULUM CELL		1/369	6/193	3/71				

SEX:	FEMALE				MALE			
DIAGNOSIS / AGE IN MONTHS:	0 < 12	12 < 18	18 < 24	24+	0 < 12	12 < 18	18 < 24	24+
*************** MECHANICAL ABNORMALITIES AND CYST FORMATION ***************								
EMBOLISM							1/1136	
THROMBUS				2/1015				
*************** INFLAMMATION AND REPAIR ***************								
INFLAMMATION			1/1356	1/1015		1/199	5/1136	
PERIVASCULITIS/VASCULITIS							1/1136	1/958
*************** DEGENERATION, NECROSIS, AND DEPOSITION ***************								
MINERALIZATION							2/1136	
SCLEROSIS				4/1015				
*************** GROWTH ALTERATIONS, NON-NEOPLASTIC, HYPERTROPHY ***************								
HYPERTROPHY				1/1015				
*************** GROWTH ALTERATIONS, NON-NEOPLASTIC, HYPERPLASIA ***************								
HYPERPLASIA			1/1356	2/1015				
*************** GROWTH ALTERATIONS, NEOPLASTIC, MALIGNANT, PRIMARY SITE NOT DETERMINED ***************								
ANGIOSARCOMA				1/1015				

SEX:	FEMALE				MALE			
DIAGNOSIS / AGE IN MONTHS:	0 < 12	12 < 18	18 < 24	24+	0 < 12	12 < 18	18 < 24	24+
*************** MECHANICAL ABNORMALITIES AND CYST FORMATION ***************								
THROMBUS	1/176	1/367						
*************** INFLAMMATION AND REPAIR ***************								
INFLAMMATION		3/367	5/192	1/75		3/300	2/161	5/102
INFLAMMATION, GRANULOMATOUS					1/131			
*************** DEGENERATION, NECROSIS, AND DEPOSITION ***************								
NECROSIS			2/192					
*************** GROWTH ALTERATIONS, NON-NEOPLASTIC, HYPERTROPHY ***************								
HYPERTROPHY			2/192			1/300		
*************** GROWTH ALTERATIONS, NEOPLASTIC, MALIGNANT, PRIMARY SITE NOT DETERMINED ***************								
LYMPHOMA, RETICULUM CELL			2/192	1/75				

SEX:	FEMALE				MALE			
DIAGNOSIS / AGE IN MONTHS:	0 < 12	12 < 18	18 < 24	24+	0 < 12	12 < 18	18 < 24	24+
*************** MECHANICAL ABNORMALITIES AND CYST FORMATION ***************								
THROMBUS			1/327	1/1575			1/437	1/1743
THROMBUS, MURAL				1/1575				
*************** INFLAMMATION AND REPAIR ***************								
INFLAMMATION					5/101			3/1743
INFLAMMATION, NONSUPPURATIVE						1/101		
PERIVASCULITIS/VASCULITIS				3/1575				3/1743
PERVASCULITIS/VASCULITIS								1/1743
*************** DEGENERATION, NECROSIS, AND DEPOSITION ***************								
ARTERIOSCLEROSIS								1/1743
MINERALIZATION			7/327	32/1575			9/437	61/1743
*************** GROWTH ALTERATIONS, NON-NEOPLASTIC, HYPERTROPHY ***************								
HYPERTROPHY				1/1575	5/101			4/1743
*************** GROWTH ALTERATIONS, NEOPLASTIC, MALIGNANT, PRIMARY SITE NOT DETERMINED ***************								
LEUKEMIA, GRANULOCYTIC								2/1743
LEUKEMIA, LYMPHOBLASTIC								1/1743
LEUKEMIA, LYMPHOCYTIC				1/1575				

SEX:	FEMALE				MALE			
DIAGNOSIS / AGE IN MONTHS:	0 < 12	12 < 18	18 < 24	24+	0 < 12	12 < 18	18 < 24	24+
*************** INFLAMMATION AND REPAIR ***************								
INFLAMMATION							1/104	
PERIVASCULITIS/VASCULITIS							1/104	
*************** DEGENERATION, NECROSIS, AND DEPOSITION ***************								
ARTERIOSCLEROSIS			2/44	1/336		4/64	10/104	9/243
MINERALIZATION			1/44	5/336		3/64	8/104	11/243
NECROSIS							1/104	
*************** GROWTH ALTERATIONS, NON-NEOPLASTIC, HYPERPLASIA ***************								
HEMATOPOIESIS							1/104	

SEX:	FEMALE				MALE			
DIAGNOSIS / AGE IN MONTHS:	0 < 12	12 < 18	18 < 24	24+	0 < 12	12 < 18	18 < 24	24+
*************** MECHANICAL ABNORMALITIES AND CYST FORMATION ***************								
CONGESTION				1/240				
RUPTURE								1/216
THROMBUS			2/420	1/240		1/97	3/392	2/216
*************** INFLAMMATION AND REPAIR ***************								
ATHEROSCLEROSIS						1/97	3/392	
INFLAMMATION			1/420					
INFLAMMATION, NONSUPPURATIVE				4/240			2/392	5/216
PERIVASCULITIS/VASCULITIS	1/181					3/97	7/392	14/216*
*************** DEGENERATION, NECROSIS, AND DEPOSITION ***************								
ARTERIOSCLEROSIS			1/420			1/97	1/392	1/216
DEGENERATION							1/392	2/216
DEGENERATION, FIBRINOID				3/240			2/392	6/216
HEMOSIDERIN DEPOSITION								1/216
MINERALIZATION		2/60	5/420	3/240		6/97	24/392*	23/216*
*************** GROWTH ALTERATIONS, NON-NEOPLASTIC, HYPERTROPHY ***************								
HYPERTROPHY								3/216

*VARIABILITY AMONG DATA SOURCES SIGNIFICANT AT THE 1% SIGNIFICANCE LEVEL

ORGAN SYSTEM: CARDIOVASCULAR SYSTEM   ORGAN: BLOOD VESSEL   STRAIN: SPRAGUE DAWLEY

SEX:	FEMALE				MALE			
DIAGNOSIS / AGE IN MONTHS:	0 < 12	12 < 18	18 < 24	24+	0 < 12	12 < 18	18 < 24	24+
*************** GROWTH ALTERATIONS, NON-NEOPLASTIC, HYPERPLASIA ***************								
HYPERPLASIA				1/240				1/216
*************** GROWTH ALTERATIONS, NEOPLASTIC, MALIGNANT, PRIMARY SITE NOT DETERMINED ***************								
LEUKEMIA			1/420	1/240				
LEUKEMIA, HISTIOCYTIC							1/392	
LYMPHOMA, RETICULUM CELL			1/420					
SARCOMA, HISTIOCYTIC								1/216

SEX:	FEMALE				MALE			
DIAGNOSIS / AGE IN MONTHS:	0 < 12	12 < 18	18 < 24	24+	0 < 12	12 < 18	18 < 24	24+
*************** MECHANICAL ABNORMALITIES AND CYST FORMATION ***************								
DILATATION	1/212	2/369	1/193		5/171	4/314	1/147	
THROMBUS		2/369		1/71	2/171	5/314		
*************** INFLAMMATION AND REPAIR ***************								
FIBROSIS		1/369		3/71	1/171	4/314	3/147	3/94
INFLAMMATION								1/94
INFLAMMATION, GRANULOMATOUS		2/369						
*************** DEGENERATION, NECROSIS, AND DEPOSITION ***************								
DEGENERATION, MYOCARDIAL				1/71		2/314		
FATTY CHANGE						1/314		
MINERALIZATION	2/212	7/369		4/71	21/171	21/314	12/147	14/94
NECROSIS		1/369	1/193		1/171	1/314		1/94
*************** GROWTH ALTERATIONS, NON-NEOPLASTIC, HYPERTROPHY ***************								
HYPERTROPHY		1/369						
*************** GROWTH ALTERATIONS, NON-NEOPLASTIC, METAPLASIA ***************								
METAPLASIA		1/369						
METAPLASIA, CARTILAGINOUS				1/71				1/94

SEX:	FEMALE				MALE			
DIAGNOSIS / AGE IN MONTHS:	0 < 12	12 < 18	18 < 24	24+	0 < 12	12 < 18	18 < 24	24+
METAPLASIA, OSSEOUS		1/369						
*************** GROWTH ALTERATIONS, NEOPLASTIC, MALIGNANT, PRIMARY ***************								
RHABDOMYOSARCOMA							1/147	
*************** GROWTH ALTERATIONS, NEOPLASTIC, MALIGNANT, METASTATIC ***************								
ADENOCARCINOMA, PULMONARY								1/94
CARCINOSARCOMA, MAMMARY		1/369						
*************** GROWTH ALTERATIONS, NEOPLASTIC, MALIGNANT, PRIMARY SITE NOT DETERMINED ***************								
LYMPHOMA	3/212	3/369	3/193	1/71	2/171			2/94
LYMPHOMA, RETICULUM CELL		1/369	5/193	1/71				

SEX:	FEMALE				MALE			
AGE IN MONTHS: / DIAGNOSIS	0 < 12	12 < 18	18 < 24	24+	0 < 12	12 < 18	18 < 24	24+
*************** MECHANICAL ABNORMALITIES AND CYST FORMATION ***************								
THROMBUS					1/124		1/1136	2/958
THROMBUS, MURAL			1/1356					
*************** INFLAMMATION AND REPAIR ***************								
ABSCESS						1/199		
FIBROSIS			1/1356			1/199	5/1136	2/958
INFLAMMATION			1/1356			2/199	5/1136	2/958
INFLAMMATION, SUPPURATIVE	1/67	1/103		1/1015		6/199	1/1136	1/958
PERIVASCULITIS/VASCULITIS			2/1356	2/1015			2/1136	2/958
*************** DEGENERATION, NECROSIS, AND DEPOSITION ***************								
DEGENERATION, MYOCARDIAL						5/199	3/1136	
MELANOSIS			4/1356				2/1136	
MINERALIZATION			4/1356	2/1015		4/199	6/1136	
*************** GROWTH ALTERATIONS, NON-NEOPLASTIC, HYPERPLASIA ***************								
HYPERPLASIA						1/199		
*************** GROWTH ALTERATIONS, NEOPLASTIC, BENIGN ***************								
ANGIOMA				1/1015				1/958

SEX:	FEMALE				MALE			
DIAGNOSIS / AGE IN MONTHS:	0 < 12	12 < 18	18 < 24	24+	0 < 12	12 < 18	18 < 24	24+
*************** GROWTH ALTERATIONS, NEOPLASTIC, MALIGNANT, PRIMARY ***************								
ANGIOSARCOMA								2/958
FIBROSARCOMA							1/1136	
*************** GROWTH ALTERATIONS, NEOPLASTIC, MALIGNANT, METASTATIC ***************								
CARCINOMA, HEPATOCELLULAR								1/958
*************** GROWTH ALTERATIONS, NEOPLASTIC, MALIGNANT, PRIMARY SITE NOT DETERMINED ***************								
LEIOMYOSARCOMA			1/1356					

SEX:	FEMALE				MALE			
DIAGNOSIS / AGE IN MONTHS:	0 < 12	12 < 18	18 < 24	24+	0 < 12	12 < 18	18 < 24	24+
*************** MECHANICAL ABNORMALITIES AND CYST FORMATION ***************								
DILATATION	1/176	1/367		1/75	1/131		3/161	
THROMBUS			1/192			1/300		
*************** INFLAMMATION AND REPAIR ***************								
FIBROSIS		1/367	1/192		1/131	3/300	1/161	
*************** DEGENERATION, NECROSIS, AND DEPOSITION ***************								
AMYLOID DEPOSITION								8/102
HEMOSIDERIN DEPOSITION						3/300	1/161	5/102
MELANOSIS					2/131			3/102
MINERALIZATION						1/300		
NECROSIS							1/161	
PIGMENT DEPOSITION				2/75				
*************** GROWTH ALTERATIONS, NON-NEOPLASTIC, METAPLASIA ***************								
METAPLASIA			1/192				1/161	1/102
METAPLASIA, CARTILAGINOUS								1/102
METAPLASIA, OSSEOUS						1/300		

SEX:	FEMALE				MALE			
DIAGNOSIS / AGE IN MONTHS:	0 < 12	12 < 18	18 < 24	24+	0 < 12	12 < 18	18 < 24	24+
*************** GROWTH ALTERATIONS, NEOPLASTIC, MALIGNANT, PRIMARY SITE NOT DETERMINED ***************								
LYMPHOMA				1/75			1/161	1/102
LYMPHOMA, RETICULUM CELL			2/192	2/75			2/161	

SEX:	FEMALE				MALE			
DIAGNOSIS / AGE IN MONTHS:	0 < 12	12 < 18	18 < 24	24+	0 < 12	12 < 18	18 < 24	24+
*************** MECHANICAL ABNORMALITIES AND CYST FORMATION ***************								
EMBOLUS, SEPTIC			1/327					
HEMORRHAGE					1/101	1/101	1/437	1/1743
THROMBUS			2/327	8/1575	1/101		6/437	11/1743
THROMBUS, MURAL				6/1575			3/437	5/1743
THROMBUS, ORGANIZED				1/1575				
*************** INFLAMMATION AND REPAIR ***************								
ABSCESS								1/1743
FIBROSIS		3/24	14/327	184/1575*	2/101	24/101	57/437*	394/1743*
INFLAMMATION			12/327	50/1575	7/101*	24/101*	48/437*	185/1743*
INFLAMMATION, GRANULOMATOUS							1/437	
INFLAMMATION, MIXED					1/101			
INFLAMMATION, NONSUPPURATIVE						3/101		1/1743
INFLAMMATION, SUPPURATIVE								2/1743
PERIVASCULITIS/VASCULITIS			1/327	5/1575			1/437	6/1743
*************** DEGENERATION, NECROSIS, AND DEPOSITION ***************								
DEGENERATION, MYOCARDIAL			3/327	52/1575	9/101*	11/101	26/437*	149/1743*
FATTY CHANGE				1/1575			1/437	
HEMOSIDERIN DEPOSITION				1/1575				

*VARIABILITY AMONG DATA SOURCES SIGNIFICANT AT THE 1% SIGNIFICANCE LEVEL

SEX:	FEMALE				MALE			
DIAGNOSIS / AGE IN MONTHS:	0 < 12	12 < 18	18 < 24	24+	0 < 12	12 < 18	18 < 24	24+
MINERALIZATION				3/1575	2/101		1/437	6/1743
NECROSIS				2/1575				1/1743
*************** GROWTH ALTERATIONS, NON-NEOPLASTIC, HYPERTROPHY ***************								
HYPERTROPHY				1/1575				
*************** GROWTH ALTERATIONS, NON-NEOPLASTIC, HYPERPLASIA ***************								
HYPERPLASIA								1/1743
HYPERPLASIA, MESOTHELIAL								1/1743
*************** GROWTH ALTERATIONS, NON-NEOPLASTIC, ATROPHY ***************								
ATROPHY								2/1743
*************** GROWTH ALTERATIONS, NON-NEOPLASTIC, METAPLASIA ***************								
METAPLASIA, OSSEOUS								1/1743
*************** GROWTH ALTERATIONS, NEOPLASTIC, BENIGN ***************								
FIBROMA							1/437	
*************** GROWTH ALTERATIONS, NEOPLASTIC, MALIGNANT, PRIMARY ***************								
NEUROFIBROSARCOMA				1/1575				
SARCOMA, ANITSCHKOW CELL								1/1743

SEX:	FEMALE				MALE			
DIAGNOSIS / AGE IN MONTHS:	0 < 12	12 < 18	18 < 24	24+	0 < 12	12 < 18	18 < 24	24+
*************** GROWTH ALTERATIONS, NEOPLASTIC, MALIGNANT, PRIMARY SITE NOT DETERMINED ***************								
LEUKEMIA	1/26						1/437	8/1743
LEUKEMIA, GRANULOCYTIC								2/1743
LEUKEMIA, LYMPHOCYTIC				1/1575			1/437	1/1743
LEUKEMIA, STEM CELL							1/437	
LYMPHOMA				1/1575				
SARCOMA							1/437	1/1743

SEX:	FEMALE				MALE			
DIAGNOSIS / AGE IN MONTHS:	0 < 12	12 < 18	18 < 24	24+	0 < 12	12 < 18	18 < 24	24+
*************** MECHANICAL ABNORMALITIES AND CYST FORMATION ***************								
CONGESTION				1/336				
*************** INFLAMMATION AND REPAIR ***************								
FIBROSIS				8/336		2/64	8/104	10/243
INFLAMMATION								1/243
PERIVASCULITIS/VASCULITIS								2/243
*************** DEGENERATION, NECROSIS, AND DEPOSITION ***************								
ARTERIOSCLEROSIS							1/104	1/243
DEGENERATION, MYOCARDIAL			2/44	18/336		4/64	12/104	31/243 *
MINERALIZATION						5/64	10/104	6/243
*************** GROWTH ALTERATIONS, NON-NEOPLASTIC, HYPERPLASIA ***************								
HYPERPLASIA		1/39	2/44	1/336	1/11			7/243
*************** GROWTH ALTERATIONS, NEOPLASTIC, BENIGN ***************								
MIXED TUMOR			1/44					
*************** GROWTH ALTERATIONS, NEOPLASTIC, MALIGNANT, PRIMARY ***************								
ANGIOSARCOMA						1/64		1/243

*VARIABILITY AMONG DATA SOURCES SIGNIFICANT AT THE 1% SIGNIFICANCE LEVEL

ORGAN SYSTEM: CARDIOVASCULAR SYSTEM   ORGAN: HEART   STRAIN: OSBORNE MENDEL

SEX:	FEMALE				MALE			
AGE IN MONTHS:	0 < 12	12 < 18	18 < 24	24+	0 < 12	12 < 18	18 < 24	24+
DIAGNOSIS								
*************** GROWTH ALTERATIONS, NEOPLASTIC, MALIGNANT, PRIMARY SITE NOT DETERMINED ***************								
SARCOMA							1/104	2/243

SEX:	FEMALE				MALE			
AGE IN MONTHS: / DIAGNOSIS	0 < 12	12 < 18	18 < 24	24+	0 < 12	12 < 18	18 < 24	24+
*************** MECHANICAL ABNORMALITIES AND CYST FORMATION ***************								
HEMORRHAGE					1/170			
THROMBUS								1/216
*************** INFLAMMATION AND REPAIR ***************								
FIBROSIS			2/420	8/240	2/170	1/97	8/392	18/216 *
INFLAMMATION	2/181	1/60	7/420	8/240	12/170	7/97	37/392*	39/216 *
INFLAMMATION, GRANULOMATOUS					2/170			
INFLAMMATION, NONSUPPURATIVE	10/181				9/170			
PERIVASCULITIS/VASCULITIS			2/420					1/216
*************** DEGENERATION, NECROSIS, AND DEPOSITION ***************								
ARTERIOSCLEROSIS								2/216
DEGENERATION, MYOCARDIAL			6/420	12/240	2/170	1/97	7/392	29/216 *
HEMOSIDERIN DEPOSITION								1/216
MINERALIZATION		1/60	5/420			4/97	14/392	16/216 *
NECROSIS								1/216
*************** GROWTH ALTERATIONS, NON-NEOPLASTIC, HYPERPLASIA ***************								
HYPERPLASIA					2/170			

*VARIABILITY AMONG DATA SOURCES SIGNIFICANT AT THE 1% SIGNIFICANCE LEVEL

SEX:	FEMALE				MALE			
DIAGNOSIS / AGE IN MONTHS:	0 < 12	12 < 18	18 < 24	24+	0 < 12	12 < 18	18 < 24	24+
*************** GROWTH ALTERATIONS, NON-NEOPLASTIC, METAPLASIA ***************								
METAPLASIA, OSSEOUS, CHONDROID							1/392	
*************** GROWTH ALTERATIONS, NEOPLASTIC, MALIGNANT, PRIMARY ***************								
RHABDOMYOSARCOMA				1/240				1/216
*************** GROWTH ALTERATIONS, NEOPLASTIC, MALIGNANT, METASTATIC ***************								
CARCINOMA, HURTHLE CELL						1/97		
*************** GROWTH ALTERATIONS, NEOPLASTIC, MALIGNANT, PRIMARY SITE NOT DETERMINED ***************								
LEIOMYOSARCOMA		1/60						
LEUKEMIA, LYMPHOCYTIC			1/420			1/97		
LYMPHOMA, RETICULUM CELL			1/420					
SARCOMA, HISTIOCYTIC			1/420					

SEX:	FEMALE				MALE			
DIAGNOSIS AGE IN MONTHS:	0 < 12	12 < 18	18 < 24	24+	0 < 12	12 < 18	18 < 24	24+
*************** INFLAMMATION AND REPAIR ***************								
INFLAMMATION			1/327					5/1743

ORGAN SYSTEM: CARDIOVASCULAR SYSTEM ORGAN: HEART; ENDOCARDIUM STRAIN: SPRAGUE DAWLEY

SEX:	FEMALE				MALE			
DIAGNOSIS / AGE IN MONTHS:	0 < 12	12 < 18	18 < 24	24+	0 < 12	12 < 18	18 < 24	24+
*************** INFLAMMATION AND REPAIR ***************								
INFLAMMATION				2/240	1/170			2/216

SEX:	FEMALE				MALE			
DIAGNOSIS / AGE IN MONTHS:	0 < 12	12 < 18	18 < 24	24+	0 < 12	12 < 18	18 < 24	24+
*************** INFLAMMATION AND REPAIR ***************								
INFLAMMATION	1/212	6/369	3/193	3/71	8/171	5/314	2/147	1/94
INFLAMMATION, NONSUPPURATIVE								2/94

ORGAN SYSTEM: CARDIOVASCULAR SYSTEM   ORGAN: HEART; PERICARDIUM   STRAIN: B6C3F1

DIAGNOSIS	FEMALE 0 < 12	FEMALE 12 < 18	FEMALE 18 < 24	FEMALE 24+	MALE 0 < 12	MALE 12 < 18	MALE 18 < 24	MALE 24+
*************** INFLAMMATION AND REPAIR ***************								
INFLAMMATION	1/67	1/103	6/1356	3/1015		3/199	4/1136	3/958

SEX:	FEMALE				MALE			
DIAGNOSIS / AGE IN MONTHS:	0 < 12	12 < 18	18 < 24	24+	0 < 12	12 < 18	18 < 24	24+
*************** INFLAMMATION AND REPAIR ***************								
INFLAMMATION		5/367	4/192	2/75	1/131	4/300	3/161	1/102

SEX:	FEMALE				MALE			
DIAGNOSIS / AGE IN MONTHS:	0 < 12	12 < 18	18 < 24	24+	0 < 12	12 < 18	18 < 24	24+
*************** INFLAMMATION AND REPAIR ***************								
INFLAMMATION			7/327	68/1575	16/101		10/437	104/1743 *
INFLAMMATION, NONSUPPURATIVE						1/101		1/1743

*VARIABILITY AMONG DATA SOURCES SIGNIFICANT AT THE 1% SIGNIFICANCE LEVEL

ORGAN SYSTEM: CARDIOVASCULAR SYSTEM ORGAN: HEART; PERICARDIUM STRAIN: OSBORNE MENDEL

SEX:	FEMALE				MALE			
DIAGNOSIS / AGE IN MONTHS:	0 < 12	12 < 18	18 < 24	24+	0 < 12	12 < 18	18 < 24	24+
*************** INFLAMMATION AND REPAIR ***************								
INFLAMMATION	2/7	1/39	1/44	3/336		7/64	8/104	24/243

SEX:	FEMALE				MALE			
DIAGNOSIS / AGE IN MONTHS:	0 < 12	12 < 18	18 < 24	24+	0 < 12	12 < 18	18 < 24	24+
*************** INFLAMMATION AND REPAIR ***************								
INFLAMMATION	9/181*	1/60	1/420		6/170	4/97	3/392	4/216
INFLAMMATION, NONSUPPURATIVE			1/420	3/240				3/216

*VARIABILITY AMONG DATA SOURCES SIGNIFICANT AT THE 1% SIGNIFICANCE LEVEL

ORGAN SYSTEM: CARDIOVASCULAR SYSTEM ORGAN: LYMPHATIC VESSEL STRAIN: FISCHER 344

SEX:	FEMALE				MALE			
DIAGNOSIS / AGE IN MONTHS:	0 < 12	12 < 18	18 < 24	24+	0 < 12	12 < 18	18 < 24	24+
*************** INFLAMMATION AND REPAIR ***************								
INFLAMMATION								5/1743

SEX:	FEMALE				MALE			
DIAGNOSIS / AGE IN MONTHS:	0 < 12	12 < 18	18 < 24	24+	0 < 12	12 < 18	18 < 24	24+
*************** GROWTH ALTERATIONS, NEOPLASTIC, BENIGN ***************								
PAPILLOMA								1/94

SEX:	FEMALE				MALE			
DIAGNOSIS / AGE IN MONTHS:	0 < 12	12 < 18	18 < 24	24+	0 < 12	12 < 18	18 < 24	24+
*************** MECHANICAL ABNORMALITIES AND CYST FORMATION ***************								
CYST							1/1136	
*************** INFLAMMATION AND REPAIR ***************								
INFLAMMATION			3/1356	1/1015			1/1136	1/958
*************** DEGENERATION, NECROSIS, AND DEPOSITION ***************								
CELLULAR CHANGE, EOSINOPHILIC			1/1356					

SEX:	FEMALE				MALE			
DIAGNOSIS / AGE IN MONTHS:	0 < 12	12 < 18	18 < 24	24+	0 < 12	12 < 18	18 < 24	24+
*************** MECHANICAL ABNORMALITIES AND CYST FORMATION ***************								
CYST		1/367						
*************** DEGENERATION, NECROSIS, AND DEPOSITION ***************								
CALCULI	4/176		4/192	2/75		2/300		6/102
*************** GROWTH ALTERATIONS, NON-NEOPLASTIC, HYPERPLASIA ***************								
HYPERPLASIA						1/300		2/102
*************** GROWTH ALTERATIONS, NEOPLASTIC, BENIGN ***************								
PAPILLOMA			1/192				1/161	

SEX:	FEMALE				MALE			
DIAGNOSIS / AGE IN MONTHS:	0 < 12	12 < 18	18 < 24	24+	0 < 12	12 < 18	18 < 24	24+
*************** MECHANICAL ABNORMALITIES AND CYST FORMATION ***************								
CYST								1/94
THROMBUS					1/171		1/147	
*************** INFLAMMATION AND REPAIR ***************								
GRANULOMA					1/171			1/94
INFLAMMATION						1/314		2/94
INFLAMMATION, GRANULOMATOUS								3/94
INFLAMMATION, NECROTIZING						2/314		3/94
INFLAMMATION, NONSUPPURATIVE			1/193	6/71		2/314	3/147	9/94
INFLAMMATION, SUPPURATIVE								1/94
*************** DEGENERATION, NECROSIS, AND DEPOSITION ***************								
CELLULAR CHANGE		1/369	1/193			3/314	1/147	1/94
CELLULAR CHANGE, VACUOLAR								3/94
FATTY CHANGE	1/212	5/369		3/71	27/171	58/314	42/147	24/94
HYALINOSIS, VASCULAR								1/94
INFARCT			1/193			1/314		
MINERALIZATION	2/212		1/193					
NECROSIS	2/212	11/369	4/193	2/71	1/171	2/314	4/147	5/94
NECROSIS, COAGULATIVE						2/314		

SEX:	FEMALE				MALE			
DIAGNOSIS / AGE IN MONTHS:	0 < 12	12 < 18	18 < 24	24+	0 < 12	12 < 18	18 < 24	24+
PIGMENT DEPOSITION	1/212		1/193			2/314	1/147	2/94
GROWTH ALTERATIONS, NON-NEOPLASTIC HYPERPLASIA								
HEMATOPOIESIS	1/212		9/193	3/71		2/314	3/147	
HYPERPLASIA								1/94
HYPERPLASIA, NODULAR								2/94
MASTOCYTOSIS								1/94
GROWTH ALTERATIONS, NON-NEOPLASTIC, ATROPHY								
ATROPHY								1/94
GROWTH ALTERATIONS, NEOPLASTIC, MALIGNANT, PRIMARY								
CARCINOMA								3/94
GROWTH ALTERATIONS, NEOPLASTIC, MALIGNANT, PRIMARY SITE NOT DETERMINED								
LYMPHOMA	4/212	7/369	10/193	3/71	3/171	1/314	4/147	4/94
LYMPHOMA, RETICULUM CELL		2/369	18/193	23/71		2/314	4/147	3/94
GROWTH ALTERATIONS, NEOPLASTIC, NOT DETERMINED								
NEOPLASM		3/369	3/193	1/71	1/171	7/314	5/147	2/94

SEX:	FEMALE				MALE			
DIAGNOSIS / AGE IN MONTHS:	0 < 12	12 < 18	18 < 24	24+	0 < 12	12 < 18	18 < 24	24+
*************** DEVELOPMENTAL ANOMALIES ***************								
HAMARTOMA				1/1015				
*************** MECHANICAL ABNORMALITIES AND CYST FORMATION ***************								
ANGIECTASIS		2/103	4/1356	8/1015		1/199	11/1136	9/958
CONGESTION			3/1356	2/1015			2/1136	
CYST, HEMORRHAGIC							1/1136	
HEMORRHAGE			1/1356				4/1136	
THROMBUS			1/1356	1/1015			2/1136	1/958
THROMBUS, ORGANIZED							1/1136	
*************** INFLAMMATION AND REPAIR ***************								
GRANULOMA			1/1356				1/1136	
INFLAMMATION, GRANULOMATOUS			1/1356	3/1015			1/1136	
INFLAMMATION, NECROTIZING				1/1015			1/1136	2/958
INFLAMMATION, NONSUPPURATIVE			11/1356	19/1015			3/1136	10/958
INFLAMMATION, SUPPURATIVE		1/103					1/1136	2/958
PERIVASCULITIS/VASCULITIS				1/1015				3/958
*************** DEGENERATION, NECROSIS, AND DEPOSITION ***************								
AMYLOID DEPOSITION					11/124*	26/199	19/1136	

SEX:	FEMALE				MALE			
AGE IN MONTHS: / DIAGNOSIS	0 < 12	12 < 18	18 < 24	24+	0 < 12	12 < 18	18 < 24	24+
CELLULAR CHANGE			2/1356	2/1015			1/1136	4/958
CELLULAR CHANGE, BASOPHILIC			1/1356				3/1136	3/958
CELLULAR CHANGE, CLEAR CELL								2/958
CELLULAR CHANGE, VACUOLAR			2/1356	9/1015			1/1136	10/958
DEGENERATION			2/1356	7/1015		7/199	7/1136	2/958
FATTY CHANGE		1/103	5/1356	7/1015	1/124	1/199	19/1136	22/958
INFARCT		1/103	1/1356	1/1015			3/1136	8/958
LIPID DEPOSITION			2/1356	4/1015			3/1136	2/958
MINERALIZATION							3/1136	1/958
NECROSIS	1/67	3/103	42/1356	16/1015	2/124	2/199	46/1136	13/958
NECROSIS, COAGULATIVE	2/67		3/1356	1/1015			8/1136	5/958
NECROSIS, HEMORRHAGIC						1/199	1/1136	
*************** GROWTH ALTERATIONS, NON-NEOPLASTIC HYPERTROPHY ***************								
CYTOMEGALY							2/1136	1/958
HYPERTROPHY							1/1136	
*************** GROWTH ALTERATIONS, NON-NEOPLASTIC HYPERPLASIA ***************								
HEMATOPOIESIS	1/67		9/1356	5/1015	1/124	1/199	12/1136	1/958
HISTIOCYTOSIS				1/1015				
HYPERPLASIA, MYELOID			1/1356				1/1136	
HYPERPLASIA, NODULAR		1/103	14/1356	6/1015		7/199	32/1136	12/958

SEX:	FEMALE				MALE			
DIAGNOSIS / AGE IN MONTHS:	0 < 12	12 < 18	18 < 24	24+	0 < 12	12 < 18	18 < 24	24+
HYPERPLASIA, RETICULUM CELL				2/1015			1/1136	
*************** GROWTH ALTERATIONS, NON-NEOPLASTIC, ATROPHY ***************								
ATROPHY								1/958
*************** GROWTH ALTERATIONS, NEOPLASTIC, BENIGN ***************								
ADENOMA		2/103	9/1356*	22/1015 *		4/199	59/1136	70/958
ANGIOMA			1/1356	1/1015			2/1136	7/958
*************** GROWTH ALTERATIONS, NEOPLASTIC, MALIGNANT, PRIMARY ***************								
ANGIOSARCOMA				6/1015		1/199	7/1136	8/958
CARCINOMA			25/1356*	50/1015 *		13/199	172/1136	222/958
HEPATOMA								1/958
*************** GROWTH ALTERATIONS, NEOPLASTIC, MALIGNANT, METASTATIC ***************								
CARCINOMA, ACINAR CELL								1/958
CARCINOMA, CORTICAL							1/1136	
*************** GROWTH ALTERATIONS, NEOPLASTIC, MALIGNANT, PRIMARY SITE NOT DETERMINED ***************								
FIBROSARCOMA			1/1356				1/1136	
LEUKEMIA, GRANULOCYTIC				1/1015			1/1136	
LEUKEMIA, UNDIFFERENTIATED			1/1356					

*VARIABILITY AMONG DATA SOURCES SIGNIFICANT AT THE 1% SIGNIFICANCE LEVEL

SEX:	FEMALE				MALE			
DIAGNOSIS / AGE IN MONTHS:	0 < 12	12 < 18	18 < 24	24+	0 < 12	12 < 18	18 < 24	24+
LYMPHOMA			3/1356	3/1015			1/1136	
LYMPHOMA, HISTIOCYTIC			4/1356	1/1015			3/1136	1/958
LYMPHOMA, LYMPHOCYTIC			1/1356				1/1136	
SARCOMA							1/1136	
*************** GROWTH ALTERATIONS, NEOPLASTIC, NOT DETERMINED ***************								
NEOPLASM						1/199	2/1136	7/958

SEX:	FEMALE				MALE			
DIAGNOSIS / AGE IN MONTHS:	0 < 12	12 < 18	18 < 24	24+	0 < 12	12 < 18	18 < 24	24+
MECHANICAL ABNORMALITIES AND CYST FORMATION								
ANGIECTASIS								1/102
THROMBUS						1/300		
INFLAMMATION AND REPAIR								
INFLAMMATION						2/300		3/102
INFLAMMATION, GRANULOMATOUS						4/300	1/161	7/102
INFLAMMATION, NECROTIZING								1/102
INFLAMMATION, NONSUPPURATIVE	2/176	12/367	6/192	3/75	1/131	13/300	6/161	24/102*
INFLAMMATION, SUPPURATIVE					2/131		1/161	
PERIVASCULITIS/VASCULITIS							1/161	
DEGENERATION, NECROSIS, AND DEPOSITION								
AMYLOID DEPOSITION							2/161	5/102
CELLULAR CHANGE		2/367	1/192	2/75		3/300	3/161	
FATTY CHANGE		11/367*	13/192	15/75	26/131	123/300	47/161	27/102*
INFARCT				1/75			1/161	
NECROSIS	5/176	7/367	6/192	1/75		2/300	5/161	1/102
NECROSIS, COAGULATIVE						2/300		1/102
PIGMENT DEPOSITION				1/75			1/161	1/102

*VARIABILITY AMONG DATA SOURCES SIGNIFICANT AT THE 1% SIGNIFICANCE LEVEL

SEX:	FEMALE				MALE			
DIAGNOSIS / AGE IN MONTHS:	0 < 12	12 < 18	18 < 24	24+	0 < 12	12 < 18	18 < 24	24+
*************** GROWTH ALTERATIONS, NON-NEOPLASTIC, HYPERPLASIA ***************								
HEMATOPOIESIS	2/176	9/367	8/192	4/75			4/161	1/102
*************** GROWTH ALTERATIONS, NEOPLASTIC, MALIGNANT, PRIMARY SITE NOT DETERMINED ***************								
LYMPHOMA	1/176	2/367	4/192	3/75		1/300	1/161	4/102
LYMPHOMA, RETICULUM CELL		6/367	17/192	21/75			8/161	11/102
*************** GROWTH ALTERATIONS, NEOPLASTIC, NOT DETERMINED ***************								
NEOPLASM		1/367	2/192	5/75		4/300	6/161	6/102

SEX:	FEMALE				MALE			
DIAGNOSIS / AGE IN MONTHS:	0 < 12	12 < 18	18 < 24	24+	0 < 12	12 < 18	18 < 24	24+
*************** MECHANICAL ABNORMALITIES AND CYST FORMATION ***************								
ANGIECTASIS			1/327	9/1575			2/437	15/1743
CONGESTION			1/327	13/1575	1/101	1/101	6/437	25/1743
CYST					1/101	2/101	1/437	28/1743
HEMORRHAGE			1/327	1/1575				
HERNIATION				3/1575				
THROMBUS			1/327				2/437	
*************** INFLAMMATION AND REPAIR ***************								
CIRRHOSIS			3/327	4/1575			1/437	5/1743
FIBROSIS			1/327			1/101	4/437	6/1743
GRANULOMA			2/327	14/1575				5/1743
INFLAMMATION			1/327	1/1575	3/101	15/101 *	9/437	26/1743
INFLAMMATION WITH FIBROSIS			1/327					
INFLAMMATION, GRANULOMATOUS				3/1575	3/101	6/101	1/437	46/1743
INFLAMMATION, NECROTIZING				3/1575		3/101	1/437	6/1743
INFLAMMATION, NONSUPPURATIVE				2/1575	2/101	11/101 *	9/437	40/1743 *
INFLAMMATION, SUPPURATIVE				1/1575				
PERIVASCULITIS/VASCULITIS			1/327	1/1575				1/1743

*VARIABILITY AMONG DATA SOURCES SIGNIFICANT AT THE 1% SIGNIFICANCE LEVEL

SEX:	FEMALE				MALE			
DIAGNOSIS / AGE IN MONTHS:	0 < 12	12 < 18	18 < 24	24+	0 < 12	12 < 18	18 < 24	24+
*************** DEGENERATION, NECROSIS, AND DEPCSITION ***************								
AMYLOID DEPOSITION								1/1743
CELLULAR CHANGE			10/327	57/1575		1/101	2/437	26/1743
CELLULAR CHANGE, BASOPHILIC			12/327	119/1575 *				25/1743
CELLULAR CHANGE, CLEAR CELL								3/1743
CELLULAR CHANGE, EOSINOPHILIC								2/1743
CELLULAR CHANGE, VACUOLAR			7/327	6/1575			2/437	50/1743
DEGENERATION		1/24	3/327	7/1575			8/437	21/1743
FATTY CHANGE		4/24	35/327	108/1575 *		7/101	31/437 *	101/1743*
LIPID DEPOSITION				2/1575		1/101	3/437	1/1743
NECROSIS		2/24	8/327	27/1575	1/101		21/437	50/1743
NECROSIS, COAGULATIVE				8/1575	2/101	2/101	4/437	12/1743
NECROSIS, FAT							1/437	
PIGMENT DEPOSITION			1/327				1/437	1/1743
*************** GROWTH ALTERATIONS, NON-NEOPLASTIC, HYPERTROPHY ***************								
CYTOMEGALY			2/327	6/1575			2/437	11/1743
HYPERTROPHY				1/1575				3/1743
*************** GROWTH ALTERATIONS, NON-NEOPLASTIC, HYPERPLASIA ***************								
HEMATOPOIESIS				10/1575		1/101	3/437	7/1743

*VARIABILITY AMONG DATA SOURCES SIGNIFICANT AT THE 1% SIGNIFICANCE LEVEL

SEX:	FEMALE				MALE			
DIAGNOSIS / AGE IN MONTHS:	0 < 12	12 < 18	18 < 24	24+	0 < 12	12 < 18	18 < 24	24+
HYPERPLASIA			2/327	2/1575				1/1743
HYPERPLASIA, BASOPHILIC				1/1575				
HYPERPLASIA, ERYTHROID				1/1575				
HYPERPLASIA, NODULAR			2/327	5/1575			1/437	18/1743
HYPERPLASIA, SCLEROSING					2/101	14/101	12/437	76/1743
*************** GROWTH ALTERATIONS, NON-NEOPLASTIC, ATROPHY ***************								
ATROPHY								5/1743
*************** GROWTH ALTERATIONS, NEOPLASTIC, BENIGN ***************								
ADENOMA				5/1575				5/1743
LIPOMA				1/1575				
NODULE				1/1575				
*************** GROWTH ALTERATIONS, NEOPLASTIC, MALIGNANT, PRIMARY ***************								
ADENOCARCINOMA			1/327					
CARCINOMA			2/327	4/1575			16/437	10/1743
HEPATOMA								1/1743
*************** GROWTH ALTERATIONS, NEOPLASTIC, MALIGNANT, PRIMARY SITE NOT DETERMINED ***************								
FIBROSARCOMA							1/437	
HISTIOCYTOMA, MALIGNANT							1/437	

SEX:	FEMALE				MALE			
DIAGNOSIS / AGE IN MONTHS:	0 < 12	12 < 18	18 < 24	24+	0 < 12	12 < 18	18 < 24	24+
LEIOMYOSARCOMA				1/1575				
LEUKEMIA	1/26		1/327	2/1575			2/437	25/1743
LEUKEMIA, GRANULOCYTIC				1/1575				5/1743
LEUKEMIA, LYMPHOBLASTIC			1/327	3/1575				2/1743
LEUKEMIA, LYMPHOCYTIC				3/1575			1/437	3/1743
LEUKEMIA, STEM CELL							1/437	
LYMPHOMA				1/1575				1/1743
NEUROFIBROSARCOMA				1/1575				
SARCOMA								1/1743
*************** GROWTH ALTERATIONS, NEOPLASTIC, NOT DETERMINED ***************								
NEOPLASM			1/327	12/1575			1/437	16/1743
*************** PARASITISM ***************								
PARASITISM				1/1575				

SEX:	FEMALE				MALE			
DIAGNOSIS / AGE IN MONTHS:	0 < 12	12 < 18	18 < 24	24+	0 < 12	12 < 18	18 < 24	24+
*************** MECHANICAL ABNORMALITIES AND CYST FORMATION ***************								
ANGIECTASIS				6/336		2/64	3/104	11/243
CONGESTION	1/7		2/44				3/104	1/243
CYST				1/336				
HEMORRHAGE							1/104	
*************** INFLAMMATION AND REPAIR ***************								
CIRRHOSIS							1/104	
GRANULOMA				1/336				1/243
*************** DEGENERATION, NECROSIS, AND DEPOSITION ***************								
CELLULAR CHANGE				5/336				1/243
CELLULAR CHANGE, VACUOLAR				6/336				3/243
DEGENERATION				2/336				3/243
FATTY CHANGE		2/39	6/44	21/336		7/64	14/104	20/243
NECROSIS	1/7		4/44	2/336			2/104	1/243
*************** GROWTH ALTERATIONS, NON-NEOPLASTIC, HYPERTROPHY ***************								
CYTOMEGALY		1/39		17/336 *		1/64	2/104	12/243

*VARIABILITY AMONG DATA SOURCES SIGNIFICANT AT THE 1% SIGNIFICANCE LEVEL

SEX:	FEMALE				MALE			
DIAGNOSIS / AGE IN MONTHS:	0 < 12	12 < 18	18 < 24	24+	0 < 12	12 < 18	18 < 24	24+
*************** GROWTH ALTERATIONS, NON-NEOPLASTIC, HYPERPLASIA ***************								
HYPERPLASIA, NODULAR								1/243
*************** GROWTH ALTERATIONS, NEOPLASTIC, BENIGN ***************								
ADENOMA								1/243
*************** GROWTH ALTERATIONS, NEOPLASTIC, MALIGNANT, PRIMARY ***************								
ANGIOSARCOMA								1/243
CARCINOMA				4/336				1/243
*************** GROWTH ALTERATIONS, NEOPLASTIC, MALIGNANT, PRIMARY SITE NOT DETERMINED ***************								
HISTIOCYTOMA, FIBROUS				1/336				
LYMPHOMA								1/243
*************** GROWTH ALTERATIONS, NEOPLASTIC, NOT DETERMINED ***************								
NEOPLASM			1/44	7/336				1/243

SEX:	FEMALE				MALE			
DIAGNOSIS / AGE IN MONTHS:	0 < 12	12 < 18	18 < 24	24+	0 < 12	12 < 18	18 < 24	24+
*************** MECHANICAL ABNORMALITIES AND CYST FORMATION ***************								
ANGIECTASIS		2/60		1/240				
CONGESTION		1/60	12/420	3/240		14/97*	15/392	14/216
CYST							1/392	10/216
CYST, HEMORRHAGIC				2/240				3/216
HEMORRHAGE					1/170		1/392	
LYMPHANGIECTASIS						1/97	1/392	2/216
THROMBUS				1/240				1/216
*************** INFLAMMATION AND REPAIR ***************								
ABSCESS		1/60		2/240		1/97		1/216
CIRRHOSIS				2/240			1/392	1/216
GRANULOMA	3/181		3/420		1/170		1/392	
INFLAMMATION	4/181	2/60	15/420	6/240	8/170		24/392 *	11/216 *
INFLAMMATION, NECROTIZING			1/420					
INFLAMMATION, NONSUPPURATIVE	19/181	1/60	2/420	10/240	23/170	1/97	1/392	14/216 *
INFLAMMATION, SUPPURATIVE			2/420			1/97		
PERIVASCULITIS/VASCULITIS				2/240				2/216
*************** DEGENERATION, NECROSIS, AND DEPOSITION ***************								
CELLULAR CHANGE, GLYCOGEN			1/420	4/240				6/216

*VARIABILITY AMONG DATA SOURCES SIGNIFICANT AT THE 1% SIGNIFICANCE LEVEL

ORGAN SYSTEM: DIGESTIVE SYSTEM; ACCESSORY ORGANS ORGAN: LIVER STRAIN: SPRAGUE DAWLEY

SEX:	FEMALE				MALE			
AGE IN MONTHS: / DIAGNOSIS	0 < 12	12 < 18	18 < 24	24+	0 < 12	12 < 18	18 < 24	24+
CELLULAR CHANGE, VACUOLAR	2/181				1/170			
DEGENERATION			1/420				2/392	15/216*
FATTY CHANGE	4/181	3/60	50/420*	40/240 *	2/170	6/97	26/392 *	51/216*
HEMOSIDERIN DEPOSITION			2/420				1/392	
LIPID DEPOSITION			2/420			1/97		
MINERALIZATION								1/216
NECROSIS		1/60	3/420	6/240		2/97	4/392	9/216
NECROSIS, COAGULATIVE		1/60	3/420	4/240	1/170	2/97	2/392	4/216
NECROSIS, HEMORRHAGIC								1/216
*************** GROWTH ALTERATIONS, NON-NEOPLASTIC, HYPERTROPHY ***************								
HYPERTROPHY								1/216
*************** GROWTH ALTERATIONS, NON-NEOPLASTIC, HYPERPLASIA ***************								
HEMATOPOIESIS	11/181 *		4/420	3/240	10/170*	1/97	1/392	5/216
HYPERPLASIA			3/420	1/240			2/392	
HYPERPLASIA, MYELOID			2/420	6/240		1/97		3/216
HYPERPLASIA, NODULAR			3/420	3/240			3/392	5/216
*************** GROWTH ALTERATIONS, NEOPLASTIC, BENIGN ***************								
ADENOMA				1/240				

*VARIABILITY AMONG DATA SOURCES SIGNIFICANT AT THE 1% SIGNIFICANCE LEVEL

SEX:	FEMALE				MALE			
AGE IN MONTHS: / DIAGNOSIS	0 < 12	12 < 18	18 < 24	24+	0 < 12	12 < 18	18 < 24	24+
*************** GROWTH ALTERATIONS, NEOPLASTIC, MALIGNANT, PRIMARY ***************								
ANGIOSARCOMA			1/420					
CARCINOMA								1/216
HEPATOMA							1/392	2/216
*************** GROWTH ALTERATIONS, NEOPLASTIC, MALIGNANT, METASTATIC ***************								
RHABDOMYOSARCOMA				1/240				
*************** GROWTH ALTERATIONS, NEOPLASTIC, MALIGNANT, PRIMARY SITE NOT DETERMINED ***************								
FIBROSARCOMA							1/392	
LEUKEMIA			1/420					1/216
LEUKEMIA, HISTIOCYTIC							1/392	
LEUKEMIA, LYMPHOCYTIC			1/420			1/97		
LYMPHOMA	1/181		1/420		1/170		1/392	
LYMPHOMA, RETICULUM CELL							1/392	
MYELOSARCOMA						1/97		
SARCOMA, HISTIOCYTIC			2/420	2/240			1/392	4/216

SEX:	FEMALE				MALE			
DIAGNOSIS / AGE IN MONTHS:	0 < 12	12 < 18	18 < 24	24+	0 < 12	12 < 18	18 < 24	24+
*************** MECHANICAL ABNORMALITIES AND CYST FORMATION ***************								
CYST	2/212	3/369	3/193	1/71	1/171	1/314	2/147	1/94
*************** INFLAMMATION AND REPAIR ***************								
FIBROSIS		1/369	3/193	1/71		2/314		
INFLAMMATION	9/212	31/369	12/193	7/71		5/314	5/147	8/94
INFLAMMATION, NONSUPPURATIVE						1/314		2/94
*************** GROWTH ALTERATIONS, NON-NEOPLASTIC, HYPERPLASIA ***************								
HYPERPLASIA		1/369	1/193					1/94

SEX:	FEMALE				MALE			
DIAGNOSIS / AGE IN MONTHS:	0 < 12	12 < 18	18 < 24	24+	0 < 12	12 < 18	18 < 24	24+
*************** MECHANICAL ABNORMALITIES AND CYST FORMATION ***************								
CYST			1/1356	1/1015				3/958
DILATATION			1/1356	1/1015				
*************** INFLAMMATION AND REPAIR ***************								
FIBROSIS			1/1356			1/199	3/1136	2/958
INFLAMMATION		2/103	25/1356	45/1015	1/124		15/1136	23/958
*************** GROWTH ALTERATIONS, NON-NEOPLASTIC, HYPERPLASIA ***************								
HYPERPLASIA			2/1356	5/1015			24/1136	21/958
*************** GROWTH ALTERATIONS, NEOPLASTIC, MALIGNANT, PRIMARY ***************								
CARCINOMA			1/1356					1/958

SEX:	FEMALE				MALE			
DIAGNOSIS / AGE IN MONTHS:	0 < 12	12 < 18	18 < 24	24+	0 < 12	12 < 18	18 < 24	24+
*************** MECHANICAL ABNORMALITIES AND CYST FORMATION ***************								
CYST	2/176	3/367	10/192	5/75	1/131	6/300	2/161	6/102
*************** INFLAMMATION AND REPAIR ***************								
FIBROSIS	1/176	2/367					1/161	
INFLAMMATION	27/176	81/367	53/192	18/75	3/131	14/300	11/161	12/102
INFLAMMATION, NONSUPPURATIVE								1/102
*************** GROWTH ALTERATIONS, NON-NEOPLASTIC, HYPERPLASIA ***************								
HYPERPLASIA		1/367	2/192				1/161	1/102

SEX:	FEMALE				MALE			
DIAGNOSIS / AGE IN MONTHS:	0 < 12	12 < 18	18 < 24	24+	0 < 12	12 < 18	18 < 24	24+
*************** MECHANICAL ABNORMALITIES AND CYST FORMATION ***************								
CHOLANGIECTASIS								2/1743
CYST				1/1575				1/1743
DILATATION						1/101		7/1743
*************** INFLAMMATION AND REPAIR ***************								
FIBROSIS			2/327	2/1575				20/1743
INFLAMMATION			10/327	26/1575			9/437	69/1743
*************** GROWTH ALTERATIONS, NON-NEOPLASTIC, HYPERPLASIA ***************								
HYPERPLASIA	1/26	2/24	55/327	363/1575*	8/101	19/101	109/437*	609/1743*
*************** GROWTH ALTERATIONS, NEOPLASTIC, MALIGNANT, PRIMARY ***************								
CARCINOMA			2/327					1/1743

*VARIABILITY AMONG DATA SOURCES SIGNIFICANT AT THE 1% SIGNIFICANCE LEVEL

ORGAN SYSTEM: DIGESTIVE SYSTEM; ACCESSORY ORGANS ORGAN: LIVER; BILE DUCT STRAIN: OSBORNE MENDEL

SEX:	FEMALE				MALE			
DIAGNOSIS / AGE IN MONTHS:	0 < 12	12 < 18	18 < 24	24+	0 < 12	12 < 18	18 < 24	24+
*************** MECHANICAL ABNORMALITIES AND CYST FORMATION ***************								
CYST				1/336			1/104	2/243
DILATATION				5/336				
*************** INFLAMMATION AND REPAIR ***************								
FIBROSIS			1/44	4/336			4/104	7/243
INFLAMMATION			2/44	15/336	1/11	1/64	8/104	11/243
*************** GROWTH ALTERATIONS, NON-NEOPLASTIC, HYPERPLASIA ***************								
HYPERPLASIA			3/44	62/336 *		3/64	15/104 *	38/243

*VARIABILITY AMONG DATA SOURCES SIGNIFICANT AT THE 1% SIGNIFICANCE LEVEL

SEX:	FEMALE				MALE			
DIAGNOSIS / AGE IN MONTHS:	0 < 12	12 < 18	18 < 24	24+	0 < 12	12 < 18	18 < 24	24+
*************** MECHANICAL ABNORMALITIES AND CYST FORMATION ***************								
CYST			1/420	2/240		1/97	2/392	5/216
DILATATION			1/420	1/240				
*************** INFLAMMATION AND REPAIR ***************								
FIBROSIS		1/60	1/420	2/240			1/392	8/216
INFLAMMATION	3/181	2/60	6/420	10/240	4/170	1/97	2/392	13/216 *
INFLAMMATION, NONSUPPURATIVE	1/181		1/420	2/240	2/170	1/97	1/392	17/216
*************** GROWTH ALTERATIONS, NON-NEOPLASTIC, HYPERPLASIA ***************								
HYPERPLASIA		1/60	7/420	23/240*		1/97	7/392	51/216 *

*VARIABILITY AMONG DATA SOURCES SIGNIFICANT AT THE 1% SIGNIFICANCE LEVEL

SEX:	FEMALE				MALE			
DIAGNOSIS / AGE IN MONTHS:	0 < 12	12 < 18	18 < 24	24+	0 < 12	12 < 18	18 < 24	24+
*************** MECHANICAL ABNORMALITIES AND CYST FORMATION ***************								
CYST						1/314		
*************** INFLAMMATION AND REPAIR ***************								
FIBROSIS						1/314		1/94
INFLAMMATION		1/369				2/314	1/147	1/94
INFLAMMATION, NONSUPPURATIVE		1/369	2/193	6/71		1/314	4/147	18/94*
*************** DEGENERATION, NECROSIS, AND DEPOSITION ***************								
NECROSIS						1/314	1/147	
*************** GROWTH ALTERATIONS, NEOPLASTIC, BENIGN ***************								
ADENOMA	1/212	1/369	2/193			14/314	7/147	7/94
MYOEPITHELIOMA	1/212							
*************** GROWTH ALTERATIONS, NEOPLASTIC, MALIGNANT, PRIMARY ***************								
ADENOCARCINOMA	1/212	1/369						
*************** GROWTH ALTERATIONS, NEOPLASTIC, MALIGNANT, PRIMARY SITE NOT DETERMINED ***************								
LYMPHOMA	2/212	1/369	2/193					2/94
LYMPHOMA, RETICULUM CELL		1/369	6/193	10/71				5/94

*VARIABILITY AMONG DATA SOURCES SIGNIFICANT AT THE 1% SIGNIFICANCE LEVEL

SEX:	FEMALE				MALE			
DIAGNOSIS / AGE IN MONTHS:	0 < 12	12 < 18	18 < 24	24+	0 < 12	12 < 18	18 < 24	24+
*************** INFLAMMATION AND REPAIR ***************								
FIBROSIS							1/1136	
INFLAMMATION			11/1356	23/1015			10/1136	13/958
INFLAMMATION, NONSUPPURATIVE			9/1356	1/1015		1/199	16/1136	
PERIVASCULITIS/VASCULITIS	1/67	2/103	9/1356	10/1015			7/1136	2/958
*************** GROWTH ALTERATIONS, NON-NEOPLASTIC, ATROPHY ***************								
ATROPHY			1/1356					
*************** GROWTH ALTERATIONS, NEOPLASTIC, MALIGNANT, PRIMARY SITE NOT DETERMINED ***************								
LEIOMYOSARCOMA			1/1356					

SEX:	FEMALE				MALE			
DIAGNOSIS / AGE IN MONTHS:	0 < 12	12 < 18	18 < 24	24+	0 < 12	12 < 18	18 < 24	24+
*************** INFLAMMATION AND REPAIR ***************								
INFLAMMATION		8/367	2/192	2/75		3/300	3/161	1/102
INFLAMMATION, NONSUPPURATIVE		26/367	19/192	27/75	1/131	19/300*	30/161	37/102*
*************** DEGENERATION, NECROSIS, AND DEPOSITION ***************								
AMYLOID DEPOSITION								10/102
FATTY CHANGE		5/367	2/192	3/75		4/300		3/102
NECROSIS						1/300		
*************** GROWTH ALTERATIONS, NON-NEOPLASTIC, ATROPHY ***************								
ATROPHY						1/300		
*************** GROWTH ALTERATIONS, NEOPLASTIC, BENIGN ***************								
ADENOMA	7/176	16/367	8/192	6/75	6/131	5/300	4/161	3/102
*************** GROWTH ALTERATIONS, NEOPLASTIC, MALIGNANT, PRIMARY SITE NOT DETERMINED ***************								
LYMPHOMA			2/192	3/75		1/300		3/102
LYMPHOMA, RETICULUM CELL			1/192	12/75				6/102

*VARIABILITY AMONG DATA SOURCES SIGNIFICANT AT THE 1% SIGNIFICANCE LEVEL

SEX:	FEMALE				MALE			
DIAGNOSIS / AGE IN MONTHS:	0 < 12	12 < 18	18 < 24	24+	0 < 12	12 < 18	18 < 24	24+
*************** MECHANICAL ABNORMALITIES AND CYST FORMATION ***************								
CYST								1/1743
*************** INFLAMMATION AND REPAIR ***************								
FIBROSIS				1/1575				
INFLAMMATION			2/327	22/1575				34/1743
INFLAMMATION, NECROTIZING			1/327	2/1575				5/1743
INFLAMMATION, NONSUPPURATIVE								1/1743
INFLAMMATION, SUPPURATIVE			1/327					
*************** DEGENERATION, NECROSIS, AND DEPOSITION ***************								
DEGENERATION				1/1575				
*************** GROWTH ALTERATIONS, NON-NEOPLASTIC, HYPERTROPHY ***************								
KARYOMEGALY								5/1743
*************** GROWTH ALTERATIONS, NON-NEOPLASTIC, HYPERPLASIA ***************								
HYPERPLASIA				8/1575				15/1743
HYPERPLASIA, INTRADUCTAL								1/1743

SEX:	FEMALE				MALE			
DIAGNOSIS / AGE IN MONTHS:	0 < 12	12 < 18	18 < 24	24+	0 < 12	12 < 18	18 < 24	24+
*************** GROWTH ALTERATIONS, NON-NEOPLASTIC, ATROPHY ***************								
ATROPHY		1/24					1/437	2/1743
*************** GROWTH ALTERATIONS, NON-NEOPLASTIC, DYSPLASIA ***************								
DYSPLASIA								9/1743
*************** GROWTH ALTERATIONS, NEOPLASTIC, BENIGN ***************								
ADENOMA				3/1575				1/1743
FIBROMA								1/1743
*************** GROWTH ALTERATIONS, NEOPLASTIC, MALIGNANT, PRIMARY ***************								
ADENOCARCINOMA							1/437	3/1743
CARCINOMA IN SITU								1/1743
FIBROADENOCARCINOMA								1/1743
*************** GROWTH ALTERATIONS, NEOPLASTIC, MALIGNANT, PRIMARY SITE NOT DETERMINED ***************								
ANGIOSARCOMA							1/437	
LEUKEMIA, LYMPHOCYTIC				1/1575				
SARCOMA								4/1743

SEX:	FEMALE				MALE			
DIAGNOSIS / AGE IN MONTHS:	0 < 12	12 < 18	18 < 24	24+	0 < 12	12 < 18	18 < 24	24+
*************** MECHANICAL ABNORMALITIES AND CYST FORMATION ***************								
CYST				1/336				
*************** INFLAMMATION AND REPAIR ***************								
INFLAMMATION				1/336				
*************** GROWTH ALTERATIONS, NON-NEOPLASTIC, HYPERPLASIA ***************								
HYPERPLASIA, INTRADUCTAL								1/243
*************** GROWTH ALTERATIONS, NEOPLASTIC, MALIGNANT, PRIMARY ***************								
CARCINOMA							1/104	
*************** GROWTH ALTERATIONS, NEOPLASTIC, MALIGNANT, PRIMARY SITE NOT DETERMINED ***************								
NEUROFIBROSARCOMA								1/243
*************** GROWTH ALTERATIONS, NEOPLASTIC, NOT DETERMINED ***************								
HISTIOCYTOMA, FIBROUS								1/243

SEX:	FEMALE				MALE			
DIAGNOSIS / AGE IN MONTHS:	0 < 12	12 < 18	18 < 24	24+	0 < 12	12 < 18	18 < 24	24+
*************** MECHANICAL ABNORMALITIES AND CYST FORMATION ***************								
EDEMA						1/97		
*************** INFLAMMATION AND REPAIR ***************								
INFLAMMATION, SUPPURATIVE	3/181				4/170			
*************** DEGENERATION, NECROSIS, AND DEPOSITION ***************								
MINERALIZATION							2/392	
*************** GROWTH ALTERATIONS, NON-NEOPLASTIC, ATROPHY ***************								
ATROPHY						1/97		
*************** GROWTH ALTERATIONS, NEOPLASTIC, MALIGNANT, PRIMARY SITE NOT DETERMINED ***************								
LEUKEMIA, LYMPHOCYTIC						1/97		
LYMPHOMA, RETICULUM CELL			1/420					

SEX:	FEMALE				MALE			
DIAGNOSIS / AGE IN MONTHS:	0 < 12	12 < 18	18 < 24	24+	0 < 12	12 < 18	18 < 24	24+
*************** INFLAMMATION AND REPAIR ***************								
INFLAMMATION						1/314		
*************** GROWTH ALTERATIONS, NEOPLASTIC, MALIGNANT, PRIMARY SITE NOT DETERMINED ***************								
LYMPHOMA		2/369	1/193					
LYMPHOMA, RETICULUM CELL			4/193	1/71				

SEX:	FEMALE				MALE			
DIAGNOSIS / AGE IN MONTHS:	0 < 12	12 < 18	18 < 24	24+	0 < 12	12 < 18	18 < 24	24+
*************** INFLAMMATION AND REPAIR ***************								
INFLAMMATION, SUPPURATIVE						1/199		
*************** GROWTH ALTERATIONS, NEOPLASTIC, BENIGN ***************								
PAPILLOMA								1/958

SEX:	FEMALE				MALE			
DIAGNOSIS / AGE IN MONTHS:	0 < 12	12 < 18	18 < 24	24+	0 < 12	12 < 18	18 < 24	24+
*************** MECHANICAL ABNORMALITIES AND CYST FORMATION ***************								
DILATATION				1/1575				
*************** INFLAMMATION AND REPAIR ***************								
INFLAMMATION								1/1743
*************** DEGENERATION, NECROSIS, AND DEPOSITION ***************								
NECROSIS, FAT			1/327					
*************** GROWTH ALTERATIONS, NON-NEOPLASTIC, DYSPLASIA ***************								
DYSPLASIA								3/1743

SEX:	FEMALE				MALE			
DIAGNOSIS / AGE IN MONTHS:	0 < 12	12 < 18	18 < 24	24+	0 < 12	12 < 18	18 < 24	24+
*************** MECHANICAL ABNORMALITIES AND CYST FORMATION ***************								
DILATATION				1/336				
*************** INFLAMMATION AND REPAIR ***************								
INFLAMMATION								1/243
INFLAMMATION, ULCERATIVE			1/44					
*************** GROWTH ALTERATIONS, NEOPLASTIC, MALIGNANT, PRIMARY SITE NOT DETERMINED ***************								
SARCOMA								1/243

SEX:	FEMALE				MALE			
DIAGNOSIS / AGE IN MONTHS:	0 < 12	12 < 18	18 < 24	24+	0 < 12	12 < 18	18 < 24	24+
*************** GROWTH ALTERATIONS, NEOPLASTIC, MALIGNANT, PRIMARY ***************								
CARCINOMA			1/420					
*************** GROWTH ALTERATIONS, NEOPLASTIC, MALIGNANT, METASTATIC ***************								
RHABDOMYOSARCOMA								1/216

SEX:	FEMALE				MALE			
AGE IN MONTHS:	0 < 12	12 < 18	18 < 24	24+	0 < 12	12 < 18	18 < 24	24+
DIAGNOSIS								
*************** MECHANICAL ABNORMALITIES AND CYST FORMATION ***************								
INTUSSUSCEPTION			2/193		1/171			
PROLAPSE, RECTAL		1/369						
*************** INFLAMMATION AND REPAIR ***************								
FIBROSIS				1/71				
INFLAMMATION	1/212	4/369	3/193	2/71	1/171			2/94
INFLAMMATION, NONSUPPURATIVE								2/94
INFLAMMATION, ULCERATIVE	1/212	2/369	1/193	5/71			1/147	
*************** DEGENERATION, NECROSIS, AND DEPOSITION ***************								
NECROSIS			1/193	1/71				
*************** GROWTH ALTERATIONS, NON-NEOPLASTIC, HYPERPLASIA ***************								
HYPERPLASIA								1/94
*************** GROWTH ALTERATIONS, NEOPLASTIC, BENIGN ***************								
ADENOMA		1/369						
POLYP		1/369	1/193	5/71				4/94

SEX:	FEMALE				MALE			
DIAGNOSIS / AGE IN MONTHS:	0 < 12	12 < 18	18 < 24	24+	0 < 12	12 < 18	18 < 24	24+
*************** GROWTH ALTERATIONS, NEOPLASTIC, MALIGNANT, PRIMARY ***************								
ADENOCARCINOMA		1/369	1/193					
*************** GROWTH ALTERATIONS, NEOPLASTIC, MALIGNANT, PRIMARY SITE NOT DETERMINED ***************								
CARCINOMA, SQUAMOUS CELL								1/94
LYMPHOMA	4/212	4/369	6/193	4/71				2/94
LYMPHOMA, RETICULUM CELL		2/369	15/193	23/71			3/147	9/94
SARCOMA, UNDIFFERENTIATED							1/147	
*************** PARASITISM ***************								
PARASITISM						2/314		2/94

SEX:	FEMALE				MALE			
DIAGNOSIS / AGE IN MONTHS:	0 < 12	12 < 18	18 < 24	24+	0 < 12	12 < 18	18 < 24	24+
*************** DEVELOPMENTAL ANOMALIES ***************								
INFLAMMATION, SUPPURATIVE			1/1356				1/1136	
*************** MECHANICAL ABNORMALITIES AND CYST FORMATION ***************								
CYST			1/1356					
HEMORRHAGE							1/1136	
PROLAPSE, RECTAL							9/1136	
*************** INFLAMMATION AND REPAIR ***************								
GRANULOMA							1/1136	
INFLAMMATION				1/1015			4/1136	1/958
INFLAMMATION, GRANULOMATOUS				2/1015				
INFLAMMATION, ULCERATIVE								2/958
PERIVASCULITIS/VASCULITIS				1/1015				
*************** DEGENERATION, NECROSIS, AND DEPOSITION ***************								
AMYLOID DEPOSITION							2/1136	2/958
*************** GROWTH ALTERATIONS, NON-NEOPLASTIC, HYPERPLASIA ***************								
HYPERPLASIA			12/1356	5/1015	1/124		9/1136	8/958
HYPERPLASIA, LYMPHOID			5/1356	8/1015			1/1136	14/958

SEX:	FEMALE				MALE			
DIAGNOSIS / AGE IN MONTHS:	0 < 12	12 < 18	18 < 24	24+	0 < 12	12 < 18	18 < 24	24+
HYPERPLASIA, RETICULUM CELL				1/1015				
*************** GROWTH ALTERATIONS, NEOPLASTIC, BENIGN ***************								
POLYP							1/1136	
POLYP, ADENOMATOUS			1/1356				1/1136	2/958
*************** GROWTH ALTERATIONS, NEOPLASTIC, MALIGNANT, PRIMARY ***************								
ADENOCARCINOMA							1/1136	2/958
LEIOMYOSARCOMA			1/1356	2/1015				
*************** GROWTH ALTERATIONS, NEOPLASTIC, MALIGNANT, PRIMARY SITE NOT DETERMINED ***************								
LYMPHOMA			2/1356	7/1015				9/958
LYMPHOMA, HISTIOCYTIC		1/103	2/1356	2/1015				1/958
LYMPHOMA, LYMPHOCYTIC								1/958
NEUROFIBROSARCOMA				1/1015				
SARCOMA				1/1015			1/1136	
*************** PARASITISM ***************								
PARASITISM		1/103	5/1356	9/1015	2/124	2/199	26/1136	42/958

DIAGNOSIS	SEX: AGE IN MONTHS:	FEMALE 0 < 12	12 < 18	18 < 24	24+	MALE 0 < 12	12 < 18	18 < 24	24+
*************** MECHANICAL ABNORMALITIES AND CYST FORMATION ***************									
INTUSSUSCEPTION			1/367	1/192					
*************** INFLAMMATION AND REPAIR ***************									
INFLAMMATION, NONSUPPURATIVE									1/102
INFLAMMATION, ULCERATIVE					1/75				
*************** DEGENERATION, NECROSIS, AND DEPOSITION ***************									
AMYLOID DEPOSITION							3/300	3/161	29/102 *
*************** GROWTH ALTERATIONS, NON-NEOPLASTIC, DYSPLASIA ***************									
DYSPLASIA, EPITHELIAL								1/161	
*************** GROWTH ALTERATIONS, NEOPLASTIC, BENIGN ***************									
POLYP				1/192	1/75			2/161	2/102
*************** GROWTH ALTERATIONS, NEOPLASTIC, MALIGNANT, PRIMARY ***************									
ADENOCARCINOMA			1/367	1/192					
*************** GROWTH ALTERATIONS, NEOPLASTIC, MALIGNANT, PRIMARY SITE NOT DETERMINED ***************									
LYMPHOMA			2/367		1/75				

*VARIABILITY AMONG DATA SOURCES SIGNIFICANT AT THE 1% SIGNIFICANCE LEVEL

SEX:	FEMALE				MALE			
DIAGNOSIS / AGE IN MONTHS:	0 < 12	12 < 18	18 < 24	24+	0 < 12	12 < 18	18 < 24	24+
LYMPHOMA, RETICULUM CELL			3/192	7/75			4/161	5/102
*************** PARASITISM ***************								
PARASITISM					1/131	1/300	2/161	6/102

DIAGNOSIS	FEMALE 0 < 12	FEMALE 12 < 18	FEMALE 18 < 24	FEMALE 24+	MALE 0 < 12	MALE 12 < 18	MALE 18 < 24	MALE 24+
*************** MECHANICAL ABNORMALITIES AND CYST FORMATION ***************								
CYST								1/1743
HEMORRHAGE								1/1743
*************** INFLAMMATION AND REPAIR ***************								
INFLAMMATION				3/1575			4/437	7/1743
INFLAMMATION, HEMORRHAGIC								1/1743
INFLAMMATION, SUPPURATIVE								2/1743
INFLAMMATION, ULCERATIVE			2/327			1/101		1/1743
PERIVASCULITIS/VASCULITIS				2/1575				1/1743
*************** DEGENERATION, NECROSIS, AND DEPOSITION ***************								
MINERALIZATION								1/1743
*************** GROWTH ALTERATIONS, NON-NEOPLASTIC, HYPERPLASIA ***************								
HYPERPLASIA			10/327	18/1575			12/437	63/1743
HYPERPLASIA, LYMPHOID		1/24	1/327	10/1575			1/437	15/1743
HYPERPLASIA, RETICULUM CELL								1/1743
*************** GROWTH ALTERATIONS, NON-NEOPLASTIC, ATROPHY ***************								
ATROPHY					1/101			

SEX:	FEMALE				MALE			
DIAGNOSIS / AGE IN MONTHS:	0 < 12	12 < 18	18 < 24	24+	0 < 12	12 < 18	18 < 24	24+
*************** GROWTH ALTERATIONS, NEOPLASTIC, BENIGN ***************								
FIBROMYXOMA								1/1743
POLYP								1/1743
POLYP, ADENOMATOUS				1/1575				2/1743
*************** GROWTH ALTERATIONS, NEOPLASTIC, MALIGNANT, PRIMARY ***************								
ADENOCARCINOMA			3/327			1/101		1/1743
ADENOCARCINOMA, MUCINOUS						1/101	2/437	
LEIOMYOSARCOMA				2/1575				2/1743
*************** GROWTH ALTERATIONS, NEOPLASTIC, MALIGNANT, METASTATIC ***************								
MESOTHELIOMA							2/437	2/1743
*************** GROWTH ALTERATIONS, NEOPLASTIC, MALIGNANT, PRIMARY SITE NOT DETERMINED ***************								
FIBROSARCOMA								1/1743
LEUKEMIA								1/1743
LEUKEMIA, LYMPHOCYTIC				1/1575				
LEUKEMIA, STEM CELL							1/437	
SARCOMA								3/1743

SEX:	FEMALE				MALE			
DIAGNOSIS / AGE IN MONTHS:	0 < 12	12 < 18	18 < 24	24+	0 < 12	12 < 18	18 < 24	24+
*************** PARASITISM ***************								
PARASITISM	4/26		6/327	94/1575*	17/101	13/101	2/437	124/1743*

*VARIABILITY AMONG DATA SOURCES SIGNIFICANT AT THE 1% SIGNIFICANCE LEVEL

SEX:	FEMALE				MALE			
DIAGNOSIS / AGE IN MONTHS:	0 < 12	12 < 18	18 < 24	24+	0 < 12	12 < 18	18 < 24	24+
*************** MECHANICAL ABNORMALITIES AND CYST FORMATION ***************								
IMPACTION			1/44					
*************** INFLAMMATION AND REPAIR ***************								
INFLAMMATION				2/336				1/243
INFLAMMATION, ULCERATIVE				1/336				
*************** DEGENERATION, NECROSIS, AND DEPOSITION ***************								
MINERALIZATION								2/243
*************** GROWTH ALTERATIONS, NEOPLASTIC, BENIGN ***************								
POLYP, ADENOMATOUS							1/104	
*************** GROWTH ALTERATIONS, NEOPLASTIC, MALIGNANT, PRIMARY SITE NOT DETERMINED ***************								
ANGIOSARCOMA			1/44					
FIBROSARCOMA								1/243
*************** PARASITISM ***************								
PARASITISM		2/39		6/336		1/64		2/243

SEX:	FEMALE				MALE			
DIAGNOSIS / AGE IN MONTHS:	0 < 12	12 < 18	18 < 24	24+	0 < 12	12 < 18	18 < 24	24+
*************** MECHANICAL ABNORMALITIES AND CYST FORMATION ***************								
CONGESTION								1/216
EDEMA				1/240				
HEMORRHAGE	3/181				1/170			1/216
IMPACTION							1/392	1/216
THROMBUS								1/216
*************** INFLAMMATION AND REPAIR ***************								
INFLAMMATION, NONSUPPURATIVE					1/170			
PERIVASCULITIS/VASCULITIS							3/392	3/216
*************** DEGENERATION, NECROSIS, AND DEPOSITION ***************								
ARTERIOLOSCLEROSIS								1/216
ARTERIOSCLEROSIS								1/216
*************** GROWTH ALTERATIONS, NON-NEOPLASTIC, HYPERPLASIA ***************								
HYPERPLASIA								1/216
*************** GROWTH ALTERATIONS, NEOPLASTIC, MALIGNANT, PRIMARY ***************								
ADENOCARCINOMA, MUCINOUS							1/392	1/216
LEIOMYOSARCOMA							1/392	

SEX:	FEMALE				MALE			
DIAGNOSIS / AGE IN MONTHS:	0 < 12	12 < 18	18 < 24	24+	0 < 12	12 < 18	18 < 24	24+
*************** GROWTH ALTERATIONS, NEOPLASTIC, MALIGNANT, METASTATIC ***************								
RHABDOMYOSARCOMA				1/240				
*************** GROWTH ALTERATIONS, NEOPLASTIC, MALIGNANT, PRIMARY SITE NOT DETERMINED ***************								
LYMPHOMA			1/420					
LYMPHOMA, RETICULUM CELL							1/392	
*************** PARASITISM ***************								
PARASITISM	4/181		7/420	8/240	3/170		10/392	2/216

SEX:	FEMALE				MALE			
DIAGNOSIS / AGE IN MONTHS:	0 < 12	12 < 18	18 < 24	24+	0 < 12	12 < 18	18 < 24	24+
*************** MECHANICAL ABNORMALITIES AND CYST FORMATION ***************								
CONGESTION								1/94
CYST						1/314		
CYST, EPIDERMAL	1/212							
*************** INFLAMMATION AND REPAIR ***************								
INFLAMMATION		1/369				1/314		
*************** GROWTH ALTERATIONS, NEOPLASTIC, MALIGNANT, PRIMARY SITE NOT DETERMINED ***************								
LYMPHOMA	1/212		1/193					
LYMPHOMA, RETICULUM CELL			1/193					

SEX:	FEMALE				MALE			
DIAGNOSIS / AGE IN MONTHS:	0 < 12	12 < 18	18 < 24	24+	0 < 12	12 < 18	18 < 24	24+
*************** MECHANICAL ABNORMALITIES AND CYST FORMATION ***************								
CYST	1/176					1/300		
*************** INFLAMMATION AND REPAIR ***************								
INFLAMMATION		2/367			1/131		1/161	2/102

SEX:	FEMALE				MALE			
DIAGNOSIS / AGE IN MONTHS:	0 < 12	12 < 18	18 < 24	24+	0 < 12	12 < 18	18 < 24	24+
*************** GROWTH ALTERATIONS, NEOPLASTIC, BENIGN ***************								
PAPILLOMA, SQUAMOUS CELL				1/1575				
*************** GROWTH ALTERATIONS, NEOPLASTIC, MALIGNANT, PRIMARY ***************								
CARCINOMA, SQUAMOUS CELL								1/1743
FIBROSARCOMA								1/1743

SEX:	FEMALE				MALE			
AGE IN MONTHS: / DIAGNOSIS	0 < 12	12 < 18	18 < 24	24+	0 < 12	12 < 18	18 < 24	24+
*************** DEVELOPMENTAL ANOMALIES ***************								
HAMARTOMA								1/94
*************** MECHANICAL ABNORMALITIES AND CYST FORMATION ***************								
CYST			2/193	2/71			3/147	2/94
DILATATION								1/94
EDEMA			1/193					
*************** INFLAMMATION AND REPAIR ***************								
INFLAMMATION	1/212		3/193	1/71	2/171	4/314		2/94
INFLAMMATION, NONSUPPURATIVE			2/193					2/94
INFLAMMATION, ULCERATIVE			1/193		2/171	3/314		
*************** GROWTH ALTERATIONS, NON-NEOPLASTIC, HYPERPLASIA ***************								
HYPERPLASIA								1/94
POLYP		2/369				2/314		
*************** GROWTH ALTERATIONS, NEOPLASTIC, BENIGN ***************								
ADENOMA		1/369					1/147	

SEX:	FEMALE				MALE			
DIAGNOSIS / AGE IN MONTHS:	0 < 12	12 < 18	18 < 24	24+	0 < 12	12 < 18	18 < 24	24+
*************** GROWTH ALTERATIONS, NEOPLASTIC, MALIGNANT, PRIMARY ***************								
ADENOCARCINOMA			1/193					
*************** GROWTH ALTERATIONS, NEOPLASTIC, MALIGNANT, PRIMARY SITE NOT DETERMINED ***************								
LYMPHOMA	2/212	3/369	3/193	1/71				1/94
LYMPHOMA, RETICULUM CELL		1/369	3/193	11/71				1/94

SEX:	FEMALE				MALE			
DIAGNOSIS / AGE IN MONTHS:	0 < 12	12 < 18	18 < 24	24+	0 < 12	12 < 18	18 < 24	24+
*************** INFLAMMATION AND REPAIR ***************								
EROSION				1/1015				
INFLAMMATION		1/103	15/1356	8/1015			23/1136	3/958
INFLAMMATION, NECROTIZING							1/1136	
INFLAMMATION, SUPPURATIVE			1/1356	1/1015				
INFLAMMATION, ULCERATIVE			4/1356	1/1015			1/1136	1/958
PERIVASCULITIS/VASCULITIS			1/1356					
*************** DEGENERATION, NECROSIS, AND DEPOSITION ***************								
AMYLOID DEPOSITION							1/1136	
MINERALIZATION					1/124	2/199	4/1136	
*************** GROWTH ALTERATIONS, NON-NEOPLASTIC, HYPERPLASIA ***************								
ACANTHOSIS			6/1356	2/1015	1/124	1/199	4/1136	5/958
HYPERKERATOSIS			4/1356	2/1015	1/124	1/199	5/1136	4/958
HYPERPLASIA			3/1356		1/124		4/1136	
HYPERPLASIA, ADENOMATOUS			1/1356				2/1136	
HYPERPLASIA, EPITHELIAL			1/1356	2/1015			3/1136	
HYPERPLASIA, MAST CELL				1/1015				
POLYP, ADENOMATOUS				1/1015				

SEX:	FEMALE				MALE			
DIAGNOSIS / AGE IN MONTHS:	0 < 12	12 < 18	18 < 24	24+	0 < 12	12 < 18	18 < 24	24+
*************** GROWTH ALTERATIONS, NEOPLASTIC, BENIGN ***************								
ADENOMA, PAPILLARY								1/958
PAPILLOMA, SQUAMOUS CELL			4/1356	2/1015			1/1136	1/958
*************** GROWTH ALTERATIONS, NEOPLASTIC, MALIGNANT, PRIMARY ***************								
CARCINOMA, ACINAR CELL								1/958
CARCINOMA, SQUAMOUS CELL							1/1136	
LEIOMYOSARCOMA				2/1015				
*************** GROWTH ALTERATIONS, NEOPLASTIC, MALIGNANT, PRIMARY SITE NOT DETERMINED ***************								
LYMPHOMA								1/958
LYMPHOMA, HISTIOCYTIC		1/103						

SEX:	FEMALE				MALE			
DIAGNOSIS / AGE IN MONTHS:	0 < 12	12 < 18	18 < 24	24+	0 < 12	12 < 18	18 < 24	24+
*************** MECHANICAL ABNORMALITIES AND CYST FORMATION ***************								
CYST		1/367	1/192	3/75			1/161	1/102
DILATATION								2/102
HEMORRHAGE			1/192					
*************** INFLAMMATION AND REPAIR ***************								
INFLAMMATION		2/367	1/192	1/75		1/300	3/161	3/102
INFLAMMATION, NONSUPPURATIVE								1/102
INFLAMMATION, ULCERATIVE							1/161	
*************** DEGENERATION, NECROSIS, AND DEPOSITION ***************								
AMYLOID DEPOSITION								4/102
MINERALIZATION						1/300		
*************** GROWTH ALTERATIONS, NON-NEOPLASTIC, HYPERPLASIA ***************								
HYPERKERATOSIS				1/75				
HYPERPLASIA, PLASMOCYTIC								1/102
POLYP							1/161	
*************** GROWTH ALTERATIONS, NON-NEOPLASTIC, DYSPLASIA ***************								
DYSPLASIA, EPITHELIAL						1/300		

SEX:	FEMALE				MALE			
DIAGNOSIS / AGE IN MONTHS:	0 < 12	12 < 18	18 < 24	24+	0 < 12	12 < 18	18 < 24	24+
*************** GROWTH ALTERATIONS, NEOPLASTIC, BENIGN ***************								
ADENOMA		1/367				1/300		2/102
*************** GROWTH ALTERATIONS, NEOPLASTIC, MALIGNANT, PRIMARY ***************								
CARCINOMA, SQUAMOUS CELL		1/367						
*************** GROWTH ALTERATIONS, NEOPLASTIC, MALIGNANT, PRIMARY SITE NOT DETERMINED ***************								
LYMPHOMA, RETICULUM CELL				2/75				
*************** GROWTH ALTERATIONS, NEOPLASTIC, NOT DETERMINED ***************								
MAST CELL TUMOR				1/75				

SEX:	FEMALE				MALE			
DIAGNOSIS / AGE IN MONTHS:	0 < 12	12 < 18	18 < 24	24+	0 < 12	12 < 18	18 < 24	24+
*************** MECHANICAL ABNORMALITIES AND CYST FORMATION ***************								
CYST				1/1575	1/101	4/101		14/1743
DILATATION				12/1575	1/101	4/101	2/437	53/1743
DIVERTICULUM								3/1743
EDEMA			1/327					4/1743
*************** INFLAMMATION AND REPAIR ***************								
DIVERTICULITIS								1/1743
EROSION							1/437	2/1743
FIBROSIS								2/1743
GRANULOMA								1/1743
INFLAMMATION			4/327	13/1575			3/437	13/1743
INFLAMMATION, SUPPURATIVE			1/327					
INFLAMMATION, ULCERATIVE			7/327	19/1575		2/101	10/437	32/1743
PERIVASCULITIS/VASCULITIS				1/1575				
*************** DEGENERATION, NECROSIS, AND DEPOSITION ***************								
INFARCT								1/1743
MINERALIZATION				4/1575			2/437	28/1743
NECROSIS		1/24						

SEX:	FEMALE				MALE			
DIAGNOSIS / AGE IN MONTHS:	0 < 12	12 < 18	18 < 24	24+	0 < 12	12 < 18	18 < 24	24+
*************** GROWTH ALTERATIONS, NON-NEOPLASTIC, HYPERPLASIA ***************								
ACANTHOSIS			1/327	1/1575			4/437	9/1743
HYPERKERATOSIS							4/437	9/1743
HYPERPLASIA				1/1575				31/1743
HYPERPLASIA, BASAL CELL			1/327	10/1575			3/437	14/1743
HYPERPLASIA, EPITHELIAL				2/1575			2/437	4/1743
HYPERPLASIA, LYMPHOID								1/1743
HYPERPLASIA, PAPILLARY				3/1575				
POLYP, ADENOMATOUS							1/437	
*************** GROWTH ALTERATIONS, NEOPLASTIC, BENIGN ***************								
PAPILLOMA								1/1743
PAPILLOMA, SQUAMOUS CELL				2/1575				1/1743
*************** GROWTH ALTERATIONS, NEOPLASTIC, MALIGNANT, PRIMARY ***************								
ADENOCARCINOMA				1/1575				
ADENOCARCINOMA, MUCINOUS							1/437	
CARCINOMA, BASAL CELL								1/1743
CARCINOMA, SQUAMOUS CELL				1/1575				2/1743

SEX:	FEMALE				MALE			
AGE IN MONTHS:	0 < 12	12 < 18	18 < 24	24+	0 < 12	12 < 18	18 < 24	24+
DIAGNOSIS								
*************** GROWTH ALTERATIONS, NEOPLASTIC, MALIGNANT, METASTATIC ***************								
MESOTHELIOMA							1/437	1/1743
*************** GROWTH ALTERATIONS, NEOPLASTIC, MALIGNANT, PRIMARY SITE NOT DETERMINED ***************								
FIBROSARCOMA								1/1743
LEUKEMIA				1/1575				1/1743

SEX:	FEMALE				MALE			
DIAGNOSIS / AGE IN MONTHS:	0 < 12	12 < 18	18 < 24	24+	0 < 12	12 < 18	18 < 24	24+
*************** INFLAMMATION AND REPAIR ***************								
INFLAMMATION			1/44	1/336		1/64	1/104	2/243
INFLAMMATION, ULCERATIVE		1/39	8/44	16/336	1/11		8/104	4/243
*************** DEGENERATION, NECROSIS, AND DEPCSITION ***************								
MINERALIZATION			3/44	1/336		6/64	16/104	16/243
*************** GROWTH ALTERATIONS, NON-NEOPLASTIC, HYPERPLASIA ***************								
ACANTHOSIS				1/336				1/243
HYPERKERATOSIS				1/336			1/104	
*************** GROWTH ALTERATIONS, NEOPLASTIC, BENIGN ***************								
PAPILLOMA							1/104	
*************** GROWTH ALTERATIONS, NEOPLASTIC, NALIGNANT, PRIMARY ***************								
CARCINOMA, SQUAMOUS CELL				1/336				
LEIOMYOSARCOMA								1/243
*************** GROWTH ALTERATIONS, NEOPLASTIC, MALIGNANT, METASTATIC ***************								
CARCINOMA, C CELL								1/243

SEX:	FEMALE				MALE			
DIAGNOSIS / AGE IN MONTHS:	0 < 12	12 < 18	18 < 24	24+	0 < 12	12 < 18	18 < 24	24+
*************** GROWTH ALTERATIONS, NEOPLASTIC, MALIGNANT, PRIMARY SITE NOT DETERMINED ***************								
SARCOMA								1/243

SEX:	FEMALE				MALE			
DIAGNOSIS / AGE IN MONTHS:	0 < 12	12 < 18	18 < 24	24+	0 < 12	12 < 18	18 < 24	24+
*************** MECHANICAL ABNORMALITIES AND CYST FORMATION ***************								
CYST	2/181		9/420	2/240			11/392	2/216
DILATATION			2/420	18/240*			1/392	15/216 *
EDEMA								1/216
*************** INFLAMMATION AND REPAIR ***************								
EROSION			1/420				1/392	
FIBROSIS								1/216
GRANULOMA			1/420					
INFLAMMATION			2/420			1/97		1/216
INFLAMMATION, GRANULOMATOUS			1/420				1/392	
INFLAMMATION, NONSUPPURATIVE					1/170			
INFLAMMATION, ULCERATIVE	1/181		4/420	8/240	2/170	3/97	4/392	8/216
*************** DEGENERATION, NECROSIS, AND DEPOSITION ***************								
MINERALIZATION		2/60	4/420	2/240		6/97	16/392	21/216 *
MINERALIZATION, VASCULAR								1/216
*************** GROWTH ALTERATIONS, NON-NEOPLASTIC, HYPERPLASIA ***************								
HYPERPLASIA			1/420					

*VARIABILITY AMONG DATA SOURCES SIGNIFICANT AT THE 1% SIGNIFICANCE LEVEL

SEX:	FEMALE				MALE			
DIAGNOSIS / AGE IN MONTHS:	0 < 12	12 < 18	18 < 24	24+	0 < 12	12 < 18	18 < 24	24+
*************** GROWTH ALTERATIONS, NON-NEOPLASTIC, ATROPHY ***************								
ATROPHY				1/240				
*************** GROWTH ALTERATIONS, NON-NEOPLASTIC, METAPLASIA ***************								
METAPLASIA							1/392	
*************** GROWTH ALTERATIONS, NEOPLASTIC, BENIGN ***************								
PAPILLOMA			1/420					
*************** GROWTH ALTERATIONS, NEOPLASTIC, MALIGNANT, PRIMARY ***************								
ADENOCARCINOMA				1/240				
CARCINOMA, SQUAMOUS CELL				2/240				1/216
LEIOMYOSARCOMA							1/392	
*************** GROWTH ALTERATIONS, NEOPLASTIC, MALIGNANT, METASTATIC ***************								
RHABDOMYOSARCOMA								1/216
*************** GROWTH ALTERATIONS, NEOPLASTIC, MALIGNANT, PRIMARY SITE NOT DETERMINED ***************								
ANGIOSARCOMA			1/420					
LYMPHOMA			1/420					
LYMPHOMA, RETICULUM CELL							2/392	

 ORGAN SYSTEM: DIGESTIVE SYSTEM; ALIMENTARY TRACT ORGAN: STOMACH STRAIN: SPRAGUE DAWLEY

SEX:	FEMALE				MALE			
DIAGNOSIS / AGE IN MONTHS:	0 < 12	12 < 18	18 < 24	24+	0 < 12	12 < 18	18 < 24	24+
SARCOMA, HISTIOCYTIC								1/216

SEX:	FEMALE				MALE			
DIAGNOSIS / AGE IN MONTHS:	0 < 12	12 < 18	18 < 24	24+	0 < 12	12 < 18	18 < 24	24+
*************** DEVELOPMENTAL ANOMALIES ***************								
ACCESSORY ADRENAL TISSUE					1/171	2/314		2/94
*************** MECHANICAL ABNORMALITIES AND CYST FORMATION ***************								
ANGIECTASIS			1/193					
CYST							1/147	1/94
HEMORRHAGE				1/71				
THROMBUS		1/369	3/193	2/71				
*************** INFLAMMATION AND REPAIR ***************								
INFLAMMATION							1/147	
*************** DEGENERATION, NECROSIS, AND DEPOSITION ***************								
MINERALIZATION		3/369	1/193					
NECROSIS				1/71				
PIGMENT DEPOSITION	1/212	8/369	4/193	5/71				
*************** GROWTH ALTERATIONS, NON-NEOPLASTIC, HYPERPLASIA ***************								
HEMATOPOIESIS		1/369	4/193	1/71			1/147	
HYPERPLASIA					3/171	3/314	1/147	7/94
HYPERPLASIA, NODULAR							1/147	1/94

SEX:	FEMALE				MALE			
DIAGNOSIS / AGE IN MONTHS:	0 < 12	12 < 18	18 < 24	24+	0 < 12	12 < 18	18 < 24	24+
HYPERPLASIA, SPINDLE CELL		4/369	9/193	13/71		5/314	8/147	14/94 *
MASTOCYTOSIS				1/71				
*************** GROWTH ALTERATIONS, NON-NEOPLASTIC, ATROPHY ***************								
ATROPHY	1/212							2/94
*************** GROWTH ALTERATIONS, NEOPLASTIC, BENIGN ***************								
ADENOMA, CORTICAL		3/369	4/193	12/71		3/314	7/147	2/94
*************** GROWTH ALTERATIONS, NEOPLASTIC, MALIGNANT, PRIMARY ***************								
CARCINOMA, CORTICAL				1/71				
*************** GROWTH ALTERATIONS, NEOPLASTIC, MALIGNANT, PRIMARY SITE NOT DETERMINED ***************								
LYMPHOMA	2/212	3/369	2/193		1/171		1/147	1/94
LYMPHOMA, RETICULUM CELL			2/193					

*VARIABILITY AMONG DATA SOURCES SIGNIFICANT AT THE 1% SIGNIFICANCE LEVEL

DIAGNOSIS	SEX: FEMALE				MALE			
AGE IN MONTHS:	0 < 12	12 < 18	18 < 24	24+	0 < 12	12 < 18	18 < 24	24+
*************** MECHANICAL ABNORMALITIES AND CYST FORMATION ***************								
ANGIECTASIS			1/1356				1/1136	
CONGESTION				1/1015				
CYST							1/1136	2/958
HEMORRHAGE			1/1356					
*************** INFLAMMATION AND REPAIR ***************								
FIBROSIS						1/199		
INFLAMMATION			2/1356					
*************** DEGENERATION, NECROSIS, AND DEPOSITION ***************								
AMYLOID DEPOSITION					1/124	5/199	8/1136	1/958
CELLULAR CHANGE, VACUOLAR				1/1015				
DEGENERATION			1/1356				1/1136	
FATTY CHANGE				1/1015				
LIPID DEPOSITION				1/1015				1/958
MINERALIZATION						1/199		
NECROSIS							1/1136	
*************** GROWTH ALTERATIONS, NON-NEOPLASTIC, HYPERTROPHY ***************								
HYPERTROPHY				1/1015			1/1136	

SEX:	FEMALE				MALE			
DIAGNOSIS / AGE IN MONTHS:	0 < 12	12 < 18	18 < 24	24+	0 < 12	12 < 18	18 < 24	24+
*************** GROWTH ALTERATIONS, NON-NEOPLASTIC, HYPERPLASIA ***************								
HEMATOPOIESIS			1/1356					
HYPERPLASIA		2/103	38/1356	101/1015 *		1/199	60/1136*	69/958*
HYPERPLASIA, NODULAR			5/1356	2/1015			1/1136	1/958
*************** GROWTH ALTERATIONS, NON-NEOPLASTIC, ATROPHY ***************								
ATROPHY			1/1356	15/1015				
*************** GROWTH ALTERATIONS, NEOPLASTIC, BENIGN ***************								
ADENOMA, CORTICAL			7/1356	4/1015			5/1136	9/958
*************** GROWTH ALTERATIONS, NEOPLASTIC, MALIGNANT, PRIMARY ***************								
CARCINOMA								1/958
CARCINOMA, CORTICAL				1/1015			1/1136	1/958
PHEOCHROMOCYTOMA, MALIGNANT		1/103						
*************** GROWTH ALTERATIONS, NEOPLASTIC, MALIGNANT, PRIMARY SITE NOT DETERMINED ***************								
CARCINOMA, ACINAR CELL								1/958
LEIOMYOSARCOMA			1/1356					
LYMPHOMA, LYMPHOCYTIC			1/1356					

*VARIABILITY AMONG DATA SOURCES SIGNIFICANT AT THE 1% SIGNIFICANCE LEVEL

ORGAN SYSTEM: ENDOCRINE SYSTEM   ORGAN: ADRENAL GLAND   STRAIN: B6C3F1

SEX:	FEMALE				MALE			
DIAGNOSIS / AGE IN MONTHS:	0 < 12	12 < 18	18 < 24	24+	0 < 12	12 < 18	18 < 24	24+
*************** GROWTH ALTERATIONS, NEOPLASTIC, NOT DETERMINED ***************								
PHEOCHROMOCYTOMA		1/103	3/1356	4/1015			2/1136	3/958

SEX:	FEMALE				MALE			
DIAGNOSIS / AGE IN MONTHS:	0 < 12	12 < 18	18 < 24	24+	0 < 12	12 < 18	18 < 24	24+
*************** MECHANICAL ABNORMALITIES AND CYST FORMATION ***************								
ANGIECTASIS					1/131			
HEMORRHAGE			1/192					
*************** DEGENERATION, NECROSIS, AND DEPOSITION ***************								
AMYLOID DEPOSITION							1/161	8/102
NECROSIS	1/176							
PIGMENT DEPOSITION			1/192	2/75				
*************** GROWTH ALTERATIONS, NON-NEOPLASTIC, HYPERPLASIA ***************								
HEMATOPOIESIS		1/367	2/192	1/75				
HYPERPLASIA						1/300		1/102
*************** GROWTH ALTERATIONS, NEOPLASTIC, MALIGNANT, PRIMARY ***************								
CARCINOMA, CORTICAL				1/75				
*************** GROWTH ALTERATIONS, NEOPLASTIC, MALIGNANT, PRIMARY SITE NOT DETERMINED ***************								
LYMPHOMA, RETICULUM CELL		1/367						

SEX:	FEMALE				MALE			
DIAGNOSIS / AGE IN MONTHS:	0 < 12	12 < 18	18 < 24	24+	0 < 12	12 < 18	18 < 24	24+
*************** MECHANICAL ABNORMALITIES AND CYST FORMATION ***************								
ANGIECTASIS			3/327	26/1575			1/437	45/1743
CONGESTION			1/327	9/1575				2/1743
CYST				3/1575				3/1743
CYST, HEMORRHAGIC				4/1575			1/437	
DILATATION, SINUSOIDAL								1/1743
HEMORRHAGE	1/26		2/327	6/1575		2/101	1/437	8/1743
INFARCT			1/327					
THROMBUS				1/1575				
*************** INFLAMMATION AND REPAIR ***************								
FIBROSIS							1/437	
INFLAMMATION, NONSUPPURATIVE				1/1575		1/101		1/1743
*************** DEGENERATION, NECROSIS, AND DEPOSITION ***************								
CELLULAR CHANGE, VACUOLAR				1/1575				4/1743
DEGENERATION	1/26		6/327	28/1575		1/101		21/1743
DEGENERATION, CYSTIC			1/327	5/1575				
FATTY CHANGE			2/327	19/1575	5/101	5/101	6/437	32/1743
HEMOSIDERIN DEPOSITION								1/1743
LIPID DEPOSITION			4/327	12/1575			1/437	6/1743

DIAGNOSIS	FEMALE 0 < 12	FEMALE 12 < 18	FEMALE 18 < 24	FEMALE 24+	MALE 0 < 12	MALE 12 < 18	MALE 18 < 24	MALE 24+
MINERALIZATION								5/1743
NECROSIS			1/327	5/1575				8/1743
PIGMENT DEPOSITION				2/1575				
PIGMENT DEPOSITION, LIPOFUSCIN								2/1743
*************** GROWTH ALTERATIONS, NON-NEOPLASTIC, HYPERTROPHY ***************								
HYPERTROPHY				3/1575			1/437	5/1743
*************** GROWTH ALTERATIONS, NON-NEOPLASTIC, HYPERPLASIA ***************								
HEMATOPOIESIS								1/1743
HYPERPLASIA	1/26		6/327	78/1575		1/101	6/437	105/1743*
HYPERPLASIA, CORTICAL								1/1743
HYPERPLASIA, NODULAR			2/327	15/1575			1/437	28/1743
*************** GROWTH ALTERATIONS, NON-NEOPLASTIC, ATROPHY ***************								
ATROPHY				2/1575				2/1743
ATROPHY, PRESSURE								1/1743
*************** GROWTH ALTERATIONS, NON-NEOPLASTIC, METAPLASIA ***************								
METAPLASIA, OSSEOUS								1/1743

*VARIABILITY AMONG DATA SOURCES SIGNIFICANT AT THE 1% SIGNIFICANCE LEVEL

SEX:	FEMALE				MALE			
DIAGNOSIS / AGE IN MONTHS:	0 < 12	12 < 18	18 < 24	24+	0 < 12	12 < 18	18 < 24	24+
*************** GROWTH ALTERATIONS, NEOPLASTIC, BENIGN ***************								
ADENOMA, CORTICAL			2/327	20/1575			1/437	14/1743
ANGIOLIPOMA				1/1575				
ANGIOMA								1/1743
GANGLIONEUROMA			1/327	2/1575				2/1743
*************** GROWTH ALTERATIONS, NEOPLASTIC, MALIGNANT, PRIMARY ***************								
CARCINOMA, CORTICAL				3/1575			2/437	8/1743
PHEOCHROMOCYTOMA, MALIGNANT				4/1575		1/101	2/437	12/1743
*************** GROWTH ALTERATIONS, NEOPLASTIC, MALIGNANT, METASTATIC ***************								
MESOTHELIOMA							1/437	1/1743
*************** GROWTH ALTERATIONS, NEOPLASTIC, MALIGNANT, PRIMARY SITE NOT DETERMINED ***************								
HISTIOCYTOMA, MALIGNANT							1/437	
LEUKEMIA	1/26		1/327				1/437	6/1743
LEUKEMIA, GRANULOCYTIC				1/1575				3/1743
LEUKEMIA, LYMPHOBLASTIC				5/1575				1/1743
LEUKEMIA, LYMPHOCYTIC				1/1575			1/437	1/1743
LEUKEMIA, STEM CELL							1/437	
NEUROFIBROSARCOMA							1/437	

SEX:	FEMALE				MALE			
DIAGNOSIS / AGE IN MONTHS:	0 < 12	12 < 18	18 < 24	24+	0 < 12	12 < 18	18 < 24	24+
*************** GROWTH ALTERATIONS, NEOPLASTIC, NOT DETERMINED ***************								
GANGLIONEUROBLASTOMA				1/1575				
PHEOCHROMOCYTOMA			1/327	53/1575		1/101	20/437	200/1743*
SYMPATHICOBLASTOMA								3/1743

*VARIABILITY AMONG DATA SOURCES SIGNIFICANT AT THE 1% SIGNIFICANCE LEVEL

SEX:	FEMALE				MALE			
DIAGNOSIS / AGE IN MONTHS:	0 < 12	12 < 18	18 < 24	24+	0 < 12	12 < 18	18 < 24	24+
*************** DEVELOPMENTAL ANOMALIES ***************								
ACCESSORY STRUCTURE				1/336				
*************** MECHANICAL ABNORMALITIES AND CYST FORMATION ***************								
ANGIECTASIS		3/39	7/44	80/336 *			4/104	12/243
CONGESTION			1/44			1/64	1/104	
CYST								1/243
*************** DEGENERATION, NECROSIS, AND DEPOSITION ***************								
DEGENERATION	1/7	1/39	3/44	35/336 *		4/64	5/104	26/243
FATTY CHANGE				1/336				1/243
MINERALIZATION				2/336		1/64		
*************** GROWTH ALTERATIONS, NON-NEOPLASTIC, HYPERTROPHY ***************								
CYTOMEGALY				7/336			2/104	6/243
*************** GROWTH ALTERATIONS, NON-NEOPLASTIC, HYPERPLASIA ***************								
HYPERPLASIA				7/336		1/64	1/104	3/243
*************** GROWTH ALTERATIONS, NEOPLASTIC, BENIGN ***************								
ADENOMA, CORTICAL				4/336			1/104	7/243

*VARIABILITY AMONG DATA SOURCES SIGNIFICANT AT THE 1% SIGNIFICANCE LEVEL

SEX:	FEMALE				MALE			
DIAGNOSIS / AGE IN MONTHS:	0 < 12	12 < 18	18 < 24	24+	0 < 12	12 < 18	18 < 24	24+
*************** GROWTH ALTERATIONS, NEOPLASTIC, MALIGNANT, PRIMARY ***************								
CARCINOMA				2/336				
CARCINOMA, CORTICAL				1/336			1/104	3/243
*************** GROWTH ALTERATIONS, NEOPLASTIC, NOT DETERMINED ***************								
PHEOCHROMOCYTOMA				1/336		2/64		4/243

SEX:	FEMALE				MALE			
DIAGNOSIS / AGE IN MONTHS:	0 < 12	12 < 18	18 < 24	24+	0 < 12	12 < 18	18 < 24	24+
*************** DEVELOPMENTAL ANOMALIES ***************								
AGENESIS							1/392	
*************** MECHANICAL ABNORMALITIES AND CYST FORMATION ***************								
ANGIECTASIS			28/420*	1/240			2/392	
CONGESTION		1/60	25/420*	20/240*			7/392	9/216
CYST		1/60		1/240				
CYST, HEMORRHAGIC		2/60	7/420	24/240*				1/216
HEMORRHAGE		2/60	23/420*	22/240*		1/97	2/392	3/216
INFARCT							1/392	
THROMBUS			6/420	13/240*			1/392	
THROMBUS, ORGANIZED				3/240				
THROMBUS, SINUSOIDAL							1/392	
*************** INFLAMMATION AND REPAIR ***************								
ABSCESS						1/97		
INFLAMMATION, NONSUPPURATIVE	1/181							
OSSIFICATION								1/216
PERIADENITIS				1/240				

*VARIABILITY AMONG DATA SOURCES SIGNIFICANT AT THE 1% SIGNIFICANCE LEVEL

SEX:	FEMALE				MALE			
AGE IN MONTHS: / DIAGNOSIS	0 < 12	12 < 18	18 < 24	24+	0 < 12	12 < 18	18 < 24	24+
*************** DEGENERATION, NECROSIS, AND DEPOSITION ***************								
CELLULAR CHANGE, VACUOLAR					1/170			
DEGENERATION		4/60	15/420	29/240 *			17/392	25/216*
FATTY CHANGE			4/420	11/240			1/392	8/216
MINERALIZATION								3/216
NECROSIS				1/240				
*************** GROWTH ALTERATIONS, NON-NEOPLASTIC, HYPERTROPHY ***************								
HYPERTROPHY				3/240				
*************** GROWTH ALTERATIONS, NON-NEOPLASTIC, HYPERPLASIA ***************								
HEMATOPOIESIS	1/181		1/420	1/240				
HYPERPLASIA		5/60	21/420	50/240 *		3/97	21/392*	41/216*
HYPERPLASIA, MEDULLARY								2/216
*************** GROWTH ALTERATIONS, NON-NEOPLASTIC, ATROPHY ***************								
ATROPHY			1/420	6/240				1/216
*************** GROWTH ALTERATIONS, NEOPLASTIC, BENIGN ***************								
ADENOMA, CORTICAL		1/60	17/420	7/240		1/97	2/392	2/216

*VARIABILITY AMONG DATA SOURCES SIGNIFICANT AT THE 1% SIGNIFICANCE LEVEL

SEX:	FEMALE				MALE			
DIAGNOSIS / AGE IN MONTHS:	0 < 12	12 < 18	18 < 24	24+	0 < 12	12 < 18	18 < 24	24+
*************** GROWTH ALTERATIONS, NEOPLASTIC, MALIGNANT, PRIMARY ***************								
CARCINOMA				1/240				
CARCINOMA, CORTICAL								1/216
PHEOCHROMOCYTOMA, MALIGNANT			1/420					
*************** GROWTH ALTERATIONS, NEOPLASTIC, MALIGNANT, PRIMARY SITE NOT DETERMINED ***************								
LEUKEMIA			1/420					
LEUKEMIA, HISTIOCYTIC							1/392	
LEUKEMIA, LYMPHOCYTIC			1/420			1/97	1/392	
LYMPHOMA			1/420					
LYMPHOMA, RETICULUM CELL			1/420				1/392	
*************** GROWTH ALTERATIONS, NEOPLASTIC, NOT DETERMINED ***************								
PHEOCHROMOCYTOMA			1/420	4/240			4/392	14/216
SYMPATHICOBLASTOMA				2/240		1/97	2/392	6/216

ORGAN SYSTEM: ENDOCRINE SYSTEM ORGAN: PANCREAS STRAIN: BALB/C

SEX:	FEMALE				MALE			
DIAGNOSIS / AGE IN MONTHS:	0 < 12	12 < 18	18 < 24	24+	0 < 12	12 < 18	18 < 24	24+
*************** MECHANICAL ABNORMALITIES AND CYST FORMATION ***************								
CYST	1/212	1/369		1/71	1/171			
EDEMA			1/193					
HEMORRHAGE		1/369						
THROMBUS		1/369						
*************** INFLAMMATION AND REPAIR ***************								
FIBROSIS			1/193					
INFLAMMATION		1/369	1/193					
*************** DEGENERATION, NECROSIS, AND DEPOSITION ***************								
CELLULAR CHANGE, VACUOLAR				1/71				
FATTY CHANGE						1/314		1/94
NECROSIS						1/314		
NECROSIS, FAT							1/147	
*************** GROWTH ALTERATIONS, NON-NEOPLASTIC, ATROPHY ***************								
ATROPHY		1/369	1/193	1/71				
*************** GROWTH ALTERATIONS, NEOPLASTIC, BENIGN ***************								
ADENOMA			1/193					

SEX:	FEMALE				MALE			
DIAGNOSIS / AGE IN MONTHS:	0 < 12	12 < 18	18 < 24	24+	0 < 12	12 < 18	18 < 24	24+
*************** GROWTH ALTERATIONS, NEOPLASTIC, MALIGNANT, PRIMARY SITE NOT DETERMINED ***************								
LYMPHOMA	2/212	2/369	7/193				1/147	3/94
LYMPHOMA, RETICULUM CELL		2/369	6/193	16/71		1/314	1/147	3/94

SEX:	FEMALE				MALE			
DIAGNOSIS / AGE IN MONTHS:	0 < 12	12 < 18	18 < 24	24+	0 < 12	12 < 18	18 < 24	24+
*************** MECHANICAL ABNORMALITIES AND CYST FORMATION ***************								
CYST			5/1356	10/1015			1/1136	5/958
CYST, EMBRYONAL DUCT			1/1356					
CYST, HEMORRHAGIC				1/1015				
DILATATION			3/1356	5/1015				1/958
HEMORRHAGE								1/958
*************** INFLAMMATION AND REPAIR ***************								
FIBROSIS								2/958
INFLAMMATION		2/103	14/1356	6/1015			18/1136	2/958
INFLAMMATION, FAT								1/958
INFLAMMATION, GRANULOMATOUS		1/103		1/1015				
INFLAMMATION, NONSUPPURATIVE		1/103	1/1356					1/958
INFLAMMATION, SUPPURATIVE				1/1015				
PERIVASCULITIS/VASCULITIS		1/103	1/1356				1/1136	2/958
*************** DEGENERATION, NECROSIS, AND DEPOSITION ***************								
AMYLOID DEPOSITION						5/199	3/1136	
DEGENERATION							1/1136	1/958
FATTY CHANGE							1/1136	
NECROSIS								2/958

SEX:	FEMALE				MALE			
DIAGNOSIS / AGE IN MONTHS:	0 < 12	12 < 18	18 < 24	24+	0 < 12	12 < 18	18 < 24	24+
NECROSIS, FAT			2/1356	1/1015			1/1136	
*************** GROWTH ALTERATIONS, NON-NEOPLASTIC, HYPERTROPHY ***************								
HYPERTROPHY							2/1136	
*************** GROWTH ALTERATIONS, NON-NEOPLASTIC, HYPERPLASIA ***************								
HYPERPLASIA		1/103		3/1015		1/199	15/1136	32/958
HYPERPLASIA, ADENOMATOUS							1/1136	
*************** GROWTH ALTERATIONS, NON-NEOPLASTIC, ATROPHY ***************								
ATROPHY			9/1356	10/1015			9/1136	8/958
*************** GROWTH ALTERATIONS, NEOPLASTIC, BENIGN ***************								
ADENOMA, ISLET CELL			2/1356	1/1015			4/1136	10/958
FIBROMA								1/958
*************** GROWTH ALTERATIONS, NEOPLASTIC, MALIGNANT, PRIMARY ***************								
CARCINOMA	1/67							
CARCINOMA, ACINAR CELL								1/958
CARCINOMA, ISLET CELL			2/1356	1/1015				

SEX:	FEMALE				MALE			
AGE IN MONTHS:	0 < 12	12 < 18	18 < 24	24+	0 < 12	12 < 18	18 < 24	24+
DIAGNOSIS								
*************** GROWTH ALTERATIONS, NEOPLASTIC, MALIGNANT, PRIMARY SITE NOT DETERMINED ***************								
ANGIOSARCOMA						1/199		
LYMPHOMA			1/1356					
*************** PARASITISM ***************								
PARASITISM								1/958

SEX:	FEMALE				MALE			
DIAGNOSIS / AGE IN MONTHS:	0 < 12	12 < 18	18 < 24	24+	0 < 12	12 < 18	18 < 24	24+
*************** MECHANICAL ABNORMALITIES AND CYST FORMATION ***************								
CYST	1/176		1/192			1/300	3/161	
*************** INFLAMMATION AND REPAIR ***************								
FIBROSIS				1/75			2/161	
INFLAMMATION	1/176	2/367	3/192	3/75	1/131	1/300	3/161	2/102
INFLAMMATION, NONSUPPURATIVE		3/367	3/192	1/75		3/300		5/102
PERIVASCULITIS/VASCULITIS								1/102
*************** DEGENERATION, NECROSIS, AND DEPOSITION ***************								
FATTY CHANGE								1/102
INFARCT		1/367						
NECROSIS		1/367			2/131			
NECROSIS, FAT				1/75				
*************** GROWTH ALTERATIONS, NON-NEOPLASTIC, HYPERPLASIA ***************								
HYPERPLASIA, INTRADUCTAL			1/192	1/75		1/300		
*************** GROWTH ALTERATIONS, NON-NEOPLASTIC, ATROPHY ***************								
ATROPHY			1/192	1/75	1/131	2/300		3/102

SEX:	FEMALE				MALE			
AGE IN MONTHS:	0 < 12	12 < 18	18 < 24	24+	0 < 12	12 < 18	18 < 24	24+
DIAGNOSIS								
*************** GROWTH ALTERATIONS, NEOPLASTIC, BENIGN ***************								
ADENOMA								1/102
*************** GROWTH ALTERATIONS, NEOPLASTIC, MALIGNANT, PRIMARY SITE NOT DETERMINED ***************								
LYMPHOMA		1/367		3/75		1/300	1/161	2/102
LYMPHOMA, RETICULUM CELL		2/367	3/192	12/75			2/161	5/102

SEX:	FEMALE				MALE			
DIAGNOSIS / AGE IN MONTHS:	0 < 12	12 < 18	18 < 24	24+	0 < 12	12 < 18	18 < 24	24+
*************** MECHANICAL ABNORMALITIES AND CYST FORMATION ***************								
CYST				1/1575			1/437	
DILATATION								1/1743
ECTASIA, DUCT					1/101	1/101		14/1743
EDEMA				1/1575				1/1743
*************** INFLAMMATION AND REPAIR ***************								
FIBROSIS			2/327	8/1575			1/437	13/1743
INFLAMMATION			6/327	35/1575		3/101	39/437*	100/1743 *
INFLAMMATION, NONSUPPURATIVE				1/1575		3/101		
PERIVASCULITIS/VASCULITIS				2/1575			1/437	20/1743
*************** DEGENERATION, NECROSIS, AND DEPOSITION ***************								
DEGENERATION				1/1575				
FATTY CHANGE								1/1743
NECROSIS, FAT			1/327				1/437	
*************** GROWTH ALTERATIONS, NON-NEOPLASTIC, HYPERTROPHY ***************								
HYPERTROPHY					1/101			

*VARIABILITY AMONG DATA SOURCES SIGNIFICANT AT THE 1% SIGNIFICANCE LEVEL

SEX:	FEMALE				MALE			
DIAGNOSIS / AGE IN MONTHS:	0 < 12	12 < 18	18 < 24	24+	0 < 12	12 < 18	18 < 24	24+
*************** GROWTH ALTERATIONS, NON-NEOPLASTIC, HYPERPLASIA ***************								
HYPERPLASIA	1/26		2/327	15/1575			5/437	42/1743
HYPERPLASIA, INTRADUCTAL							5/437	
HYPERPLASIA, NODULAR							2/437	2/1743
*************** GROWTH ALTERATIONS, NON-NEOPLASTIC, ATROPHY ***************								
ATROPHY			7/327	48/1575	2/101	4/101	32/437*	113/1743 *
*************** GROWTH ALTERATIONS, NEOPLASTIC BENIGN ***************								
ADENOMA							2/437	11/1743
ADENOMA, ACINAR CELL							1/437	4/1743
ADENOMA, ISLET CELL			1/327	15/1575			15/437	69/1743
*************** GROWTH ALTERATIONS, NEOPLASTIC, MALIGNANT, PRIMARY ***************								
ADENOCARCINOMA, MUCINOUS							1/437	
CARCINOMA								1/1743
CARCINOMA, ISLET CELL			1/327	1/1575			1/437	6/1743
*************** GROWTH ALTERATIONS, NEOPLASTIC, MALIGNANT, METASTATIC ***************								
MESOTHELIOMA							2/437	3/1743

*VARIABILITY AMONG DATA SOURCES SIGNIFICANT AT THE 1% SIGNIFICANCE LEVEL

SEX:	FEMALE				MALE			
DIAGNOSIS / AGE IN MONTHS:	0 < 12	12 < 18	18 < 24	24+	0 < 12	12 < 18	18 < 24	24+
*************** GROWTH ALTERATIONS, NEOPLASTIC, MALIGNANT, PRIMARY SITE NOT DETERMINED ***************								
LEUKEMIA								3/1743
LEUKEMIA, GRANULOCYTIC								2/1743
LEUKEMIA, LYMPHOBLASTIC				2/1575				
LEUKEMIA, LYMPHOCYTIC				1/1575			1/437	
NEUROFIBROSARCOMA							1/437	
*************** GROWTH ALTERATIONS, NEOPLASTIC, NOT DETERMINED ***************								
ISLET CELL TUMOR			1/327					1/1743

SEX:	FEMALE				MALE			
DIAGNOSIS / AGE IN MONTHS:	0 < 12	12 < 18	18 < 24	24+	0 < 12	12 < 18	18 < 24	24+
*************** MECHANICAL ABNORMALITIES AND CYST FORMATION ***************								
DILATATION								2/243
THROMBUS							1/104	
*************** INFLAMMATION AND REPAIR ***************								
FIBROSIS				1/336				1/243
INFLAMMATION								1/243
PERIVASCULITIS/VASCULITIS			2/44	7/336			7/104	22/243
*************** DEGENERATION, NECROSIS, AND DEPOSITION ***************								
ARTERIOSCLEROSIS							4/104	
*************** GROWTH ALTERATIONS, NON-NEOPLASTIC, HYPERPLASIA ***************								
HYPERPLASIA				1/336				
*************** GROWTH ALTERATIONS, NON-NEOPLASTIC, ATROPHY ***************								
ATROPHY				1/336			1/104	5/243
*************** GROWTH ALTERATIONS, NEOPLASTIC, BENIGN ***************								
ADENOMA, ISLET CELL				5/336		2/64	1/104	7/243

SEX:	FEMALE				MALE			
DIAGNOSIS / AGE IN MONTHS:	0 < 12	12 < 18	18 < 24	24+	0 < 12	12 < 18	18 < 24	24+
*************** GROWTH ALTERATIONS, NEOPLASTIC, MALIGNANT, PRIMARY ***************								
CARCINOMA, ISLET CELL								1/243
*************** GROWTH ALTERATIONS, NEOPLASTIC, MALIGNANT, PRIMARY SITE NOT DETERMINED ***************								
ANGIOSARCOMA			1/44					
FIBROSARCOMA				1/336				
HISTIOCYTOMA, FIBROUS								1/243

ORGAN SYSTEM: ENDOCRINE SYSTEM   ORGAN: PANCREAS   STRAIN: SPRAGUE DAWLEY

SEX:	FEMALE				MALE			
AGE IN MONTHS: / DIAGNOSIS	0 < 12	12 < 18	18 < 24	24+	0 < 12	12 < 18	18 < 24	24+
*************** MECHANICAL ABNORMALITIES AND CYST FORMATION ***************								
EDEMA				2/240			1/392	1/216
THROMBUS								1/216
*************** INFLAMMATION AND REPAIR ***************								
ABSCESS							1/392	
FIBROSIS				4/240	1/170		3/392	11/216 *
INFILTRATION, FATTY	1/181		1/420	3/240	1/170			
INFLAMMATION		1/60	3/420	5/240	2/170	1/97	7/392	15/216 *
INFLAMMATION, EOSINOPHILIC					1/170			
INFLAMMATION, NONSUPPURATIVE	2/181			1/240	6/170		1/392	3/216
PERIVASCULITIS/VASCULITIS			2/420			3/97	3/392	4/216
*************** DEGENERATION, NECROSIS, AND DEPOSITION ***************								
ARTERIOSCLEROSIS				1/240			1/392	1/216
CAST, PROTEINACEOUS	1/181							
CELLULAR CHANGE, VACUOLAR					1/170			
DEGENERATION					1/170	1/97	1/392	1/216
FATTY CHANGE								1/216
FIBROCYSTIC CHANGE								2/216
HEMOSIDERIN DEPOSITION			1/420		2/170		1/392	1/216

*VARIABILITY AMONG DATA SOURCES SIGNIFICANT AT THE 1% SIGNIFICANCE LEVEL

SEX:	FEMALE				MALE			
DIAGNOSIS / AGE IN MONTHS:	0 < 12	12 < 18	18 < 24	24+	0 < 12	12 < 18	18 < 24	24+
INFARCT				1/240				
MINERALIZATION			1/420					3/216
MINERALIZATION, VASCULAR							1/392	1/216
NECROSIS		1/60			1/170			
NECROSIS, FAT			1/420					
*************** GROWTH ALTERATIONS, NON-NEOPLASTIC, HYPERTROPHY ***************								
HYPERTROPHY	3/181			1/240	3/170			
*************** GROWTH ALTERATIONS, NON-NEOPLASTIC, HYPERPLASIA ***************								
HYPERPLASIA	2/181			5/240	6/170	1/97	1/392	2/216
HYPERPLASIA, ISLET CELL							1/392	1/216
HYPERPLASIA, LYMPHOID	1/181							
HYPERPLASIA, NODULAR	1/181							
*************** GROWTH ALTERATIONS, NON-NEOPLASTIC, ATROPHY ***************								
ATROPHY		1/60		4/240		1/97	2/392	11/216*
*************** GROWTH ALTERATIONS, NEOPLASTIC, BENIGN ***************								
ADENOMA			1/420	3/240				3/216
ADENOMA, ISLET CELL			1/420	1/240			4/392	3/216

*VARIABILITY AMONG DATA SOURCES SIGNIFICANT AT THE 1% SIGNIFICANCE LEVEL

SEX:	FEMALE				MALE			
DIAGNOSIS / AGE IN MONTHS:	0 < 12	12 < 18	18 < 24	24+	0 < 12	12 < 18	18 < 24	24+
*************** GROWTH ALTERATIONS, NEOPLASTIC, MALIGNANT, PRIMARY ***************								
CARCINOMA, ISLET CELL				1/240			2/392	
*************** GROWTH ALTERATIONS, NEOPLASTIC, MALIGNANT, PRIMARY SITE NOT DETERMINED ***************								
LEIOMYOSARCOMA							1/392	
LEUKEMIA, LYMPHOCYTIC						1/97		
LYMPHOMA, RETICULUM CELL			1/420				1/392	
RHABDOMYOSARCOMA								1/216
SARCOMA, HISTIOCYTIC							1/392	
*************** GROWTH ALTERATIONS, NEOPLASTIC, NOT DETERMINED ***************								
ISLET CELL TUMOR								1/216

DIAGNOSIS	FEMALE 0 < 12	FEMALE 12 < 18	FEMALE 18 < 24	FEMALE 24+	MALE 0 < 12	MALE 12 < 18	MALE 18 < 24	MALE 24+
MECHANICAL ABNORMALITIES AND CYST FORMATION								
CYST							2/1136	1/958
DEGENERATION, NECROSIS, AND DEPOSITION								
AMYLOID DEPOSITION							1/1136	
GROWTH ALTERATIONS, NON-NEOPLASTIC, HYPERPLASIA								
HYPERPLASIA							1/1136	1/958
GROWTH ALTERATIONS, NEOPLASTIC, BENIGN								
ADENOMA				1/1015				

SEX:	FEMALE				MALE			
AGE IN MONTHS:	0 < 12	12 < 18	18 < 24	24+	0 < 12	12 < 18	18 < 24	24+
DIAGNOSIS								
*************** MECHANICAL ABNORMALITIES AND CYST FORMATION ***************								
CYST			1/192					1/102
*************** DEGENERATION, NECROSIS, AND DEPOSITION ***************								
AMYLOID DEPOSITION								6/102
*************** GROWTH ALTERATIONS, NON-NEOPLASTIC, HYPERPLASIA ***************								
HYPERPLASIA			1/192					
*************** GROWTH ALTERATIONS, NEOPLASTIC, BENIGN ***************								
ADENOMA				1/75				

SEX:	FEMALE				MALE			
DIAGNOSIS / AGE IN MONTHS:	0 < 12	12 < 18	18 < 24	24+	0 < 12	12 < 18	18 < 24	24+
*************** GROWTH ALTERATIONS, NON-NEOPLASTIC, HYPERPLASIA ***************								
HYPERPLASIA				3/1575		1/101		18/1743
HYPERPLASIA, ADENOMATOUS								1/1743
HYPERPLASIA, NODULAR				3/1575				2/1743
HYPERPLASIA, SECONDARY								1/1743
*************** GROWTH ALTERATIONS, NEOPLASTIC, BENIGN ***************								
ADENOMA							1/437	7/1743
*************** GROWTH ALTERATIONS, NEOPLASTIC, MALIGNANT, PRIMARY SITE NOT DETERMINED ***************								
LEUKEMIA, GRANULOCYTIC								1/1743

SEX:	FEMALE				MALE			
DIAGNOSIS / AGE IN MONTHS:	0 < 12	12 < 18	18 < 24	24+	0 < 12	12 < 18	18 < 24	24+
*************** GROWTH ALTERATIONS, NON-NEOPLASTIC, HYPERPLASIA ***************								
HYPERPLASIA			2/44	13/336		5/64	19/104	21/243
*************** GROWTH ALTERATIONS, NEOPLASTIC, BENIGN ***************								
ADENOMA		1/39					1/104	2/243

SEX:	FEMALE				MALE			
DIAGNOSIS / AGE IN MONTHS:	0 < 12	12 < 18	18 < 24	24+	0 < 12	12 < 18	18 < 24	24+
*************** DEGENERATION, NECROSIS, AND DEPOSITION ***************								
MINERALIZATION							1/392	
*************** GROWTH ALTERATIONS, NON-NEOPLASTIC, HYPERPLASIA ***************								
HYPERPLASIA			1/420	2/240		7/97	12/392	15/216 *
HYPERPLASIA, ADENOMATOUS			5/420	15/240*		1/97	7/392	13/216 *
*************** GROWTH ALTERATIONS, NEOPLASTIC, BENIGN ***************								
ADENOMA				1/240			1/392	3/216
ADENOMA, CHIEF CELL							1/392	
*************** GROWTH ALTERATIONS, NEOPLASTIC, MALIGNANT, PRIMARY SITE NOT DETERMINED ***************								
LEUKEMIA, LYMPHOCYTIC						1/97		

*VARIABILITY AMONG DATA SOURCES SIGNIFICANT AT THE 1% SIGNIFICANCE LEVEL

SEX:	FEMALE				MALE			
AGE IN MONTHS:	0 < 12	12 < 18	18 < 24	24+	0 < 12	12 < 18	18 < 24	24+
DIAGNOSIS								
*************** DEGENERATION, NECROSIS, AND DEPOSITION ***************								
MINERALIZATION							1/392	

SEX:	FEMALE				MALE			
DIAGNOSIS / AGE IN MONTHS:	0 < 12	12 < 18	18 < 24	24+	0 < 12	12 < 18	18 < 24	24+
*************** MECHANICAL ABNORMALITIES AND CYST FORMATION ***************								
CYST	1/212	7/369	7/193	5/71		5/314	G/147	2/94
*************** DEGENERATION, NECROSIS, AND DEPOSITION ***************								
CELLULAR CHANGE, VACUOLAR				1/71				
MINERALIZATION						1/314		
*************** GROWTH ALTERATIONS, NEOPLASTIC, MALIGNANT, PRIMARY SITE NOT DETERMINED ***************								
LYMPHOMA	3/212	3/369	1/193				1/147	
LYMPHOMA, RETICULUM CELL			1/193					

SEX:	FEMALE				MALE			
AGE IN MONTHS: / DIAGNOSIS	0 < 12	12 < 18	18 < 24	24+	0 < 12	12 < 18	18 < 24	24+
*************** MECHANICAL ABNORMALITIES AND CYST FORMATION ***************								
ANGIECTASIS			2/1356	4/1015				
CYST			2/1356				5/1136	2/958
HEMORRHAGE	1/67							
*************** GROWTH ALTERATIONS, NON-NEOPLASTIC, HYPERPLASIA ***************								
HYPERPLASIA			6/1356	5/1015			7/1136	
*************** GROWTH ALTERATIONS, NEOPLASTIC, BENIGN ***************								
ADENOMA	1/67		25/1356	15/1015			1/1136	2/958
ADENOMA, BASOPHIL				1/1015				
ADENOMA, CHROMOPHOBE		1/103	11/1356	29/1015				
*************** GROWTH ALTERATIONS, NEOPLASTIC, MALIGNANT, PRIMARY ***************								
CARCINOMA			1/1356	1/1015				
CARCINOMA, CHROMOPHOBE			1/1356				1/1136	

DIAGNOSIS	FEMALE 0 < 12	FEMALE 12 < 18	FEMALE 18 < 24	FEMALE 24+	MALE 0 < 12	MALE 12 < 18	MALE 18 < 24	MALE 24+
*************** MECHANICAL ABNORMALITIES AND CYST FORMATION ***************								
CONGESTION				2/75				
CYST		8/367	1/192	1/75	3/131	14/300	13/161	12/102
HEMORRHAGE			1/192	2/75				1/102
*************** INFLAMMATION AND REPAIR ***************								
INFLAMMATION				1/75				
*************** DEGENERATION, NECROSIS, AND DEPOSITION ***************								
PIGMENT DEPOSITION								1/102
*************** GROWTH ALTERATIONS, NEOPLASTIC, BENIGN ***************								
ADENOMA			18/192	11/75			1/161	1/102
*************** GROWTH ALTERATIONS, NEOPLASTIC, MALIGNANT, PRIMARY ***************								
ADENOCARCINOMA				1/75				
*************** GROWTH ALTERATIONS, NEOPLASTIC, MALIGNANT, PRIMARY SITE NOT DETERMINED ***************								
LYMPHOMA				1/75		1/300		
LYMPHOMA, RETICULUM CELL			1/192					

SEX:	FEMALE				MALE			
DIAGNOSIS / AGE IN MONTHS:	0 < 12	12 < 18	18 < 24	24+	0 < 12	12 < 18	18 < 24	24+
MECHANICAL ABNORMALITIES AND CYST FORMATION								
ANGIECTASIS			10/327	48/1575				14/1743
CONGESTION			1/327	16/1575	1/101	1/101		
CYST		1/24	1/327	64/1575			2/437	17/1743
CYST, COLLOID					1/101			
CYST, HEMORRHAGIC		1/24	2/327	16/1575		1/101		3/1743
HEMORRHAGE	1/26		8/327	9/1575			2/437	2/1743
INFLAMMATION AND REPAIR								
PERIVASCULITIS/VASCULITIS				1/1575				
DEGENERATION, NECROSIS, AND DEPOSITION								
HEMOSIDERIN DEPOSITION				4/1575				
MINERALIZATION				3/1575				
NECROSIS				1/1575				1/1743
GROWTH ALTERATIONS, NON-NEOPLASTIC, HYPERTROPHY								
HYPERTROPHY				1/1575				
GROWTH ALTERATIONS, NON-NEOPLASTIC, HYPERPLASIA								
HYPERPLASIA			3/327	26/1575			13/437	39/1743

SEX:	FEMALE				MALE			
DIAGNOSIS AGE IN MONTHS:	0 < 12	12 < 18	18 < 24	24+	0 < 12	12 < 18	18 < 24	24+
HYPERPLASIA, BASOPHILIC							1/437	1/1743
HYPERPLASIA, CHROMOPHOBE CELL		1/24	2/327	2/1575				13/1743
HYPERPLASIA, NODULAR							1/437	4/1743
NODULE								1/1743
*************** GROWTH ALTERATIONS, NEOPLASTIC, BENIGN ***************								
ADENOMA			43/327	257/1575*			23/437*	150/1743 *
ADENOMA, ACIDOPHIL			2/327	3/1575				4/1743
ADENOMA, BASOPHIL								2/1743
ADENOMA, CHROMOPHOBE	1/26	3/24	64/327	348/1575*			33/437*	187/1743 *
*************** GROWTH ALTERATIONS, NEOPLASTIC, MALIGNANT, PRIMARY ***************								
ADENOCARCINOMA			1/327	3/1575				
CARCINOMA		2/24	8/327	10/1575		1/101	3/437	9/1743
CARCINOMA, CHROMOPHOBE			2/327	5/1575				3/1743
*************** GROWTH ALTERATIONS, NEOPLASTIC, MALIGNANT, PRIMARY SITE NOT DETERMINED ***************								
LEUKEMIA								1/1743
LEUKEMIA, STEM CELL							1/437	

*VARIABILITY AMONG DATA SOURCES SIGNIFICANT AT THE 1% SIGNIFICANCE LEVEL

SEX:	FEMALE				MALE			
DIAGNOSIS / AGE IN MONTHS:	0 < 12	12 < 18	18 < 24	24+	0 < 12	12 < 18	18 < 24	24+
*************** MECHANICAL ABNORMALITIES AND CYST FORMATION ***************								
ANGIECTASIS				1/336				1/243
CONGESTION				1/336				
CYST			1/44	5/336				2/243
*************** DEGENERATION, NECROSIS, AND DEPOSITION ***************								
CELLULAR CHANGE, VACUOLAR				1/336				
*************** GROWTH ALTERATIONS, NON-NEOPLASTIC, HYPERPLASIA ***************								
HYPERPLASIA				1/336			1/104	2/243
HYPERPLASIA, CHROMOPHOBE CELL				10/336			1/104	3/243
*************** GROWTH ALTERATIONS, NEOPLASTIC, BENIGN ***************								
ADENOMA			1/44	2/336			1/104	4/243
ADENOMA, CHROMOPHOBE		1/39	10/44	134/336			12/104	45/243
*************** GROWTH ALTERATIONS, NEOPLASTIC, MALIGNANT, PRIMARY ***************								
CARCINOMA				2/336				
CARCINOMA, CHROMOPHOBE							2/104	

SEX:	FEMALE				MALE			
DIAGNOSIS / AGE IN MONTHS:	0 < 12	12 < 18	18 < 24	24+	0 < 12	12 < 18	18 < 24	24+
*************** MECHANICAL ABNORMALITIES AND CYST FORMATION ***************								
CONGESTION				1/240				
CYST				2/240				5/216
HEMORRHAGE			2/420	2/240				
INFARCT, HEMORRHAGIC				1/240				
*************** GROWTH ALTERATIONS, NON-NEOPLASTIC, HYPERTROPHY ***************								
HYPERTROPHY			2/420					
*************** GROWTH ALTERATIONS, NON-NEOPLASTIC, HYPERPLASIA ***************								
HYPERPLASIA			9/420	16/240			9/392	9/216
HYPERPLASIA, ADENOMATOUS				1/240				
HYPERPLASIA, CHROMOPHOBE CELL			4/420	1/240			2/392	5/216
HYPERPLASIA, EPITHELIAL				1/240				
*************** GROWTH ALTERATIONS, NEOPLASTIC, BENIGN ***************								
ADENOMA			3/420	18/240 *			2/392	
ADENOMA, ACIDOPHIL							2/392	
ADENOMA, CHROMOPHOBE	1/181	15/60	118/420*	92/240 *		2/97	40/392	41/216
EPENDYMOMA			1/420					

*VARIABILITY AMONG DATA SOURCES SIGNIFICANT AT THE 1% SIGNIFICANCE LEVEL

DIAGNOSIS	FEMALE 0 < 12	FEMALE 12 < 18	FEMALE 18 < 24	FEMALE 24+	MALE 0 < 12	MALE 12 < 18	MALE 18 < 24	MALE 24+
*************** GROWTH ALTERATIONS, NEOPLASTIC MALIGNANT, PRIMARY ***************								
ADENOCARCINOMA								1/216
CARCINOMA, CHROMOPHOBE		2/60	13/420	4/240			15/392	
*************** GROWTH ALTERATIONS, NEOPLASTIC, MALIGNANT, PRIMARY SITE NOT DETERMINED ***************								
LEUKEMIA, LYMPHOCYTIC						1/97		
LYMPHOMA			1/420		1/170			

SEX:	FEMALE				MALE			
DIAGNOSIS / AGE IN MONTHS:	0 < 12	12 < 18	18 < 24	24+	0 < 12	12 < 18	18 < 24	24+
*************** MECHANICAL ABNORMALITIES AND CYST FORMATION ***************								
CYST	1/212	1/369	1/193	1/71		3/314	1/147	5/94
*************** GROWTH ALTERATIONS, NON-NEOPLASTIC, ATROPHY ***************								
ATROPHY		1/369						
*************** GROWTH ALTERATIONS, NEOPLASTIC, BENIGN ***************								
ADENOMA						1/314		
ADENOMA, PAPILLARY								1/94
*************** GROWTH ALTERATIONS, NEOPLASTIC, MALIGNANT, PRIMARY SITE NOT DETERMINED ***************								
LYMPHOMA			1/193					
LYMPHOMA, RETICULUM CELL			1/193					

SEX:	FEMALE				MALE			
DIAGNOSIS / AGE IN MONTHS:	0 < 12	12 < 18	18 < 24	24+	0 < 12	12 < 18	18 < 24	24+
*************** MECHANICAL ABNORMALITIES AND CYST FORMATION ***************								
CYST								1/958
CYST, COLLOID							1/1136	1/958
CYST, FOLLICULAR			24/1356	5/1015	1/124	1/199	19/1136	6/958
CYST, ULTIMOBRANCHIAL DUCT				1/1015				
*************** INFLAMMATION AND REPAIR ***************								
INFLAMMATION			3/1356	3/1015			2/1136	
INFLAMMATION, NONSUPPURATIVE							1/1136	
PERIVASCULITIS/VASCULITIS						1/199		
*************** DEGENERATION, NECROSIS, AND DEPOSITION ***************								
AMYLOID DEPOSITION						2/199	3/1136	
*************** GROWTH ALTERATIONS, NON-NEOPLASTIC, HYPERTROPHY ***************								
HYPERTROPHY								1/958
*************** GROWTH ALTERATIONS, NON-NEOPLASTIC, HYPERPLASIA ***************								
HYPERPLASIA, ADENOMATOUS			1/1356					
HYPERPLASIA, C CELL			2/1356					
HYPERPLASIA, CYSTIC			1/1356	1/1015				

SEX:	FEMALE				MALE			
DIAGNOSIS / AGE IN MONTHS:	0 < 12	12 < 18	18 < 24	24+	0 < 12	12 < 18	18 < 24	24+
HYPERPLASIA, FOLLICULAR CELL			5/1356	22/1015			2/1136	13/958
HYPERPLASIA, PAPILLARY			3/1356				1/1136	
*************** GROWTH ALTERATIONS, NON-NEOPLASTIC, ATROPHY ***************								
ATROPHY								1/958
*************** GROWTH ALTERATIONS, NEOPLASTIC, BENIGN ***************								
ADENOMA				2/1015				
ADENOMA, FOLLICULAR CELL			12/1356	24/1015			6/1136	6/958
ADENOMA, PAPILLARY				3/1015				2/958
*************** GROWTH ALTERATIONS, NEOPLASTIC, MALIGNANT, PRIMARY ***************								
ADENOCARCINOMA							3/1136	
ADENOCARCINOMA, PAPILLARY			1/1356					
CARCINOMA, FOLLICULAR CELL			1/1356	4/1015			2/1136	2/958

SEX:	FEMALE				MALE			
AGE IN MONTHS: / DIAGNOSIS	0 < 12	12 < 18	18 < 24	24+	0 < 12	12 < 18	18 < 24	24+
*************** MECHANICAL ABNORMALITIES AND CYST FORMATION ***************								
CYST			1/192	2/75		1/300		1/102
*************** INFLAMMATION AND REPAIR ***************								
INFLAMMATION, NONSUPPURATIVE								1/102
*************** DEGENERATION, NECROSIS, AND DEPOSITION ***************								
AMYLOID DEPOSITION							1/161	11/102
*************** GROWTH ALTERATIONS, NEOPLASTIC, BENIGN ***************								
ADENOMA			2/192	1/75				

SEX:	FEMALE				MALE			
DIAGNOSIS / AGE IN MONTHS:	0 < 12	12 < 18	18 < 24	24+	0 < 12	12 < 18	18 < 24	24+
*************** MECHANICAL ABNORMALITIES AND CYST FORMATION ***************								
CYST				1/1575			1/437	2/1743
CYST, COLLOID				1/1575				
CYST, FOLLICULAR			3/327	4/1575		5/101 *		12/1743
CYST, THYROGLOSSAL DUCT								1/1743
*************** INFLAMMATION AND REPAIR ***************								
FIBROSIS				2/1575				
INFLAMMATION				4/1575				1/1743
INFLAMMATION, NECROTIZING				1/1575				
*************** DEGENERATION, NECROSIS, AND DEPOSITION ***************								
PIGMENT DEPOSITION							1/437	9/1743
*************** GROWTH ALTERATIONS, NON-NEOPLASTIC, HYPERPLASIA ***************								
HYPERPLASIA				4/1575			1/437	8/1743
HYPERPLASIA, ADENOMATOUS								5/1743
HYPERPLASIA, C CELL			13/327	172/1575*			9/437	122/1743 *
HYPERPLASIA, CYSTIC	1/26						1/437	1/1743
HYPERPLASIA, FOLLICULAR CELL				5/1575			1/437	6/1743
HYPERPLASIA, NODULAR								1/1743

*VARIABILITY AMONG DATA SOURCES SIGNIFICANT AT THE 1% SIGNIFICANCE LEVEL

SEX:	FEMALE				MALE			
AGE IN MONTHS: / DIAGNOSIS	0 < 12	12 < 18	18 < 24	24+	0 < 12	12 < 18	18 < 24	24+
*************** GROWTH ALTERATIONS, NON-NEOPLASTIC, ATROPHY ***************								
ATROPHY								2/1743
*************** GROWTH ALTERATIONS, NON-NEOPLASTIC, METAPLASIA ***************								
METAPLASIA, SQUAMOUS								1/1743
*************** GROWTH ALTERATIONS, NEOPLASTIC, BENIGN ***************								
ADENOMA			1/327	2/1575				8/1743
ADENOMA, C CELL			3/327	65/1575		1/101	7/437	97/1743*
ADENOMA, FOLLICULAR CELL			1/327	5/1575				8/1743
CYSTADENOMA				1/1575				1/1743
*************** GROWTH ALTERATIONS, NEOPLASTIC, MALIGNANT, PRIMARY ***************								
ADENOCARCINOMA				2/1575				12/1743
ADENOCARCINOMA, PAPILLARY								2/1743
CARCINOMA								1/1743
CARCINOMA, C CELL			1/327	31/1575		1/101	4/437	34/1743
CARCINOMA, FOLLICULAR CELL			3/327	5/1575			1/437	15/1743
CARCINOMA, PAPILLARY			1/327				1/437	
CARCINOMA, UNDIFFERENTIATED			1/327					
CYSTADENOCARCINOMA, PAPILLARY				1/1575				

*VARIABILITY AMONG DATA SOURCES SIGNIFICANT AT THE 1% SIGNIFICANCE LEVEL

SEX:	FEMALE				MALE			
DIAGNOSIS / AGE IN MONTHS:	0 < 12	12 < 18	18 < 24	24+	0 < 12	12 < 18	18 < 24	24+
*************** GROWTH ALTERATIONS, NEOPLASTIC, MALIGNANT, PRIMARY SITE NOT DETERMINED ***************								
LEUKEMIA, GRANULOCYTIC								1/1743

SEX:	FEMALE				MALE			
DIAGNOSIS / AGE IN MONTHS:	0 < 12	12 < 18	18 < 24	24+	0 < 12	12 < 18	18 < 24	24+
*************** MECHANICAL ABNORMALITIES AND CYST FORMATION ***************								
CYST								1/243
CYST, COLLOID								1/243
CYST, FOLLICULAR			2/44	11/336			2/104	13/243
CYST, ULTIMOBRANCHIAL DUCT						1/64	2/104	2/243
*************** INFLAMMATION AND REPAIR ***************								
INFLAMMATION							1/104	
*************** DEGENERATION, NECROSIS, AND DEPOSITION ***************								
DEGENERATION, CYSTIC				1/336				
*************** GROWTH ALTERATIONS, NON-NEOPLASTIC, HYPERPLASIA ***************								
HYPERPLASIA, C CELL		1/39		32/336*		2/64	3/104	21/243*
HYPERPLASIA, FOLLICULAR CELL				6/336			4/104	8/243
*************** GROWTH ALTERATIONS, NON-NEOPLASTIC, ATROPHY ***************								
ATROPHY				1/336				
*************** GROWTH ALTERATIONS, NEOPLASTIC, BENIGN ***************								
ADENOMA, C CELL		1/39		20/336				9/243

*VARIABILITY AMONG DATA SOURCES SIGNIFICANT AT THE 1% SIGNIFICANCE LEVEL

SEX:	FEMALE				MALE			
DIAGNOSIS / AGE IN MONTHS:	0 < 12	12 < 18	18 < 24	24+	0 < 12	12 < 18	18 < 24	24+
ADENOMA, FOLLICULAR CELL			1/44	7/336			3/104	20/243
*************** GROWTH ALTERATIONS, NEOPLASTIC, MALIGNANT, PRIMARY ***************								
CARCINOMA, C CELL		1/39	1/44	9/336				3/243
CARCINOMA, FOLLICULAR CELL				7/336			2/104	13/243
*************** GROWTH ALTERATIONS, NEOPLASTIC, MALIGNANT, PRIMARY SITE NOT DETERMINED ***************								
NEUROFIBROSARCOMA								1/243

SEX	FEMALE				MALE			
DIAGNOSIS / AGE IN MONTHS	0 < 12	12 < 18	18 < 24	24+	0 < 12	12 < 18	18 < 24	24+
*************** MECHANICAL ABNORMALITIES AND CYST FORMATION ***************								
CYST					2/170			
CYST, FOLLICULAR							1/392	
CYST, THYROGLOSSAL DUCT		1/60	2/420	1/240			3/392	
CYST, ULTIMOBRANCHIAL DUCT			5/420	2/240		1/97	2/392	1/216
*************** INFLAMMATION AND REPAIR ***************								
INFLAMMATION, NONSUPPURATIVE			1/420		1/170			
PERIVASCULITIS/VASCULITIS								1/216
*************** DEGENERATION, NECROSIS, AND DEPOSITION ***************								
PIGMENT DEPOSITION					1/170			
*************** GROWTH ALTERATIONS, NON-NEOPLASTIC, HYPERPLASIA ***************								
HYPERPLASIA								3/216
HYPERPLASIA, NODULAR					1/170			
*************** GROWTH ALTERATIONS, NON-NEOPLASTIC, ATROPHY ***************								
ATROPHY							2/392	

SEX:	FEMALE				MALE			
DIAGNOSIS / AGE IN MONTHS:	0 < 12	12 < 18	18 < 24	24+	0 < 12	12 < 18	18 < 24	24+
*************** GROWTH ALTERATIONS, NEOPLASTIC, BENIGN ***************								
ADENOMA				3/240				
ADENOMA, C CELL			17/420	11/240		1/97	6/392	9/216
ADENOMA, FOLLICULAR CELL				1/240				1/216
ADENOMA, HURTHLE CELL							2/392	
CYSTADENOMA		1/60						
*************** GROWTH ALTERATIONS, NEOPLASTIC, MALIGNANT, PRIMARY ***************								
CARCINOMA			1/420					
CARCINOMA, C CELL			1/420					1/216
CARCINOMA, HURTHLE CELL						1/97		
*************** GROWTH ALTERATIONS, NEOPLASTIC, MALIGNANT, METASTATIC ***************								
RHABDOMYOSARCOMA								1/216
*************** GROWTH ALTERATIONS, NEOPLASTIC, MALIGNANT, PRIMARY SITE NOT DETERMINED ***************								
LEUKEMIA, LYMPHOCYTIC						1/97		
LYMPHOMA			1/420					
LYMPHOMA, RETICULUM CELL			1/420					

SEX:	FEMALE				MALE			
DIAGNOSIS / AGE IN MONTHS:	0 < 12	12 < 18	18 < 24	24+	0 < 12	12 < 18	18 < 24	24+
*************** INFLAMMATION AND REPAIR ***************								
INFLAMMATION, SUPPURATIVE				1/1015				
*************** DEGENERATION, NECROSIS, AND DEPOSITION ***************								
CHOLESTEROL DEPOSITION				1/1015				

SEX:	FEMALE				MALE			
DIAGNOSIS / AGE IN MONTHS:	0 < 12	12 < 18	18 < 24	24+	0 < 12	12 < 18	18 < 24	24+
***************								
MECHANICAL ABNORMALITIES AND CYST FORMATION								
***************								
CYST				4/1575				
***************								
INFLAMMATION AND REPAIR								
***************								
INFLAMMATION				2/1575				
POLYP, INFLAMMATORY				1/1575				
***************								
GROWTH ALTERATIONS, NON-NEOPLASTIC, HYPERPLASIA								
***************								
ACANTHOSIS				1/1575				
HYPERPLASIA, BASAL CELL				1/1575				

ORGAN SYSTEM: GENITAL SYSTEM; FEMALES ORGAN: CERVIX STRAIN: SPRAGUE DAWLEY

SEX:	FEMALE				MALE			
AGE IN MONTHS:	0 < 12	12 < 18	18 < 24	24+	0 < 12	12 < 18	18 < 24	24+
DIAGNOSIS								
*************** INFLAMMATION AND REPAIR ***************								
INFLAMMATION			1/420					
*************** GROWTH ALTERATIONS, NEOPLASTIC, BENIGN ***************								
PAPILLOMA, SQUAMOUS CELL			1/420					

SEX:	FEMALE				MALE			
DIAGNOSIS / AGE IN MONTHS:	0 < 12	12 < 18	18 < 24	24+	0 < 12	12 < 18	18 < 24	24+
*************** INFLAMMATION AND REPAIR ***************								
ABSCESS				1/1575				
INFLAMMATION, GRANULOMATOUS				1/1575				
*************** GROWTH ALTERATIONS, NON-NEOPLASTIC, METAPLASIA ***************								
METAPLASIA, SQUAMOUS			1/327					
*************** GROWTH ALTERATIONS, NEOPLASTIC, BENIGN ***************								
ADENOMA				3/1575				
PAPILLOMA, SQUAMOUS CELL				1/1575				
*************** GROWTH ALTERATIONS, NEOPLASTIC, MALIGNANT, PRIMARY ***************								
CARCINOMA			3/327	2/1575				

ORGAN SYSTEM: GENITAL SYSTEM; FEMALES   ORGAN: CLITORIS   STRAIN: OSBORNE MENDEL

SEX:	FEMALE				MALE			
AGE IN MONTHS: / DIAGNOSIS	0 < 12	12 < 18	18 < 24	24+	0 < 12	12 < 18	18 < 24	24+
*************** GROWTH ALTERATIONS, NEOPLASTIC, MALIGNANT, PRIMARY ***************								
ADENOCARCINOMA				1/336				

SEX:	FEMALE				MALE			
DIAGNOSIS / AGE IN MONTHS:	0 < 12	12 < 18	18 < 24	24+	0 < 12	12 < 18	18 < 24	24+
*************** DEVELOPMENTAL ANOMALIES ***************								
HAMARTOMA				1/71				
*************** MECHANICAL ABNORMALITIES AND CYST FORMATION ***************								
CONGESTION				2/71				
CYST	2/212	18/369	25/193	12/71				
HEMORRHAGE		2/369	3/193	5/71				
THROMBUS		1/369	1/193	1/71				
*************** INFLAMMATION AND REPAIR ***************								
FIBROSIS			1/193					
INFLAMMATION		1/369						
*************** DEGENERATION, NECROSIS, AND DEPOSITION ***************								
PIGMENT DEPOSITION	1/212	2/369	8/193	18/71				
*************** GROWTH ALTERATIONS, NON-NEOPLASTIC, HYPERPLASIA ***************								
HYPERPLASIA		1/369						
*************** GROWTH ALTERATIONS, NON-NEOPLASTIC, ATROPHY ***************								
ATROPHY			1/193	3/71				

SEX:	FEMALE				MALE			
DIAGNOSIS / AGE IN MONTHS:	0 < 12	12 < 18	18 < 24	24+	0 < 12	12 < 18	18 < 24	24+
*************** GROWTH ALTERATIONS, NEOPLASTIC, BENIGN ***************								
ADENOMA		1/369	1/193	1/71				
ANGIOMA			5/193	5/71				
LUTEOMA			1/193	2/71				
*************** GROWTH ALTERATIONS, NEOPLASTIC, MALIGNANT, PRIMARY SITE NOT DETERMINED ***************								
LYMPHOMA	4/212	2/369	6/193					
LYMPHOMA, RETICULUM CELL			5/193	3/71				
*************** GROWTH ALTERATIONS, NEOPLASTIC, NOT DETERMINED ***************								
GRANULOSA CELL TUMOR		2/369	3/193	1/71				

SEX:	FEMALE				MALE			
AGE IN MONTHS: / DIAGNOSIS	0 < 12	12 < 18	18 < 24	24+	0 < 12	12 < 18	18 < 24	24+
*************** MECHANICAL ABNORMALITIES AND CYST FORMATION ***************								
ANGIECTASIS			1/1356					
CYST		6/103	109/1356*	107/1015*				
CYST, FOLLICULAR		1/103	27/1356	11/1015				
CYST, HEMORRHAGIC			2/1356	14/1015				
CYST, LUTEAL			1/1356					
CYST, PARAOVARIAN			7/1356	21/1015				
HEMORRHAGE				1/1015				
POLYCYSTIC OVARY				2/1015				
*************** INFLAMMATION AND REPAIR ***************								
ABSCESS			9/1356	4/1015				
INFLAMMATION		1/103	52/1356	6/1015				
INFLAMMATION, NONSUPPURATIVE			1/1356	2/1015				
INFLAMMATION, SUPPURATIVE		1/103	34/1356	5/1015				
PERIVASCULITIS/VASCULITIS			2/1356					
*************** DEGENERATION, NECROSIS, AND DEPOSITION ***************								
CORPORA AMYLACEA				1/1015				
CYST, DEGENERATIVE			4/1356					
INFARCT				2/1015				

*VARIABILITY AMONG DATA SOURCES SIGNIFICANT AT THE 1% SIGNIFICANCE LEVEL

SEX:	FEMALE				MALE			
DIAGNOSIS / AGE IN MONTHS:	0 < 12	12 < 18	18 < 24	24+	0 < 12	12 < 18	18 < 24	24+
MINERALIZATION			1/1356					
NECROSIS, FAT			4/1356	1/1015				
*************** GROWTH ALTERATIONS, NON-NEOPLASTIC, ATROPHY ***************								
ATROPHY			19/1356					
*************** GROWTH ALTERATIONS, NEOPLASTIC, BENIGN ***************								
ADENOMA			1/1356	4/1015				
ADENOMA, FOLLICULAR CELL				1/1015				
ADENOMA, TUBULAR			1/1356	1/1015				
CYSTADENOMA			4/1356	4/1015				
LUTEOMA			1/1356	1/1015				
*************** GROWTH ALTERATIONS, NEOPLASTIC, MALIGNANT, PRIMARY ***************								
ADENOCARCINOMA		1/103						
CARCINOMA	1/67							
*************** GROWTH ALTERATIONS, NEOPLASTIC, MALIGNANT, PRIMARY SITE NOT DETERMINED ***************								
LEIOMYOSARCOMA			1/1356					
LYMPHOMA			2/1356	1/1015				
LYMPHOMA, HISTIOCYTIC				1/1015				

SEX:	FEMALE				MALE			
DIAGNOSIS / AGE IN MONTHS:	0 < 12	12 < 18	18 < 24	24+	0 < 12	12 < 18	18 < 24	24+
*************** GROWTH ALTERATIONS, NEOPLASTIC, NOT DETERMINED ***************								
GRANULOSA CELL TUMOR				3/1015				

SEX:	FEMALE				MALE			
DIAGNOSIS / AGE IN MONTHS:	0 < 12	12 < 18	18 < 24	24+	0 < 12	12 < 18	18 < 24	24+
*************** MECHANICAL ABNORMALITIES AND CYST FORMATION ***************								
CYST	6/176	47/367	51/192	23/75				
HEMORRHAGE		6/367	7/192	1/75				
THROMBUS		1/367		1/75				
*************** INFLAMMATION AND REPAIR ***************								
FIBROSIS		1/367						
*************** DEGENERATION, NECROSIS, AND DEPOSITION ***************								
PIGMENT DEPOSITION	1/176	42/367	97/192	56/75				
*************** GROWTH ALTERATIONS, NON-NEOPLASTIC, ATROPHY ***************								
ATROPHY		3/367	16/192	14/75				
*************** GROWTH ALTERATIONS, NEOPLASTIC, BENIGN ***************								
ADENOMA	1/176		4/192	7/75				
ANGIOMA		1/367		1/75				
*************** GROWTH ALTERATIONS, NEOPLASTIC, MALIGNANT, PRIMARY SITE NOT DETERMINED ***************								
LYMPHOMA	1/176	3/367	1/192					
LYMPHOMA, RETICULUM CELL		1/367	3/192	3/75				

SEX:	FEMALE				MALE			
DIAGNOSIS / AGE IN MONTHS:	0 < 12	12 < 18	18 < 24	24+	0 < 12	12 < 18	18 < 24	24+
*************** MECHANICAL ABNORMALITIES AND CYST FORMATION ***************								
CYST			4/327	75/1575				
CYST, FOLLICULAR			2/327	9/1575				
CYST, PARAOVARIAN			2/327	16/1575				
*************** INFLAMMATION AND REPAIR ***************								
ABSCESS			1/327	5/1575				
INFLAMMATION			1/327	6/1575				
INFLAMMATION, GRANULOMATOUS	1/26							
INFLAMMATION, SUPPURATIVE			2/327	4/1575				
*************** DEGENERATION, NECROSIS, AND DEPOSITION ***************								
CYST, DEGENERATIVE				6/1575				
NECROSIS				1/1575				
NECROSIS, FAT			1/327	1/1575				
*************** GROWTH ALTERATIONS, NON-NEOPLASTIC, HYPERPLASIA ***************								
HYPERPLASIA			1/327	3/1575				
HYPERPLASIA, INTERSTITIAL CELL				1/1575				

SEX:	FEMALE				MALE			
AGE IN MONTHS: / DIAGNOSIS	0 < 12	12 < 18	18 < 24	24+	0 < 12	12 < 18	18 < 24	24+
*************** GROWTH ALTERATIONS, NEOPLASTIC, BENIGN ***************								
ADENOMA, TUBULAR				2/1575				
FIBROMA				1/1575				
THECOMA			1/327					
*************** GROWTH ALTERATIONS, NEOPLASTIC, MALIGNANT, PRIMARY ***************								
ADENOCARCINOMA				1/1575				
CARCINOMA, GRANULOSA CELL			1/327					
DYSGERMINOMA			1/327					
*************** GROWTH ALTERATIONS, NEOPLASTIC, MALIGNANT, PRIMARY SITE NOT DETERMINED ***************								
LEUKEMIA				1/1575				
LEUKEMIA, GRANULOCYTIC				1/1575				
LEUKEMIA, LYMPHOBLASTIC				1/1575				
LEUKEMIA, LYMPHOCYTIC				1/1575				
*************** GROWTH ALTERATIONS, NEOPLASTIC, NOT DETERMINED ***************								
GRANULOSA CELL TUMOR				3/1575				
SERTOLI CELL TUMOR				1/1575				

DIAGNOSIS	FEMALE 0 < 12	FEMALE 12 < 18	FEMALE 18 < 24	FEMALE 24+	MALE 0 < 12	MALE 12 < 18	MALE 18 < 24	MALE 24+
*************** MECHANICAL ABNORMALITIES AND CYST FORMATION ***************								
CYST		1/39		11/336				
CYST, FOLLICULAR				3/336				
*************** GROWTH ALTERATIONS, NON-NEOPLASTIC, HYPERPLASIA ***************								
HYPERPLASIA, GRANULOSA CELL				1/336				
*************** GROWTH ALTERATIONS, NEOPLASTIC, BENIGN ***************								
CYSTADENOMA				1/336				
LIPOMA				1/336				
*************** GROWTH ALTERATIONS, NEOPLASTIC, MALIGNANT, PRIMARY ***************								
ADENOCARCINOMA				1/336				
CARCINOMA				1/336				
*************** GROWTH ALTERATIONS, NEOPLASTIC, NOT DETERMINED ***************								
GRANULOSA CELL TUMOR				4/336				

SEX:	FEMALE				MALE			
DIAGNOSIS / AGE IN MONTHS:	0 < 12	12 < 18	18 < 24	24+	0 < 12	12 < 18	18 < 24	24+
*************** MECHANICAL ABNORMALITIES AND CYST FORMATION ***************								
CONGESTION	1/181							
CYST	1/181		14/420	10/240				
CYST, FOLLICULAR			1/420	7/240				
CYST, MESOTHELIAL				1/240				
*************** INFLAMMATION AND REPAIR ***************								
ABSCESS			1/420					
INFLAMMATION, HEMORRHAGIC		1/60						
INFLAMMATION, SUPPURATIVE			4/420					
*************** GROWTH ALTERATIONS, NON-NEOPLASTIC, HYPERPLASIA ***************								
HYPERPLASIA				1/240				
*************** GROWTH ALTERATIONS, NON-NEOPLAS¯IC, ATROPHY ***************								
ATROPHY				1/240				
*************** GROWTH ALTERATIONS, NEOPLASTIC, BENIGN ***************								
PAPILLOMA				1/240				
THECOMA				1/240				

SEX:	FEMALE				MALE			
DIAGNOSIS / AGE IN MONTHS:	0 < 12	12 < 18	18 < 24	24+	0 < 12	12 < 18	18 < 24	24+
*************** GROWTH ALTERATIONS, NEOPLASTIC, MALIGNANT, PRIMARY ***************								
ARRHENOBLASTOMA				1/240				
*************** GROWTH ALTERATIONS, NEOPLASTIC, MALIGNANT, PRIMARY SITE NOT DETERMINED ***************								
LYMPHOMA			1/420					
*************** GROWTH ALTERATIONS, NEOPLASTIC, NOT DETERMINED ***************								
GRANULOSA CELL TUMOR			1/420	1/240				
GRANULOSA THECAL CELL TUMOR			1/420					

SEX:	FEMALE				MALE			
AGE IN MONTHS:	0 < 12	12 < 18	18 < 24	24+	0 < 12	12 < 18	18 < 24	24+
DIAGNOSIS								
*************** GROWTH ALTERATIONS, NEOPLASTIC, MALIGNANT, PRIMARY SITE NOT DETERMINED ***************								
LYMPHOMA	3/212	2/369	5/193					
LYMPHOMA, RETICULUM CELL		1/369	4/193	4/71				

SEX:	FEMALE				MALE			
DIAGNOSIS / AGE IN MONTHS:	0 < 12	12 < 18	18 < 24	24+	0 < 12	12 < 18	18 < 24	24+
*************** MECHANICAL ABNORMALITIES AND CYST FORMATION ***************								
CYST				2/1015				
*************** INFLAMMATION AND REPAIR ***************								
ABSCESS			2/1356					
INFLAMMATION			13/1356					
INFLAMMATION, NONSUPPURATIVE				2/1015				
INFLAMMATION, SUPPURATIVE			4/1356	3/1015				
*************** DEGENERATION, NECROSIS, AND DEPOSITION ***************								
DEGENERATION			1/1356					
NECROSIS				1/1015				
*************** GROWTH ALTERATIONS, NEOPLASTIC, BENIGN ***************								
ADENOMA, PAPILLARY			3/1356					
PAPILLOMA, INTRADUCTAL			1/1356					

SEX:	FEMALE				MALE			
DIAGNOSIS / AGE IN MONTHS:	0 < 12	12 < 18	18 < 24	24+	0 < 12	12 < 18	18 < 24	24+
*************** MECHANICAL ABNORMALITIES AND CYST FORMATION ***************								
CYST		1/367	2/192	3/75				
*************** INFLAMMATION AND REPAIR ***************								
FIBROSIS			2/192	5/75				
INFLAMMATION				1/75				
*************** GROWTH ALTERATIONS, NEOPLASTIC, MALIGNANT, PRIMARY SITE NOT DETERMINED ***************								
LYMPHOMA		2/367	1/192	2/75				
LYMPHOMA, RETICULUM CELL		1/367	2/192	10/75				
*************** GROWTH ALTERATIONS, NEOPLASTIC, NOT DETERMINED ***************								
NEOPLASM			1/192					

SEX:	FEMALE				MALE			
DIAGNOSIS / AGE IN MONTHS:	0 < 12	12 < 18	18 < 24	24+	0 < 12	12 < 18	18 < 24	24+
*************** MECHANICAL ABNORMALITIES AND CYST FORMATION ***************								
CYST				2/1575				
RETENTION FLUID				1/1575				
*************** INFLAMMATION AND REPAIR ***************								
ABSCESS			1/327	1/1575				
INFLAMMATION				21/1575				
INFLAMMATION, SUPPURATIVE	1/26		7/327	41/1575				

SEX:	FEMALE				MALE			
DIAGNOSIS / AGE IN MONTHS:	0 < 12	12 < 18	18 < 24	24+	0 < 12	12 < 18	18 < 24	24+
*************** INFLAMMATION AND REPAIR ***************								
INFLAMMATION, SUPPURATIVE				1/336				

SEX:	FEMALE				MALE			
DIAGNOSIS / AGE IN MONTHS:	0 < 12	12 < 18	18 < 24	24+	0 < 12	12 < 18	18 < 24	24+
*************** INFLAMMATION AND REPAIR ***************								
GRANULOMA			1/420					
INFLAMMATION, SUPPURATIVE			2/420					

SEX:	FEMALE				MALE			
DIAGNOSIS / AGE IN MONTHS:	0 < 12	12 < 18	18 < 24	24+	0 < 12	12 < 18	18 < 24	24+
*************** MECHANICAL ABNORMALITIES AND CYST FORMATION ***************								
CYST				1/71				
HEMORRHAGE				2/71				
HYDROMETRA	1/212	15/369	9/193	10/71				
THROMBUS	1/212	4/369	2/193	1/71				
*************** INFLAMMATION AND REPAIR ***************								
FIBROSIS		2/369	3/193	7/71				
INFLAMMATION		3/369	1/193					
INFLAMMATION, NONSUPPURATIVE			1/193					
*************** DEGENERATION, NECROSIS, AND DEPOSITION ***************								
MINERALIZATION	48/212	89/369	37/193					
NECROSIS		1/369						
PIGMENT DEPOSITION	47/212	58/369	14/193	1/71				
*************** GROWTH ALTERATIONS, NON-NEOPLASTIC, HYPERPLASIA ***************								
HYPERPLASIA, ENDOMETRIAL, CYSTIC	3/212	3/369	3/193	6/71				
HYPERPLASIA, NODULAR			1/193					
POLYP	3/212	67/369	53/193	34/71				

SEX:	FEMALE				MALE			
DIAGNOSIS / AGE IN MONTHS:	0 < 12	12 < 18	18 < 24	24+	0 < 12	12 < 18	18 < 24	24+
*************** GROWTH ALTERATIONS, NEOPLASTIC, BENIGN ***************								
ANGIOMA	1/212	5/369	3/193	3/71				
LEIOMYOMA			1/193					
*************** GROWTH ALTERATIONS, NEOPLASTIC, MALIGNANT, PRIMARY ***************								
ADENOCARCINOMA		1/369						
*************** GROWTH ALTERATIONS, NEOPLASTIC, MALIGNANT, PRIMARY SITE NOT DETERMINED ***************								
ANGIOSARCOMA			1/193					
LYMPHOMA	4/212	4/369	6/193					
LYMPHOMA, RETICULUM CELL		1/369	3/193	6/71				

SEX:	FEMALE				MALE			
DIAGNOSIS / AGE IN MONTHS:	0 < 12	12 < 18	18 < 24	24+	0 < 12	12 < 18	18 < 24	24+
*************** MECHANICAL ABNORMALITIES AND CYST FORMATION ***************								
ANGIECTASIS			2/1356					
CYST		2/103	12/1356	97/1015*				
CYST, HEMORRHAGIC				2/1015				
DILATATION			1/1356	12/1015				
EDEMA			1/1356	1/1015				
HEMORRHAGE			1/1356	2/1015				
HYDROMETRA		5/103	79/1356*	22/1015				
*************** INFLAMMATION AND REPAIR ***************								
ABSCESS			7/1356	1/1015				
FIBROSIS			1/1356					
INFLAMMATION	1/67	2/103	70/1356*	16/1015				
INFLAMMATION, GRANULOMATOUS			1/1356	1/1015				
INFLAMMATION, NONSUPPURATIVE				1/1015				
INFLAMMATION, SUPPURATIVE			38/1356	13/1015				
POLYP, INFLAMMATORY			1/1356	1/1015				
PYOMETRA		2/103	6/1356	12/1015				
*************** DEGENERATION, NECROSIS, AND DEPOSITION ***************								
MINERALIZATION			1/1356					

*VARIABILITY AMONG DATA SOURCES SIGNIFICANT AT THE 1% SIGNIFICANCE LEVEL

SEX:	FEMALE				MALE			
DIAGNOSIS / AGE IN MONTHS:	0 < 12	12 < 18	18 < 24	24+	0 < 12	12 < 18	18 < 24	24+
NECROSIS, FAT			2/1356					
*************** GROWTH ALTERATIONS, NON-NEOPLASTIC, HYPERPLASIA ***************								
ENDOMETRIAL STROMAL POLYP		1/103	8/1356	5/1015				
HYPERPLASIA		2/103	21/1356	21/1015				
HYPERPLASIA, ADENOMATOUS			1/1356					
HYPERPLASIA, CYSTIC	4/67	21/103	590/1356*	270/1015				
HYPERPLASIA, EPITHELIAL		1/103						
*************** GROWTH ALTERATIONS, NON-NEOPLASTIC, METAPLASIA ***************								
METAPLASIA, SQUAMOUS			2/1356	1/1015				
*************** GROWTH ALTERATIONS, NEOPLASTIC, BENIGN ***************								
ANGIOMA			3/1356	1/1015				
LEIOMYOMA			2/1356	1/1015				
NEURILEMMOMA				1/1015				
*************** GROWTH ALTERATIONS, NEOPLASTIC, MALIGNANT, PRIMARY ***************								
ADENOCARCINOMA			2/1356	1/1015				
CARCINOMA			1/1356					
LEIOMYOSARCOMA			1/1356	5/1015				

*VARIABILITY AMONG DATA SOURCES SIGNIFICANT AT THE 1% SIGNIFICANCE LEVEL

SEX:	FEMALE				MALE			
DIAGNOSIS / AGE IN MONTHS:	0 < 12	12 < 18	18 < 24	24+	0 < 12	12 < 18	18 < 24	24+
*************** GROWTH ALTERATIONS, NEOPLASTIC, MALIGNANT, METASTATIC ***************								
CARCINOMA, TRANSITIONAL CELL				1/1015				
*************** GROWTH ALTERATIONS, NEOPLASTIC, MALIGNANT, PRIMARY SITE NOT DETERMINED ***************								
ANGIOSARCOMA			1/1356	2/1015				
LYMPHOMA				1/1015				
LYMPHOMA, HISTIOCYTIC			1/1356					
NEOPLASM, MALIGNANT			1/1356					
SARCOMA				2/1015				

SEX:	FEMALE				MALE			
DIAGNOSIS / AGE IN MONTHS:	0 < 12	12 < 18	18 < 24	24+	0 < 12	12 < 18	18 < 24	24+
*************** MECHANICAL ABNORMALITIES AND CYST FORMATION ***************								
CYST		1/367	1/192					
HYDROMETRA	3/176	27/367	7/192	2/75				
THROMBUS	1/176	1/367	2/192					
*************** INFLAMMATION AND REPAIR ***************								
FIBROSIS	1/176	24/367	29/192	15/75				
INFLAMMATION	3/176	7/367	5/192	3/75				
INFLAMMATION, NONSUPPURATIVE				1/75				
*************** DEGENERATION, NECROSIS, AND DEPOSITION ***************								
AMYLOID DEPOSITION			2/192					
MINERALIZATION	13/176	12/367	4/192					
PIGMENT DEPOSITION	23/176	25/367	6/192					
*************** GROWTH ALTERATIONS, NON-NEOPLASTIC, HYPERPLASIA ***************								
HYPERPLASIA, ADENOMATOUS				1/75				
HYPERPLASIA, ENDOMETRIAL, CYSTIC	15/176	90/367	61/192	14/75				
POLYP		1/367		2/75				

SEX:	FEMALE				MALE			
DIAGNOSIS / AGE IN MONTHS:	0 < 12	12 < 18	18 < 24	24+	0 < 12	12 < 18	18 < 24	24+
*************** GROWTH ALTERATIONS, NON-NEOPLASTIC, METAPLASIA ***************								
METAPLASIA	1/176							
METAPLASIA, OSSEOUS		1/367						
*************** GROWTH ALTERATIONS, NEOPLASTIC, BENIGN ***************								
ANGIOMA		1/367	2/192	1/75				
*************** GROWTH ALTERATIONS, NEOPLASTIC, MALIGNANT, PRIMARY SITE NOT DETERMINED ***************								
LYMPHOMA		2/367	1/192	2/75				
LYMPHOMA, RETICULUM CELL		3/367	8/192	4/75				

SEX:	FEMALE				MALE			
DIAGNOSIS / AGE IN MONTHS:	0 < 12	12 < 18	18 < 24	24+	0 < 12	12 < 18	18 < 24	24+
*************** MECHANICAL ABNORMALITIES AND CYST FORMATION ***************								
CYST			2/327	20/1575				
DILATATION				3/1575				
HEMATOMA			1/327	2/1575				
HEMORRHAGE				2/1575				
HYDROMETRA	2/26		5/327	29/1575				
THROMBUS				1/1575				
THROMBUS, ORGANIZED			1/327	1/1575				
*************** INFLAMMATION AND REPAIR ***************								
ABSCESS		1/24	2/327	8/1575				
FIBROSIS			5/327	15/1575				
INFLAMMATION		1/24	11/327	97/1575*				
INFLAMMATION, GRANULOMATOUS				1/1575				
INFLAMMATION, NECROTIZING				1/1575				
INFLAMMATION, NONSUPPURATIVE			1/327					
INFLAMMATION, SUPPURATIVE	1/26	1/24	8/327	94/1575*				
INFLAMMATION, ULCERATIVE				1/1575				
POLYP, INFLAMMATORY			2/327	10/1575				
PYOMETRA			3/327	35/1575				

*VARIABILITY AMONG DATA SOURCES SIGNIFICANT AT THE 1% SIGNIFICANCE LEVEL

SEX:	FEMALE				MALE			
DIAGNOSIS / AGE IN MONTHS:	0 < 12	12 < 18	18 < 24	24+	0 < 12	12 < 18	18 < 24	24+
*************** DEGENERATION, NECROSIS, AND DEPOSITION ***************								
DEGENERATION, CYSTIC			1/327					
HEMOSIDERIN DEPOSITION				1/1575				
NECROSIS			1/327	5/1575				
NECROSIS, FAT				2/1575				
*************** GROWTH ALTERATIONS, NON-NEOPLASTIC, HYPERTROPHY ***************								
HYPERTROPHY				1/1575				
*************** GROWTH ALTERATIONS, NON-NEOPLASTIC, HYPERPLASIA ***************								
ENDOMETRIAL STROMAL POLYP	1/26	2/24	38/327	266/1575*				
HYPERPLASIA			5/327	37/1575				
HYPERPLASIA, ADENOMATOUS				8/1575				
HYPERPLASIA, CYSTIC			6/327	51/1575				
HYPERPLASIA, STROMAL			2/327					
POLYP			3/327	5/1575				
POLYP, INFLAMMATORY				1/1575				
*************** GROWTH ALTERATIONS, NON-NEOPLASTIC, METAPLASIA ***************								
METAPLASIA, SQUAMOUS				3/1575				

*VARIABILITY AMONG DATA SOURCES SIGNIFICANT AT THE 1% SIGNIFICANCE LEVEL

DIAGNOSIS — SEX:	FEMALE				MALE			
AGE IN MONTHS:	0 < 12	12 < 18	18 < 24	24+	0 < 12	12 < 18	18 < 24	24+
*************** GROWTH ALTERATIONS, NEOPLASTIC, BENIGN ***************								
ADENOMA				1/1575				
ADENOMA, PAPILLARY			1/327					
ANGIOMA			1/327					
FIBROMA				6/1575				
LEIOMYOMA				4/1575				
*************** GROWTH ALTERATIONS, NEOPLASTIC, MALIGNANT, PRIMARY ***************								
ADENOCARCINOMA			4/327	21/1575				
CARCINOMA			1/327	2/1575				
LEIOMYOSARCOMA			2/327	2/1575				
SARCOMA, ENDOMETRIAL STROMAL				17/1575				
*************** GROWTH ALTERATIONS, NEOPLASTIC, MALIGNANT, PRIMARY SITE NOT DETERMINED ***************								
LEUKEMIA				1/1575				
LEUKEMIA, LYMPHOCYTIC				2/1575				
SARCOMA				2/1575				

SEX:	FEMALE				MALE			
DIAGNOSIS / AGE IN MONTHS:	0 < 12	12 < 18	18 < 24	24+	0 < 12	12 < 18	18 < 24	24+
*************** MECHANICAL ABNORMALITIES AND CYST FORMATION ***************								
CYST			1/44	1/336				
CYST, HEMORRHAGIC				1/336				
HYDROMETRA		7/39	3/44	33/336				
*************** INFLAMMATION AND REPAIR ***************								
INFLAMMATION		1/39	1/44	18/336				
*************** GROWTH ALTERATIONS, NON-NEOPLASTIC, HYPERPLASIA ***************								
ENDOMETRIAL STROMAL POLYP		1/39		17/336				
HYPERPLASIA				1/336				
HYPERPLASIA, CYSTIC		1/39	1/44	11/336				
*************** GROWTH ALTERATIONS, NEOPLASTIC, BENIGN ***************								
ADENOMA, PAPILLARY				1/336				
ANGIOMA				1/336				
LIPOMA				1/336				
*************** GROWTH ALTERATIONS, NEOPLASTIC, MALIGNANT, PRIMARY ***************								
ADENOCARCINOMA				2/336				
LEIOMYOSARCOMA			1/44	1/336				

ORGAN SYSTEM: GENITAL SYSTEM; FEMALES ORGAN: UTERUS STRAIN: OSBORNE MENDEL

SEX:	FEMALE				MALE			
DIAGNOSIS / AGE IN MONTHS:	0 < 12	12 < 18	18 < 24	24+	0 < 12	12 < 18	18 < 24	24+
SARCOMA, ENDOMETRIAL STROMAL			1/44					

SEX:	FEMALE				MALE			
DIAGNOSIS / AGE IN MONTHS:	0 < 12	12 < 18	18 < 24	24+	0 < 12	12 < 18	18 < 24	24+
*************** MECHANICAL ABNORMALITIES AND CYST FORMATION ***************								
CYST			1/420	1/240				
DILATATION		1/60	1/420	1/240				
HEMORRHAGE		1/60						
HYDROMETRA		1/60	1/420	4/240				
*************** INFLAMMATION AND REPAIR ***************								
ABSCESS				1/240				
EROSION				1/240				
INFLAMMATION			11/420	4/240				
INFLAMMATION, HEMORRHAGIC				1/240				
INFLAMMATION, SUPPURATIVE		2/60	48/420*	8/240				
PYOMETRA			5/420					
*************** DEGENERATION, NECROSIS, AND DEPOSITION ***************								
HEMOSIDERIN DEPOSITION				1/240				
*************** GROWTH ALTERATIONS, NON-NEOPLASTIC, HYPERPLASIA ***************								
ENDOMETRIAL STROMAL POLYP			9/420	2/240				
HYPERPLASIA			2/420	6/240				
HYPERPLASIA, ADENOMATOUS	1/181		12/420	16/240*				

*VARIABILITY AMONG DATA SOURCES SIGNIFICANT AT THE 1% SIGNIFICANCE LEVEL

SEX:	FEMALE				MALE			
DIAGNOSIS / AGE IN MONTHS:	0 < 12	12 < 18	18 < 24	24+	0 < 12	12 < 18	18 < 24	24+
HYPERPLASIA, CYSTIC			3/420	5/240				
HYPERPLASIA, ENDOMETRIAL, CYSTIC			1/420					
POLYP			2/420	2/240				
*************** GROWTH ALTERATIONS, NON-NEOPLASTIC, METAPLASIA ***************								
METAPLASIA, SQUAMOUS			1/420	3/240				
*************** GROWTH ALTERATIONS, NEOPLASTIC, BENIGN ***************								
LEIOMYOMA			2/420	1/240				
*************** GROWTH ALTERATIONS, NEOPLASTIC, MALIGNANT, PRIMARY ***************								
FIBROSARCOMA			1/420					
*************** GROWTH ALTERATIONS, NEOPLASTIC, MALIGNANT, PRIMARY SITE NOT DETERMINED ***************								
ANGIOSARCOMA				1/240				
LYMPHOMA			1/420					

SEX:	FEMALE				MALE			
DIAGNOSIS / AGE IN MONTHS:	0 < 12	12 < 18	18 < 24	24+	0 < 12	12 < 18	18 < 24	24+
*************** MECHANICAL ABNORMALITIES AND CYST FORMATION ***************								
EDEMA				1/71				
PROLAPSE, UTERINE			1/193					
PROLAPSE, VAGINAL				1/71				
*************** INFLAMMATION AND REPAIR ***************								
FIBROSIS			1/193	1/71				
INFLAMMATION		1/369	1/193	1/71				
INFLAMMATION, NONSUPPURATIVE				2/71				
*************** DEGENERATION, NECROSIS, AND DEPOSITION ***************								
INFARCT		1/369						
*************** GROWTH ALTERATIONS, NON-NEOPLASTIC, DYSPLASIA ***************								
DYSPLASIA, EPIDERMOID			6/193	14/71				
*************** GROWTH ALTERATIONS, NEOPLASTIC, BENIGN ***************								
PAPILLOMA		3/369	1/193					
*************** GROWTH ALTERATIONS, NEOPLASTIC, MALIGNANT, PRIMARY ***************								
MYOSARCOMA			1/193					

SEX:	FEMALE				MALE			
AGE IN MONTHS: / DIAGNOSIS	0 < 12	12 < 18	18 < 24	24+	0 < 12	12 < 18	18 < 24	24+
*************** GROWTH ALTERATIONS, NEOPLASTIC, MALIGNANT, PRIMARY SITE NOT DETERMINED ***************								
LYMPHOMA	2/212	2/369	4/193					
LYMPHOMA, RETICULUM CELL		1/369	4/193	6/71				

SEX:	FEMALE				MALE			
DIAGNOSIS / AGE IN MONTHS:	0 < 12	12 < 18	18 < 24	24+	0 < 12	12 < 18	18 < 24	24+
*************** INFLAMMATION AND REPAIR ***************								
INFLAMMATION			1/1356					
INFLAMMATION, SUPPURATIVE				1/1015				
*************** GROWTH ALTERATIONS, NON-NEOPLASTIC, HYPERPLASIA ***************								
HYPERPLASIA, EPITHELIAL				1/1015				
*************** GROWTH ALTERATIONS, NEOPLASTIC, MALIGNANT, PRIMARY ***************								
CARCINOMA, SQUAMOUS CELL			1/1356					

DIAGNOSIS	SEX: FEMALE				MALE			
AGE IN MONTHS:	0 < 12	12 < 18	18 < 24	24+	0 < 12	12 < 18	18 < 24	24+
*************** MECHANICAL ABNORMALITIES AND CYST FORMATION ***************								
CYST			1/192					
*************** INFLAMMATION AND REPAIR ***************								
FIBROSIS		1/367						
INFLAMMATION			2/192					
*************** GROWTH ALTERATIONS, NON-NEOPLASTIC, DYSPLASIA ***************								
DYSPLASIA, EPIDERMOID		1/367	1/192	1/75				
*************** GROWTH ALTERATIONS, NEOPLASTIC, BENIGN ***************								
PAPILLOMA	1/176							
*************** GROWTH ALTERATIONS, NEOPLASTIC, MALIGNANT, PRIMARY ***************								
MYOSARCOMA		1/367						
*************** GROWTH ALTERATIONS, NEOPLASTIC, MALIGNANT, PRIMARY SITE NOT DETERMINED ***************								
LYMPHOMA				1/75				
LYMPHOMA, RETICULUM CELL		2/367	3/192	2/75				

SEX:	FEMALE				MALE			
DIAGNOSIS / AGE IN MONTHS:	0 < 12	12 < 18	18 < 24	24+	0 < 12	12 < 18	18 < 24	24+
*************** MECHANICAL ABNORMALITIES AND CYST FORMATION ***************								
CYST, EPIDERMAL INCLUSION				1/1575				
*************** INFLAMMATION AND REPAIR ***************								
INFLAMMATION				2/1575				
*************** DEGENERATION, NECROSIS, AND DEPOSITION ***************								
NECROSIS, FAT				1/1575				
*************** GROWTH ALTERATIONS, NON-NEOPLASTIC, HYPERPLASIA ***************								
POLYP				1/1575				
*************** GROWTH ALTERATIONS, NEOPLASTIC, MALIGNANT, PRIMARY ***************								
CARCINOMA, SQUAMOUS CELL			1/327					
FIBROSARCOMA				1/1575				
*************** GROWTH ALTERATIONS, NEOPLASTIC, MALIGNANT, PRIMARY SITE NOT DETERMINED ***************								
LYMPHANGIOSARCOMA				1/1575				
SARCOMA			1/327					

SEX:	FEMALE				MALE			
DIAGNOSIS / AGE IN MONTHS:	0 < 12	12 < 18	18 < 24	24+	0 < 12	12 < 18	18 < 24	24+
*************** INFLAMMATION AND REPAIR ***************								
INFLAMMATION		1/39		8/336				
INFLAMMATION, SUPPURATIVE			1/44					
POLYP, INFLAMMATORY			1/44	1/336				
*************** GROWTH ALTERATIONS, NEOPLASTIC, MALIGNANT, PRIMARY ***************								
LEIOMYOSARCOMA			1/44					
*************** GROWTH ALTERATIONS, NEOPLASTIC, MALIGNANT, METASTATIC ***************								
SARCOMA, ENDOMETRIAL STROMAL			1/44					

SEX:	FEMALE				MALE			
DIAGNOSIS / AGE IN MONTHS:	0 < 12	12 < 18	18 < 24	24+	0 < 12	12 < 18	18 < 24	24+
*************** INFLAMMATION AND REPAIR ***************								
INFLAMMATION, SUPPURATIVE			1/420					
INFLAMMATION, ULCERATIVE			1/420					
*************** DEGENERATION, NECROSIS, AND DEPOSITION ***************								
NECROSIS			1/420					
*************** GROWTH ALTERATIONS, NEOPLASTIC, MALIGNANT, PRIMARY ***************								
FIBROSARCOMA			1/420					

SEX:	FEMALE				MALE			
DIAGNOSIS / AGE IN MONTHS:	0 < 12	12 < 18	18 < 24	24+	0 < 12	12 < 18	18 < 24	24+
*************** GROWTH ALTERATIONS, NON-NEOPLASTIC, HYPERTROPHY ***************								
KARYOMEGALY						1/314		
*************** GROWTH ALTERATIONS, NON-NEOPLASTIC, DYSPLASIA ***************								
DYSPLASIA						1/314	1/147	2/94

DIAGNOSIS	FEMALE 0 < 12	FEMALE 12 < 18	FEMALE 18 < 24	FEMALE 24+	MALE 0 < 12	MALE 12 < 18	MALE 18 < 24	MALE 24+
*************** INFLAMMATION AND REPAIR ***************								
INFLAMMATION, NONSUPPURATIVE								1/102
*************** GROWTH ALTERATIONS, NON-NEOPLASTIC, HYPERTROPHY ***************								
KARYOMEGALY						1/300	2/161	1/102
*************** GROWTH ALTERATIONS, NON-NEOPLASTIC, DYSPLASIA ***************								
DYSPLASIA								1/102

SEX:	FEMALE				MALE			
DIAGNOSIS / AGE IN MONTHS:	0 < 12	12 < 18	18 < 24	24+	0 < 12	12 < 18	18 < 24	24+
*************** INFLAMMATION AND REPAIR ***************								
ABSCESS								1/1743
GRANULOMA								1/1743
INFLAMMATION								1/1743

ORGAN SYSTEM: GENITAL SYSTEM; MALES   ORGAN: EPIDIDYMIS   STRAIN: BALB/C

SEX:	FEMALE				MALE			
DIAGNOSIS / AGE IN MONTHS:	0 < 12	12 < 18	18 < 24	24+	0 < 12	12 < 18	18 < 24	24+
*************** MECHANICAL ABNORMALITIES AND CYST FORMATION ***************								
CYST								2/94
HEMORRHAGE							1/147	
SPERMATOCELE					4/171	20/314	9/147	3/94
*************** INFLAMMATION AND REPAIR ***************								
GRANULOMA						4/314	1/147	
INFLAMMATION						5/314	3/147	
INFLAMMATION, NONSUPPURATIVE								2/94*
*************** DEGENERATION, NECROSIS, AND DEPOSITION ***************								
CELLULAR CHANGE, VACUOLAR								1/94
MINERALIZATION								2/94
PIGMENT DEPOSITION						1/314		1/94
*************** GROWTH ALTERATIONS, NON-NEOPLASTIC, DYSPLASIA ***************								
DYSPLASIA					2/171	1/314		1/94
*************** GROWTH ALTERATIONS, NEOPLASTIC, MALIGNANT, PRIMARY ***************								
ADENOCARCINOMA							1/147	

*VARIABILITY AMONG DATA SOURCES SIGNIFICANT AT THE 1% SIGNIFICANCE LEVEL

SEX:	FEMALE				MALE			
DIAGNOSIS / AGE IN MONTHS:	0 < 12	12 < 18	18 < 24	24+	0 < 12	12 < 18	18 < 24	24+
*************** GROWTH ALTERATIONS, NEOPLASTIC, MALIGNANT, PRIMARY SITE NOT DETERMINED ***************								
LYMPHOMA								1/94
LYMPHOMA, RETICULUM CELL						1/314		3/94
*************** GROWTH ALTERATIONS, NEOPLASTIC, NOT DETERMINED ***************								
MYOBLASTOMA, GRANULAR CELL							4/147	

SEX:	FEMALE				MALE			
DIAGNOSIS / AGE IN MONTHS:	0 < 12	12 < 18	18 < 24	24+	0 < 12	12 < 18	18 < 24	24+
*************** MECHANICAL ABNORMALITIES AND CYST FORMATION ***************								
DILATATION, TUBULAR							1/1136	
*************** INFLAMMATION AND REPAIR ***************								
ABSCESS							1/1136	
GRANULOMA						1/199	5/1136	1/958
INFLAMMATION							1/1136	2/958
INFLAMMATION, FAT								1/958
INFLAMMATION, NONSUPPURATIVE								4/958
INFLAMMATION, SUPPURATIVE					1/124		1/1136	
*************** DEGENERATION, NECROSIS, AND DEPOSITION ***************								
NECROSIS, FAT							1/1136	2/958
*************** GROWTH ALTERATIONS, NEOPLASTIC, BENIGN ***************								
LIPOMA							1/1136	

SEX:	FEMALE				MALE			
DIAGNOSIS / AGE IN MONTHS:	0 < 12	12 < 18	18 < 24	24+	0 < 12	12 < 18	18 < 24	24+
*************** MECHANICAL ABNORMALITIES AND CYST FORMATION ***************								
CYST						1/300		1/102
*************** INFLAMMATION AND REPAIR ***************								
FIBROSIS								3/102
INFLAMMATION							1/161	3/102
INFLAMMATION, NONSUPPURATIVE							1/161	6/102
*************** GROWTH ALTERATIONS, NON-NEOPLASTIC, HYPERTROPHY ***************								
KARYOMEGALY								1/102
*************** GROWTH ALTERATIONS, NON-NEOPLASTIC, DYSPLASIA ***************								
DYSPLASIA							1/161	4/102
*************** GROWTH ALTERATIONS, NEOPLASTIC, MALIGNANT, PRIMARY SITE NOT DETERMINED ***************								
LYMPHOMA								2/102
LYMPHOMA, RETICULUM CELL								1/102

SEX:	FEMALE				MALE			
DIAGNOSIS / AGE IN MONTHS:	0 < 12	12 < 18	18 < 24	24+	0 < 12	12 < 18	18 < 24	24+
*************** INFLAMMATION AND REPAIR ***************								
ABSCESS						1/101		
FIBROSIS								2/1743
GRANULOMA								1/1743
*************** DEGENERATION, NECROSIS, AND DEPOSITION ***************								
CELLULAR CHANGE, VACUOLAR								1/1743
NECROSIS, FAT						1/101		2/1743
*************** GROWTH ALTERATIONS, NON-NEOPLASTIC, HYPERPLASIA ***************								
HYPERPLASIA								1/1743
*************** GROWTH ALTERATIONS, NON-NEOPLASTIC, DYSPLASIA ***************								
DYSPLASIA							2/437	31/1743
*************** GROWTH ALTERATIONS, NEOPLASTIC, MALIGNANT, METASTATIC ***************								
INTERSTITIAL CELL TUMOR								1/1743
*************** GROWTH ALTERATIONS, NEOPLASTIC, MALIGNANT, PRIMARY SITE NOT DETERMINED ***************								
MESOTHELIOMA							2/437	3/1743
NEUROFIBROSARCOMA							1/437	

SEX:	FEMALE				MALE			
DIAGNOSIS / AGE IN MONTHS:	0 < 12	12 < 18	18 < 24	24+	0 < 12	12 < 18	18 < 24	24+
*************** INFLAMMATION AND REPAIR ***************								
INFLAMMATION							1/104	
PERIVASCULITIS/VASCULITIS								1/243
*************** DEGENERATION, NECROSIS, AND DEPOSITION ***************								
MINERALIZATION								1/243
NECROSIS, FAT						2/64		3/243
*************** GROWTH ALTERATIONS, NON-NEOPLASTIC, ATROPHY ***************								
ATROPHY						1/64	5/104	7/243
*************** GROWTH ALTERATIONS, NEOPLASTIC, BENIGN ***************								
LIPOMA								1/243

SEX:	FEMALE				MALE			
AGE IN MONTHS:	0 < 12	12 < 18	18 < 24	24+	0 < 12	12 < 18	18 < 24	24+
DIAGNOSIS								
*************** INFLAMMATION AND REPAIR ***************								
FIBROSIS								1/216
GRANULOMA							1/392	
INFLAMMATION					1/170			2/216
*************** DEGENERATION, NECROSIS, AND DEPOSITION ***************								
MINERALIZATION								1/216
*************** GROWTH ALTERATIONS, NON-NEOPLASTIC, HYPERTROPHY ***************								
HYPERTROPHY								1/216

DIAGNOSIS	FEMALE 0 < 12	FEMALE 12 < 18	FEMALE 18 < 24	FEMALE 24+	MALE 0 < 12	MALE 12 < 18	MALE 18 < 24	MALE 24+
*************** MECHANICAL ABNORMALITIES AND CYST FORMATION ***************								
CYST							1/147	1/94
DILATATION								1/94
*************** INFLAMMATION AND REPAIR ***************								
FIBROSIS					2/171			1/94
INFLAMMATION					5/171	2/314	1/147	
INFLAMMATION WITH FIBROSIS, PROSTATE								1/94
INFLAMMATION, NONSUPPURATIVE							1/147	1/94
INFLAMMATION, NONSUPPURATIVE, PROSTATE							1/147	1/94
*************** GROWTH ALTERATIONS, NON-NEOPLASTIC, ATROPHY ***************								
ATROPHY						1/314	1/147	1/94
*************** GROWTH ALTERATIONS, NON-NEOPLASTIC, METAPLASIA ***************								
METAPLASIA, SQUAMOUS							1/147	
*************** GROWTH ALTERATIONS, NEOPLASTIC, MALIGNANT, PRIMARY SITE NOT DETERMINED ***************								
LYMPHOMA					3/171		1/147	2/94
LYMPHOMA, RETICULUM CELL							2/147	2/94

SEX:	FEMALE				MALE			
DIAGNOSIS / AGE IN MONTHS:	0 < 12	12 < 18	18 < 24	24+	0 < 12	12 < 18	18 < 24	24+
*************** MECHANICAL ABNORMALITIES AND CYST FORMATION ***************								
CAST						1/199	1/1136	3/958
CYST							1/1136	1/958
DILATATION							6/1136	4/958
*************** INFLAMMATION AND REPAIR ***************								
INFLAMMATION					7/124 *	7/199	7/1136	2/958
INFLAMMATION, NONSUPPURATIVE								1/958
INFLAMMATION, SUPPURATIVE					4/124		2/1136	1/958
*************** DEGENERATION, NECROSIS, AND DEPOSITION ***************								
CHOLESTEROL DEPOSITION								1/958
*************** GROWTH ALTERATIONS, NON-NEOPLASTIC, HYPERTROPHY ***************								
CYTOMEGALY								1/958
HYPERTROPHY					1/124			
*************** GROWTH ALTERATIONS, NON-NEOPLASTIC, HYPERPLASIA ***************								
HYPERPLASIA, CYSTIC							1/1136	1/958
HYPERPLASIA, EPITHELIAL							1/1136	
HYPERPLASIA, LYMPHOID							1/1136	

*VARIABILITY AMONG DATA SOURCES SIGNIFICANT AT THE 1% SIGNIFICANCE LEVEL

SEX:	FEMALE				MALE			
DIAGNOSIS / AGE IN MONTHS:	0 < 12	12 < 18	18 < 24	24+	0 < 12	12 < 18	18 < 24	24+
*************** GROWTH ALTERATIONS, NEOPLASTIC, BENIGN ***************								
ANGIOMA								1/958
*************** GROWTH ALTERATIONS, NEOPLASTIC, MALIGNANT, PRIMARY ***************								
CARCINOMA, ACINAR CELL								1/958
*************** GROWTH ALTERATIONS, NEOPLASTIC, MALIGNANT, PRIMARY SITE NOT DETERMINED ***************								
LYMPHOMA, HISTIOCYTIC							2/1136	

SEX:	FEMALE				MALE			
DIAGNOSIS / AGE IN MONTHS:	0 < 12	12 < 18	18 < 24	24+	0 < 12	12 < 18	18 < 24	24+
*************** MECHANICAL ABNORMALITIES AND CYST FORMATION ***************								
CYST						1/300		3/102
DILATATION								10/102
HEMORRHAGE								1/102
HERNIATION, SCROTAL								1/102
*************** INFLAMMATION AND REPAIR ***************								
FIBROSIS					1/131			
INFLAMMATION						1/300	3/161	7/102
INFLAMMATION, GRANULOMATOUS						1/300		
INFLAMMATION, NONSUPPURATIVE								1/102
INFLAMMATION, NONSUPPURATIVE, PROSTATE						3/300		1/102
INFLAMMATION, NONSUPPURATIVE, SEM. VESIC								1/102
INFLAMMATION, SUPPURATIVE, PROSTATE							1/161	
INFLAMMATION, SUPPURATIVE, SEMINAL VESIC							1/161	
*************** DEGENERATION, NECROSIS, AND DEPOSITION ***************								
NECROSIS							1/161	
*************** GROWTH ALTERATIONS, NON-NEOPLASTIC, HYPERPLASIA ***************								
HYPERPLASIA						1/300		

SEX:	FEMALE				MALE			
DIAGNOSIS / AGE IN MONTHS:	0 < 12	12 < 18	18 < 24	24+	0 < 12	12 < 18	18 < 24	24+
HYPERPLASIA, LYMPHOID								1/102
*************** GROWTH ALTERATIONS, NON-NEOPLASTIC, ATROPHY ***************								
ATROPHY								4/102
*************** GROWTH ALTERATIONS, NEOPLASTIC, BENIGN ***************								
LEIOMYOMA								1/102
PAPILLOMA								1/102
*************** GROWTH ALTERATIONS, NEOPLASTIC, MALIGNANT, METASTATIC ***************								
MYOSARCOMA								1/102
*************** GROWTH ALTERATIONS, NEOPLASTIC, MALIGNANT, PRIMARY SITE NOT DETERMINED ***************								
LYMPHOMA						1/300		3/102
LYMPHOMA, RETICULUM CELL								1/102

SEX:	FEMALE				MALE			
DIAGNOSIS / AGE IN MONTHS:	0 < 12	12 < 18	18 < 24	24+	0 < 12	12 < 18	18 < 24	24+
*************** INFLAMMATION AND REPAIR ***************								
ABSCESS							1/437	1/1743
FIBROSIS							1/437	
INFLAMMATION						5/101*	61/437*	185/1743 *
INFLAMMATION, NONSUPPURATIVE, PROSTATE								4/1743
INFLAMMATION, PROSTATE							2/437	8/1743
INFLAMMATION, SEMINAL VESICLE								1/1743
INFLAMMATION, SUPPURATIVE					1/101		9/437	35/1743
INFLAMMATION, SUPPURATIVE, PROSTATE						3/101	3/437	11/1743
INFLAMMATION, SUPPURATIVE, SEMINAL VESIC								1/1743
PERIVASCULITIS/VASCULITIS								1/1743
*************** DEGENERATION, NECROSIS, AND DEPOSITION ***************								
DEGENERATION							1/437	12/1743
MINERALIZATION								1/1743
*************** GROWTH ALTERATIONS, NON-NEOPLASTIC, HYPERPLASIA ***************								
HYPERPLASIA								32/1743
HYPERPLASIA, ADENOMATOUS								1/1743
HYPERPLASIA, CYSTIC								1/1743
HYPERPLASIA, EPITHELIAL								16/1743

*VARIABILITY AMONG DATA SOURCES SIGNIFICANT AT THE 1% SIGNIFICANCE LEVEL

SEX:	FEMALE				MALE			
DIAGNOSIS / AGE IN MONTHS:	0 < 12	12 < 18	18 < 24	24+	0 < 12	12 < 18	18 < 24	24+
HYPERPLASIA, NODULAR								1/1743
*************** GROWTH ALTERATIONS, NON-NEOPLASTIC, ATROPHY ***************								
ATROPHY							5/437	54/1743
*************** GROWTH ALTERATIONS, NON-NEOPLASTIC, METAPLASIA ***************								
METAPLASIA, SQUAMOUS							5/437	20/1743
*************** GROWTH ALTERATIONS, NEOPLASTIC, BENIGN ***************								
ADENOMA, ACINAR CELL								2/1743
ADENOMA, PAPILLARY								1/1743
PARAGANGLIOMA								5/1743
*************** GROWTH ALTERATIONS, NEOPLASTIC, MALIGNANT, PRIMARY ***************								
ADENOCARCINOMA, MUCINOUS							1/437	
CARCINOMA								1/1743
*************** GROWTH ALTERATIONS, NEOPLASTIC, MALIGNANT, PRIMARY SITE NOT DETERMINED ***************								
LEUKEMIA								2/1743
LEUKEMIA, STEM CELL							1/437	
LYMPHOMA								1/1743
MESOTHELIOMA							4/437	4/1743

SEX:	FEMALE				MALE			
DIAGNOSIS / AGE IN MONTHS:	0 < 12	12 < 18	18 < 24	24+	0 < 12	12 < 18	18 < 24	24+
NEUROFIBROSARCOMA							1/437	

SEX:	FEMALE				MALE			
DIAGNOSIS / AGE IN MONTHS:	0 < 12	12 < 18	18 < 24	24+	0 < 12	12 < 18	18 < 24	24+
*************** INFLAMMATION AND REPAIR ***************								
INFLAMMATION						8/64	6/104	26/243
INFLAMMATION, GRANULOMATOUS								2/243
INFLAMMATION, SUPPURATIVE								1/243
*************** DEGENERATION, NECROSIS, AND DEPOSITION ***************								
CALCULI								1/243
MINERALIZATION							1/104	
*************** GROWTH ALTERATIONS, NON-NEOPLASTIC, HYPERPLASIA ***************								
HYPERPLASIA, EPITHELIAL							1/104	
*************** GROWTH ALTERATIONS, NEOPLASTIC, BENIGN ***************								
FIBROMA								1/243
*************** GROWTH ALTERATIONS, NEOPLASTIC, MALIGNANT, PRIMARY SITE NOT DETERMINED ***************								
ANGIOSARCOMA								2/243

SEX:	FEMALE				MALE			
DIAGNOSIS / AGE IN MONTHS:	0 < 12	12 < 18	18 < 24	24+	0 < 12	12 < 18	18 < 24	24+
*************** INFLAMMATION AND REPAIR ***************								
INFLAMMATION							5/392	
INFLAMMATION, NONSUPPURATIVE							1/392	9/216
INFLAMMATION, NONSUPPURATIVE, PROSTATE					2/170			
INFLAMMATION, PROSTATE						1/97	4/392	6/216
INFLAMMATION, SUPPURATIVE						3/97	8/392	4/216
INFLAMMATION, SUPPURATIVE, PROSTATE						1/97	4/392	5/216
PERIVASCULITIS/VASCULITIS						1/97		
*************** DEGENERATION, NECROSIS, AND DEPOSITION ***************								
CALCULI							4/392	12/216 *
*************** GROWTH ALTERATIONS, NON-NEOPLASTIC, ATROPHY ***************								
ATROPHY							1/392	
*************** GROWTH ALTERATIONS, NON-NEOPLASTIC, METAPLASIA ***************								
METAPLASIA, SQUAMOUS							1/392	
*************** GROWTH ALTERATIONS, NEOPLASTIC, MALIGNANT, PRIMARY SITE NOT DETERMINED ***************								
LEIOMYOSARCOMA							1/392	
LEUKEMIA								1/216

*VARIABILITY AMONG DATA SOURCES SIGNIFICANT AT THE 1% SIGNIFICANCE LEVEL

SEX:	FEMALE				MALE			
DIAGNOSIS / AGE IN MONTHS:	0 < 12	12 < 18	18 < 24	24+	0 < 12	12 < 18	18 < 24	24+
LYMPHOMA, RETICULUM CELL							1/392	

SEX:	FEMALE				MALE			
DIAGNOSIS / AGE IN MONTHS:	0 < 12	12 < 18	18 < 24	24+	0 < 12	12 < 18	18 < 24	24+
*************** MECHANICAL ABNORMALITIES AND CYST FORMATION ***************								
THROMBUS						1/314		
*************** INFLAMMATION AND REPAIR ***************								
INFLAMMATION						1/314	1/147	
INFLAMMATION, ULCERATIVE							1/147	
*************** GROWTH ALTERATIONS, NON-NEOPLASTIC, HYPERPLASIA ***************								
HYPERPLASIA						1/314		
*************** GROWTH ALTERATIONS, NEOPLASTIC, MALIGNANT, PRIMARY SITE NOT DETERMINED ***************								
LYMPHOMA								1/94

SEX:	FEMALE				MALE			
DIAGNOSIS / AGE IN MONTHS:	0 < 12	12 < 18	18 < 24	24+	0 < 12	12 < 18	18 < 24	24+
*************** MECHANICAL ABNORMALITIES AND CYST FORMATION ***************								
PROLAPSE					1/124			
*************** INFLAMMATION AND REPAIR ***************								
INFLAMMATION							1/1136	

SEX:	FEMALE				MALE			
AGE IN MONTHS:	0 < 12	12 < 18	18 < 24	24+	0 < 12	12 < 18	18 < 24	24+
DIAGNOSIS								
***************								
INFLAMMATION AND REPAIR								
***************								
INFLAMMATION								1/102

SEX:	FEMALE				MALE			
DIAGNOSIS / AGE IN MONTHS:	0 < 12	12 < 18	18 < 24	24+	0 < 12	12 < 18	18 < 24	24+
*************** MECHANICAL ABNORMALITIES AND CYST FORMATION ***************								
CYST						3/314	1/147	1/94
ECTASIA					16/171	26/314	12/147	14/94*
*************** INFLAMMATION AND REPAIR ***************								
ABSCESS						2/314		1/94
INFLAMMATION					2/171	5/314	3/147	7/94
INFLAMMATION, NONSUPPURATIVE								1/94
INFLAMMATION, SUPPURATIVE								2/94
*************** DEGENERATION, NECROSIS, AND DEPOSITION ***************								
NECROSIS								2/94
*************** GROWTH ALTERATIONS, NON-NEOPLASTIC, HYPERPLASIA ***************								
HYPERKERATOSIS					1/171			1/94
HYPERPLASIA						1/314		
*************** GROWTH ALTERATIONS, NON-NEOPLASTIC, ATROPHY ***************								
ATROPHY					1/171	2/314	2/147	

*VARIABILITY AMONG DATA SOURCES SIGNIFICANT AT THE 1% SIGNIFICANCE LEVEL

SEX:	FEMALE				MALE			
DIAGNOSIS / AGE IN MONTHS:	0 < 12	12 < 18	18 < 24	24+	0 < 12	12 < 18	18 < 24	24+
*************** GROWTH ALTERATIONS, NON-NEOPLASTIC, METAPLASIA ***************								
METAPLASIA								1/94
*************** GROWTH ALTERATIONS, NEOPLASTIC, MALIGNANT, PRIMARY SITE NOT DETERMINED ***************								
LYMPHOMA								1/94
LYMPHOMA, RETICULUM CELL								1/94

SEX:	FEMALE				MALE			
DIAGNOSIS / AGE IN MONTHS:	0 < 12	12 < 18	18 < 24	24+	0 < 12	12 < 18	18 < 24	24+
*************** MECHANICAL ABNORMALITIES AND CYST FORMATION ***************								
CYST								1/958
*************** INFLAMMATION AND REPAIR ***************								
ABSCESS						1/199	4/1136	
INFLAMMATION						2/199	1/1136	2/958
INFLAMMATION, SUPPURATIVE					1/124			3/958
*************** DEGENERATION, NECROSIS, AND DEPOSITION ***************								
CALCULI								1/958
*************** GROWTH ALTERATIONS, NON-NEOPLASTIC, HYPERPLASIA ***************								
ACANTHOSIS					1/124			
HYPERKERATOSIS					1/124			
HYPERPLASIA								2/958

SEX:	FEMALE				MALE			
DIAGNOSIS / AGE IN MONTHS:	0 < 12	12 < 18	18 < 24	24+	0 < 12	12 < 18	18 < 24	24+
*************** MECHANICAL ABNORMALITIES AND CYST FORMATION ***************								
ECTASIA					1/131	2/300	8/161	7/102
*************** INFLAMMATION AND REPAIR ***************								
INFLAMMATION					1/131		3/161	2/102
INFLAMMATION, GRANULOMATOUS							1/161	
INFLAMMATION, NONSUPPURATIVE								1/102
*************** GROWTH ALTERATIONS, NON-NEOPLASTIC, HYPERPLASIA ***************								
HYPERKERATOSIS								1/102
HYPERPLASIA								1/102
*************** GROWTH ALTERATIONS, NON-NEOPLASTIC, ATROPHY ***************								
ATROPHY							1/161	
*************** GROWTH ALTERATIONS, NON-NEOPLASTIC, METAPLASIA ***************								
METAPLASIA, SQUAMOUS								1/102

SEX:	FEMALE				MALE			
DIAGNOSIS / AGE IN MONTHS:	0 < 12	12 < 18	18 < 24	24+	0 < 12	12 < 18	18 < 24	24+
*************** MECHANICAL ABNORMALITIES AND CYST FORMATION ***************								
CYST, INCLUSION								1/1743
DILATATION							1/437	
*************** INFLAMMATION AND REPAIR ***************								
ABSCESS							1/437	8/1743
INFLAMMATION								5/1743
INFLAMMATION, SUPPURATIVE								4/1743
INFLAMMATION, ULCERATIVE								1/1743
*************** DEGENERATION, NECROSIS, AND DEPOSITION ***************								
NECROSIS								3/1743
*************** GROWTH ALTERATIONS, NON-NEOPLASTIC, HYPERPLASIA ***************								
HYPERPLASIA								6/1743
*************** GROWTH ALTERATIONS, NEOPLASTIC, BENIGN ***************								
ADENOMA						1/101	6/437	17/1743
ADENOMA, SEBACEOUS							1/437	

SEX:	FEMALE				MALE			
AGE IN MONTHS:	0 < 12	12 < 18	18 < 24	24+	0 < 12	12 < 18	18 < 24	24+
DIAGNOSIS								
*************** GROWTH ALTERATIONS, NEOPLASTIC, MALIGNANT, PRIMARY ***************								
CARCINOMA						1/101	1/437	14/1743
CARCINOMA, SQUAMOUS CELL						1/101	1/437	1/1743

SEX:	FEMALE				MALE			
DIAGNOSIS / AGE IN MONTHS:	0 < 12	12 < 18	18 < 24	24+	0 < 12	12 < 18	18 < 24	24+
*************** INFLAMMATION AND REPAIR ***************								
ABSCESS								1/216
INFLAMMATION, NONSUPPURATIVE						1/97		

SEX:	FEMALE				MALE			
DIAGNOSIS / AGE IN MONTHS:	0 < 12	12 < 18	18 < 24	24+	0 < 12	12 < 18	18 < 24	24+
*************** GROWTH ALTERATIONS, NEOPLASTIC, BENIGN ***************								
FIBROMA								1/1743

SEX:	FEMALE				MALE			
DIAGNOSIS / AGE IN MONTHS:	0 < 12	12 < 18	18 < 24	24+	0 < 12	12 < 18	18 < 24	24+
*************** MECHANICAL ABNORMALITIES AND CYST FORMATION ***************								
CYST					2/171	8/314	6/147	7/94
SPERMATOCELE								1/94
*************** INFLAMMATION AND REPAIR ***************								
FIBROSIS							1/147	
INFLAMMATION, NONSUPPURATIVE								1/94
PERIVASCULITIS/VASCULITIS								1/94
*************** DEGENERATION, NECROSIS, AND DEPOSITION ***************								
DEGENERATION								2/94
MINERALIZATION					1/171	2/314	1/147	13/94
NECROSIS, COAGULATIVE					1/171			
PIGMENT DEPOSITION								1/94
*************** GROWTH ALTERATIONS, NON-NEOPLASTIC, HYPERTROPHY ***************								
KARYOMEGALY						1/314		1/94
*************** GROWTH ALTERATIONS, NON-NEOPLASTIC, HYPERPLASIA ***************								
HYPERPLASIA, INTERSTITIAL CELL					1/171	9/314	2/147	3/94

SEX:	FEMALE				MALE			
DIAGNOSIS / AGE IN MONTHS:	O < 12	12 < 18	18 < 24	24+	O < 12	12 < 18	18 < 24	24+
*************** GROWTH ALTERATIONS, NON-NEOPLASTIC, ATROPHY ***************								
ATROPHY					3/171	22/314*	18/147	68/94
*************** GROWTH ALTERATIONS, NON-NEOPLASTIC, HYPOPLASIA ***************								
HYPOSPERMIA					3/171	35/314	38/147	27/94 *
*************** GROWTH ALTERATIONS, NEOPLASTIC, BENIGN ***************								
INTERSTITIAL CELL TUMOR							1/147	3/94
*************** GROWTH ALTERATIONS, NEOPLASTIC, MALIGNANT, PRIMARY SITE NOT DETERMINED ***************								
LYMPHOMA					1/171			1/94

*VARIABILITY AMONG DATA SOURCES SIGNIFICANT AT THE 1% SIGNIFICANCE LEVEL

SEX:	FEMALE				MALE			
DIAGNOSIS / AGE IN MONTHS:	0 < 12	12 < 18	18 < 24	24+	0 < 12	12 < 18	18 < 24	24+
*************** INFLAMMATION AND REPAIR ***************								
FIBROSIS								1/958
GRANULOMA						1/199		1/958
INFLAMMATION, SUPPURATIVE							1/1136	
*************** DEGENERATION, NECROSIS, AND DEPOSITION ***************								
DEGENERATION							9/1136	
INFARCT								1/958
MINERALIZATION						2/199	2/1136	2/958
*************** GROWTH ALTERATIONS, NON-NEOPLASTIC, ATROPHY ***************								
ATROPHY						1/199	11/1136	10/958
*************** GROWTH ALTERATIONS, NON-NEOPLASTIC, HYPOPLASIA ***************								
ASPERMATOGENESIS								1/958
*************** GROWTH ALTERATIONS, NEOPLASTIC, BENIGN ***************								
INTERSTITIAL CELL TUMOR							2/1136	1/958
*************** GROWTH ALTERATIONS, NEOPLASTIC, MALIGNANT, PRIMARY ***************								
CARCINOMA, EMBRYONAL							1/1136	1/958

ORGAN SYSTEM: GENITAL SYSTEM; MALES ORGAN: TESTIS STRAIN: B6C3F1

SEX:	FEMALE				MALE			
DIAGNOSIS / AGE IN MONTHS:	0 < 12	12 < 18	18 < 24	24+	0 < 12	12 < 18	18 < 24	24+
SEMINOMA							1/1136	

SEX:	FEMALE				MALE			
DIAGNOSIS / AGE IN MONTHS:	0 < 12	12 < 18	18 < 24	24+	0 < 12	12 < 18	18 < 24	24+
*************** MECHANICAL ABNORMALITIES AND CYST FORMATION ***************								
CYST					1/131	1/300	1/161	5/102
SPERMATOCELE								1/102
*************** INFLAMMATION AND REPAIR ***************								
INFLAMMATION, GRANULOMATOUS							1/161	1/102
INFLAMMATION, SUPPURATIVE							1/161	
*************** DEGENERATION, NECROSIS, AND DEPOSITION ***************								
DEGENERATION, VACUOLAR					2/131	2/300		
MINERALIZATION								2/102
NECROSIS								1/102
*************** GROWTH ALTERATIONS, NON-NEOPLASTIC, HYPERPLASIA ***************								
HYPERPLASIA, INTERSTITIAL CELL						1/300	1/161	1/102
*************** GROWTH ALTERATIONS, NON-NEOPLASTIC, ATROPHY ***************								
ATROPHY					2/131	16/300*	6/161	29/102*
*************** GROWTH ALTERATIONS, NON-NEOPLASTIC, HYPOPLASIA ***************								
HYPOSPERMIA					2/131	2/300	1/161	3/102

*VARIABILITY AMONG DATA SOURCES SIGNIFICANT AT THE 1% SIGNIFICANCE LEVEL

SEX:	FEMALE				MALE			
DIAGNOSIS / AGE IN MONTHS:	0 < 12	12 < 18	18 < 24	24+	0 < 12	12 < 18	18 < 24	24+
*************** GROWTH ALTERATIONS, NEOPLASTIC, BENIGN ***************								
ANGIOMA						1/300		
INTERSTITIAL CELL TUMOR						1/300		2/102
*************** GROWTH ALTERATIONS, NEOPLASTIC, MALIGNANT, PRIMARY SITE NOT DETERMINED ***************								
LYMPHOMA							1/161	

SEX:	FEMALE				MALE			
DIAGNOSIS / AGE IN MONTHS:	0 < 12	12 < 18	18 < 24	24+	0 < 12	12 < 18	18 < 24	24+
*************** MECHANICAL ABNORMALITIES AND CYST FORMATION ***************								
THROMBUS							1/437	
*************** INFLAMMATION AND REPAIR ***************								
GRANULOMA								2/1743
INFLAMMATION, FAT							1/437	
PERIVASCULITIS/VASCULITIS							3/437	1/1743
*************** DEGENERATION, NECROSIS, AND DEPOSITION ***************								
DEGENERATION						1/101	19/437	66/1743
INFARCT								1/1743
MINERALIZATION					2/101		5/437	35/1743
NECROSIS							1/437	
NECROSIS, FAT							1/437	
NECROSIS, VASCULAR, FIBRINOID								1/1743
*************** GROWTH ALTERATIONS, NON-NEOPLASTIC, HYPERPLASIA ***************								
HYPERPLASIA						2/101	3/437	7/1743
HYPERPLASIA, INTERSTITIAL CELL						6/101*	46/437 *	79/1743
HYPERPLASIA, NODULAR					1/101	15/101*	7/437	7/1743

*VARIABILITY AMONG DATA SOURCES SIGNIFICANT AT THE 1% SIGNIFICANCE LEVEL

ORGAN SYSTEM: GENITAL SYSTEM; MALES ORGAN: TESTIS STRAIN: FISCHER 344

SEX:	FEMALE				MALE			
DIAGNOSIS / AGE IN MONTHS:	0 < 12	12 < 18	18 < 24	24+	0 < 12	12 < 18	18 < 24	24+
*************** GROWTH ALTERATIONS, NON-NEOPLASTIC, ATROPHY ***************								
ATROPHY					6/101	11/101	44/437*	273/1743*
ATROPHY, TUBULAR							1/437	
*************** GROWTH ALTERATIONS, NON-NEOPLASTIC, HYPOPLASIA ***************								
ASPERMATOGENESIS					5/101		2/437	3/1743
*************** GROWTH ALTERATIONS, NEOPLASTIC, BENIGN ***************								
INTERSTITIAL CELL TUMOR					1/101	18/101	257/437*	1586/1743*
*************** GROWTH ALTERATIONS, NEOPLASTIC, MALIGNANT, PRIMARY ***************								
INTERSTITIAL CELL TUMOR, MALIGNANT							2/437	42/1743
SEMINOMA							1/437	2/1743
*************** GROWTH ALTERATIONS, NEOPLASTIC, MALIGNANT, PRIMARY SITE NOT DETERMINED ***************								
LEUKEMIA								1/1743
LEUKEMIA, GRANULOCYTIC								1/1743
LEUKEMIA, LYMPHOCYTIC								1/1743
MESOTHELIOMA					1/101	2/101	6/437	14/1743

*VARIABILITY AMONG DATA SOURCES SIGNIFICANT AT THE 1% SIGNIFICANCE LEVEL

SEX:	FEMALE				MALE			
DIAGNOSIS / AGE IN MONTHS:	0 < 12	12 < 18	18 < 24	24+	0 < 12	12 < 18	18 < 24	24+
*************** MECHANICAL ABNORMALITIES AND CYST FORMATION ***************								
EDEMA								1/243
*************** INFLAMMATION AND REPAIR ***************								
GRANULOMA						1/64		
PERIVASCULITIS/VASCULITIS							1/104	4/243
*************** DEGENERATION, NECROSIS, AND DEPOSITION ***************								
ARTERIOSCLEROSIS							1/104	
MINERALIZATION							2/104	9/243
*************** GROWTH ALTERATIONS, NON-NEOPLASTIC, ATROPHY ***************								
ATROPHY						12/64	39/104	70/243
*************** GROWTH ALTERATIONS, NON-NEOPLASTIC, HYPOPLASIA ***************								
ASPERMATOGENESIS								2/243
HYPOPLASIA						1/64	5/104	9/243

SEX:	FEMALE				MALE			
DIAGNOSIS / AGE IN MONTHS:	0 < 12	12 < 18	18 < 24	24+	0 < 12	12 < 18	18 < 24	24+
DEVELOPMENTAL ANOMALIES								
CRYPTORCHID						1/97		
INFLAMMATION AND REPAIR								
FIBROSIS								1/216
GRANULOMA					1/170			
INFLAMMATION							1/392	
INFLAMMATION, NONSUPPURATIVE					4/170			
PERIVASCULITIS/VASCULITIS						3/97	13/392	29/216 *
DEGENERATION, NECROSIS, AND DEPOSITION								
ARTERIOLOSCLEROSIS								3/216
ARTERIOSCLEROSIS							1/392	
DEGENERATION						1/97	3/392	2/216
MINERALIZATION						1/97	1/392	2/216
MINERALIZATION, VASCULAR							1/392	
GROWTH ALTERATIONS, NON-NEOPLASTIC HYPERTROPHY								
HYPERTROPHY								2/216

*VARIABILITY AMONG DATA SOURCES SIGNIFICANT AT THE 1% SIGNIFICANCE LEVEL

SEX:	FEMALE				MALE			
DIAGNOSIS / AGE IN MONTHS:	0 < 12	12 < 18	18 < 24	24+	0 < 12	12 < 18	18 < 24	24+
*************** GROWTH ALTERATIONS, NON-NEOPLASTIC, HYPERPLASIA ***************								
HYPERPLASIA						1/97		1/216
HYPERPLASIA, INTERSTITIAL CELL								3/216
*************** GROWTH ALTERATIONS, NON-NEOPLASTIC, ATROPHY ***************								
ATROPHY					2/170	4/97	10/392	19/216 *
*************** GROWTH ALTERATIONS, NON-NEOPLASTIC, HYPOPLASIA ***************								
ASPERMATOGENESIS							1/392	1/216
HYPOPLASIA							1/392	2/216
*************** GROWTH ALTERATIONS, NEOPLASTIC, BENIGN ***************								
INTERSTITIAL CELL TUMOR						1/97	4/392	10/216
*************** GROWTH ALTERATIONS, NEOPLASTIC, NOT DETERMINED ***************								
NEOPLASM							1/392	

*VARIABILITY AMONG DATA SOURCES SIGNIFICANT AT THE 1% SIGNIFICANCE LEVEL

SEX:	FEMALE				MALE			
DIAGNOSIS / AGE IN MONTHS:	0 < 12	12 < 18	18 < 24	24+	0 < 12	12 < 18	18 < 24	24+
*************** INFLAMMATION AND REPAIR ***************								
FIBROSIS				1/71				
*************** DEGENERATION, NECROSIS, AND DEPOSITION ***************								
NECROSIS								1/94
PIGMENT DEPOSITION			1/193	1/71				
*************** GROWTH ALTERATIONS, NON-NEOPLASTIC, HYPERPLASIA ***************								
HYPERPLASIA, EOSINOPHILIC		2/369	1/193					
HYPERPLASIA, MEGAKARYOCYTIC		1/369		1/71				
HYPERPLASIA, MYELOID								1/94
LEUKOCYTOSIS, LYMPHOCYTIC	1/212	4/369	2/193			1/314		
LEUKOCYTOSIS, NEUTROPHILIC	10/212	31/369	24/193	28/71	21/171	55/314	36/147	28/94 *
*************** GROWTH ALTERATIONS, NON-NEOPLASTIC, DYSPLASIA ***************								
ATYPICAL RED BLOOD CELLS		1/369	1/193		1/171			
ATYPICAL WHITE BLOOD CELLS	2/212	3/369	5/193			1/314	2/147	
*************** GROWTH ALTERATIONS, NON-NEOPLASTIC, HYPOPLASIA ***************								
HYPOPLASIA, HEMATOPOIETIC				1/71	1/171			
NEUTROPENIA			1/193					

*VARIABILITY AMONG DATA SOURCES SIGNIFICANT AT THE 1% SIGNIFICANCE LEVEL

SEX:	FEMALE				MALE			
DIAGNOSIS / AGE IN MONTHS:	0 < 12	12 < 18	18 < 24	24+	0 < 12	12 < 18	18 < 24	24+
*************** GROWTH ALTERATIONS, NEOPLASTIC, MALIGNANT, PRIMARY ***************								
LEUKEMIA			1/193					1/94
*************** GROWTH ALTERATIONS, NEOPLASTIC, MALIGNANT, PRIMARY SITE NOT DETERMINED ***************								
LYMPHOMA	3/212	5/369	4/193		3/171			2/94
LYMPHOMA, HISTIOCYTIC								1/94
LYMPHOMA, RETICULUM CELL		2/369	5/193	8/71			1/147	2/94

SEX:	FEMALE				MALE			
DIAGNOSIS / AGE IN MONTHS:	0 < 12	12 < 18	18 < 24	24+	0 < 12	12 < 18	18 < 24	24+
*************** GROWTH ALTERATIONS, NON-NEOPLASTIC, HYPERPLASIA ***************								
LEUKOCYTOSIS, NEUTROPHILIC			1/1356	1/1015				1/958
*************** GROWTH ALTERATIONS, NON-NEOPLASTIC, ATROPHY ***************								
LEUKOPENIA								1/958
*************** GROWTH ALTERATIONS, NON-NEOPLASTIC, HYPOPLASIA ***************								
HYPOPLASIA, HEMATOPOIETIC			1/1356					
*************** GROWTH ALTERATIONS, NEOPLASTIC, MALIGNANT, PRIMARY ***************								
LEUKEMIA				1/1015				
LEUKEMIA, GRANULOCYTIC		1/103						
LEUKEMIA, LYMPHOCYTIC				1/1015				1/958
*************** GROWTH ALTERATIONS, NEOPLASTIC, MALIGNANT, PRIMARY SITE NOT DETERMINED ***************								
ANGIOSARCOMA								1/958
LYMPHOMA			1/1356					1/958

SEX:	FEMALE				MALE			
DIAGNOSIS / AGE IN MONTHS:	0 < 12	12 < 18	18 < 24	24+	0 < 12	12 < 18	18 < 24	24+
*************** MECHANICAL ABNORMALITIES AND CYST FORMATION ***************								
HEMORRHAGE				1/75				
*************** INFLAMMATION AND REPAIR ***************								
FIBROSIS				5/75				2/102
*************** DEGENERATION, NECROSIS, AND DEPOSITION ***************								
PIGMENT DEPOSITION			1/192	1/75				1/102
*************** GROWTH ALTERATIONS, NON-NEOPLASTIC, HYPERPLASIA ***************								
HEMATOPOIESIS	1/176							
HYPERPLASIA							1/161	
HYPERPLASIA, EOSINOPHILIC			1/192			1/300		
HYPERPLASIA, MEGAKARYOCYTIC							2/161	
HYPERPLASIA, MYELOID							1/161	
LEUKOCYTOSIS, LYMPHOCYTIC	11/176	7/367			8/131	8/300	6/161	
LEUKOCYTOSIS, NEUTROPHILIC	2/176	16/367	12/192	34/75	3/131	14/300	20/161	25/102
*************** GROWTH ALTERATIONS, NON-NEOPLASTIC, DYSPLASIA ***************								
ATYPICAL RED BLOOD CELLS		1/367	1/192					
ATYPICAL WHITE BLOOD CELLS	1/176	1/367	2/192	1/75			2/161	1/102

SEX:	FEMALE				MALE			
DIAGNOSIS / AGE IN MONTHS:	0 < 12	12 < 18	18 < 24	24+	0 < 12	12 < 18	18 < 24	24+
*************** GROWTH ALTERATIONS, NON-NEOPLASTIC, HYFOPLASIA ***************								
HYPOPLASIA, HEMATOPOIETIC							1/161	
NEUTROPENIA	1/176				1/131			
*************** GROWTH ALTERATIONS, NEOPLASTIC, MALIGNANT, PRIMARY SITE NOT DETERMINED ***************								
LYMPHOMA	1/176			1/75		1/300		1/102
LYMPHOMA, RETICULUM CELL		2/367	7/192	11/75			1/161	2/102

SEX:	FEMALE				MALE			
DIAGNOSIS / AGE IN MONTHS:	0 < 12	12 < 18	18 < 24	24+	0 < 12	12 < 18	18 < 24	24+
*************** INFLAMMATION AND REPAIR ***************								
PERIVASCULITIS/VASCULITIS								1/1743
*************** DEGENERATION, NECROSIS, AND DEPOSITION ***************								
CELLULAR CHANGE, VACUOLAR			1/327					
MYELOFIBROSIS								3/1743
NECROSIS								1/1743
OSTEODYSTROPHY, FIBROUS								1/1743
OSTEOSCLEROSIS				2/1575				
*************** GROWTH ALTERATIONS, NON-NEOPLASTIC, HYPERTROPHY ***************								
CYTOMEGALY			1/327					
*************** GROWTH ALTERATIONS, NON-NEOPLASTIC, HYPERPLASIA ***************								
HISTIOCYTOSIS				1/1575				
HYPERPLASIA								1/1743
HYPERPLASIA, ERYTHROID							1/437	1/1743
HYPERPLASIA, MEGAKARYOCYTIC								1/1743
HYPERPLASIA, MYELOID					1/101	1/101		
LEUKOCYTOSIS				1/1575				
LEUKOCYTOSIS, NEUTROPHILIC			1/327	2/1575				

SEX:	FEMALE				MALE			
DIAGNOSIS / AGE IN MONTHS:	0 < 12	12 < 18	18 < 24	24+	0 < 12	12 < 18	18 < 24	24+
MYELOHISTIOCYTOSIS								1/1743
*************** GROWTH ALTERATIONS, NON-NEOPLASTIC, ATROPHY ***************								
LEUKOPENIA							1/437	
*************** GROWTH ALTERATIONS, NON-NEOPLASTIC, HYPOPLASIA ***************								
HYPOPLASIA, HEMATOPOIETIC				1/1575				
*************** GROWTH ALTERATIONS, NEOPLASTIC, MALIGNANT, PRIMARY ***************								
LEUKEMIA						1/101	1/437	10/1743
LEUKEMIA, GRANULOCYTIC				1/1575				4/1743
LEUKEMIA, LYMPHOBLASTIC				2/1575				2/1743
LEUKEMIA, LYMPHOCYTIC				1/1575				5/1743
LEUKEMIA, MONONUCLEAR							1/437	8/1743
*************** GROWTH ALTERATIONS, NEOPLASTIC, NOT DETERMINED ***************								
NEOPLASM				6/1575			3/437	1/1743

SEX:	FEMALE				MALE			
DIAGNOSIS / AGE IN MONTHS:	0 < 12	12 < 18	18 < 24	24+	0 < 12	12 < 18	18 < 24	24+
*************** DEGENERATION, NECROSIS, AND DEPOSITION ***************								
FATTY CHANGE					2/170			
*************** GROWTH ALTERATIONS, NON-NEOPLASTIC, HYPERPLASIA ***************								
HYPERPLASIA			4/420	5/240		1/97	1/392	3/216
HYPERPLASIA, ERYTHROID							3/392	1/216
HYPERPLASIA, MEGAKARYOCYTIC			2/420	1/240				
HYPERPLASIA, MYELOID							2/392	
*************** GROWTH ALTERATIONS, NON-NEOPLASTIC, HYPOPLASIA ***************								
HYPOPLASIA, HEMATOPOIETIC			1/420					2/216
*************** GROWTH ALTERATIONS, NEOPLASTIC, MALIGNANT, PRIMARY ***************								
LEUKEMIA								1/216
LEUKEMIA, LYMPHOCYTIC							1/392	
*************** GROWTH ALTERATIONS, NEOPLASTIC, MALIGNANT, PRIMARY SITE NOT DETERMINED ***************								
LYMPHOMA			1/420	1/240	1/170			
LYMPHOMA, RETICULUM CELL							1/392	

SEX:	FEMALE				MALE			
DIAGNOSIS / AGE IN MONTHS:	0 < 12	12 < 18	18 < 24	24+	0 < 12	12 < 18	18 < 24	24+
*************** MECHANICAL ABNORMALITIES AND CYST FORMATION ***************								
EDEMA							1/147	
*************** INFLAMMATION AND REPAIR ***************								
INFLAMMATION		1/369						
*************** DEGENERATION, NECROSIS, AND DEPOSITION ***************								
NECROSIS						1/314		
*************** GROWTH ALTERATIONS, NON-NEOPLASTIC, HYPERPLASIA ***************								
HYPERPLASIA, LYMPHOID	1/212	1/369	4/193	3/71		1/314	3/147	1/94
HYPERPLASIA, RETICULOENDOTHELIAL		1/369	1/193	2/71		3/314	1/147	3/94
*************** GROWTH ALTERATIONS, NEOPLASTIC, MALIGNANT, PRIMARY SITE NOT DETERMINED ***************								
LYMPHOMA	4/212	8/369	10/193	4/71	3/171		4/147	2/94
LYMPHOMA, HISTIOCYTIC								3/94
LYMPHOMA, RETICULUM CELL		3/369	23/193	25/71		1/314	5/147	10/94

SEX:	FEMALE				MALE			
DIAGNOSIS / AGE IN MONTHS:	0 < 12	12 < 18	18 < 24	24+	0 < 12	12 < 18	18 < 24	24+
*************** MECHANICAL ABNORMALITIES AND CYST FORMATION ***************								
ANGIECTASIS			2/1356			3/199	3/1136	
CONGESTION		2/103	8/1356	3/1015		1/199	28/1136	37/958
CYST			1/1356	2/1015			1/1136	3/958
CYST, HEMORRHAGIC							1/1136	
EDEMA							3/1136	
ERYTHROPHAGOCYTOSIS								1/958
HEMORRHAGE			10/1356				28/1136	8/958
LYMPHANGIECTASIS			1/1356				1/1136	3/958
THROMBUS							2/1136	
*************** INFLAMMATION AND REPAIR ***************								
ABSCESS			1/1356					
FIBROSIS								1/958
INFLAMMATION	1/67		30/1356	4/1015	1/124	8/199	52/1136	1/958
INFLAMMATION, GRANULOMATOUS			1/1356	5/1015				3/958
INFLAMMATION, HEMORRHAGIC				1/1015				
INFLAMMATION, NECROTIZING							1/1136	
INFLAMMATION, SUPPURATIVE			2/1356	1/1015			5/1136	

SEX:	FEMALE				MALE			
DIAGNOSIS AGE IN MONTHS:	0 < 12	12 < 18	18 < 24	24+	0 < 12	12 < 18	18 < 24	24+
*************** DEGENERATION, NECROSIS, AND DEPOSITION ***************								
DEGENERATION, CYSTIC							1/1136	
FATTY CHANGE								1/958
HEMOSIDERIN DEPOSITION							1/1136	2/958
NECROSIS			1/1356	1/1015			1/1136	
*************** GROWTH ALTERATIONS, NON-NEOPLASTIC, HYPERPLASIA ***************								
HEMATOPOIESIS			7/1356	2/1015			17/1136	13/958
HYPERPLASIA	1/67	1/103	8/1356	14/1015			5/1136	10/958
HYPERPLASIA, GRANULOCYTIC	1/67							
HYPERPLASIA, HISTIOCYTIC							1/1136	1/958
HYPERPLASIA, LYMPHOID	2/67		23/1356	21/1015	1/124		20/1136	30/958
HYPERPLASIA, MEGAKARYOCYTIC								1/958
HYPERPLASIA, PLASMA CELL			4/1356	2/1015			2/1136	3/958
HYPERPLASIA, RETICULOENDOTHELIAL			2/1356				2/1136	
HYPERPLASIA, RETICULUM CELL			1/1356	3/1015			5/1136	4/958
*************** GROWTH ALTERATIONS, NON-NEOPLASTIC, ATROPHY ***************								
ATROPHY			1/1356	1/1015	1/124		3/1136	2/958

SEX:	FEMALE				MALE			
DIAGNOSIS / AGE IN MONTHS:	0 < 12	12 < 18	18 < 24	24+	0 < 12	12 < 18	18 < 24	24+
*************** GROWTH ALTERATIONS, NEOPLASTIC, BENIGN ***************								
ANGIOMA								2/958
*************** GROWTH ALTERATIONS, NEOPLASTIC, MALIGNANT, METASTATIC ***************								
ADENOCARCINOMA			1/1356					
CARCINOMA, ACINAR CELL								1/958
CARCINOMA, ALVEOLAR BRONCHIOLAR							2/1136	
CARCINOMA, BASAL CELL				1/1015				
CARCINOMA, HEPATOCELLULAR							3/1136	1/958
LEIOMYOSARCOMA		1/103						
*************** GROWTH ALTERATIONS, NEOPLASTIC, MALIGNANT, PRIMARY SITE NOT DETERMINED ***************								
ANGIOSARCOMA				3/1015			1/1136	1/958
LYMPHOMA			5/1356	10/1015			3/1136	11/958
LYMPHOMA, HISTIOCYTIC			3/1356	4/1015			12/1136	2/958
LYMPHOMA, LYMPHOCYTIC			1/1356	3/1015			3/1136	3/958
SARCOMA				1/1015			1/1136	

SEX:	FEMALE				MALE			
DIAGNOSIS / AGE IN MONTHS:	0 < 12	12 < 18	18 < 24	24+	0 < 12	12 < 18	18 < 24	24+
*************** MECHANICAL ABNORMALITIES AND CYST FORMATION ***************								
CYST		1/367	1/192	1/75		1/300		
EDEMA			1/192	2/75				
HEMORRHAGE				1/75				
*************** INFLAMMATION AND REPAIR ***************								
INFLAMMATION	1/176	5/367						
INFLAMMATION, GRANULOMATOUS	1/176							
*************** DEGENERATION, NECROSIS, AND DEPOSITION ***************								
PIGMENT DEPOSITION								1/102
*************** GROWTH ALTERATIONS, NON-NEOPLASTIC, HYPERPLASIA ***************								
HEMATOPOIESIS		1/367	1/192				1/161	
HYPERPLASIA, LYMPHOID	7/176	2/367	9/192	3/75		1/300	4/161	3/102
HYPERPLASIA, PLASMA CELL	3/176	4/367	3/192					
HYPERPLASIA, RETICULOENDOTHELIAL	2/176		8/192	4/75		2/300	4/161	1/102
*************** GROWTH ALTERATIONS, NEOPLASTIC, BENIGN ***************								
ANGIOMA								1/102

SEX:	FEMALE				MALE			
DIAGNOSIS / AGE IN MONTHS:	0 < 12	12 < 18	18 < 24	24+	0 < 12	12 < 18	18 < 24	24+
*************** GROWTH ALTERATIONS, NEOPLASTIC, MALIGNANT, PRIMARY SITE NOT DETERMINED ***************								
LYMPHOMA	1/176	8/367	6/192	3/75		1/300	2/161	5/102
LYMPHOMA, HISTIOCYTIC								1/102
LYMPHOMA, RETICULUM CELL		13/367	20/192	28/75		4/300	15/161	14/102

SEX:	FEMALE				MALE			
DIAGNOSIS / AGE IN MONTHS:	0 < 12	12 < 18	18 < 24	24+	0 < 12	12 < 18	18 < 24	24+
*************** MECHANICAL ABNORMALITIES AND CYST FORMATION ***************								
CONGESTION			3/327	2/1575			2/437	4/1743
CYST				2/1575			1/437	2/1743
DILATATION								3/1743
EDEMA								5/1743
ERYTHROPHAGOCYTOSIS				1/1575				
HEMORRHAGE				1/1575		1/101	2/437	7/1743
LYMPHANGIECTASIS				8/1575			1/437	19/1743
LYMPHEDEMA								1/1743
*************** INFLAMMATION AND REPAIR ***************								
GRANULOMA								1/1743
INFLAMMATION			1/327	10/1575		1/101	1/437	6/1743
INFLAMMATION, FAT								1/1743
INFLAMMATION, GRANULOMATOUS			1/327	1/1575				
INFLAMMATION, NECROTIZING				1/1575				
INFLAMMATION, NONSUPPURATIVE								1/1743
INFLAMMATION, SUPPURATIVE								1/1743
*************** DEGENERATION, NECROSIS, AND DEPOSITION ***************								
CELLULAR CHANGE				1/1575				

SEX:	FEMALE				MALE			
DIAGNOSIS / AGE IN MONTHS:	0 < 12	12 < 18	18 < 24	24+	0 < 12	12 < 18	18 < 24	24+
HEMOSIDERIN DEPOSITION				13/1575				6/1743
LIPOCHROMATOSIS								3/1743
PIGMENT DEPOSITION			1/327					2/1743
*************** GROWTH ALTERATIONS, NON-NEOPLASTIC, HYPERPLASIA ***************								
HYPERPLASIA			1/327	2/1575			2/437	11/1743
HYPERPLASIA, HISTIOCYTIC								5/1743
HYPERPLASIA, LYMPHOID				8/1575		1/101	2/437	14/1743
HYPERPLASIA, PLASMA CELL			1/327	18/1575			1/437	19/1743
HYPERPLASIA, RETICULUM CELL				1/1575				16/1743
*************** GROWTH ALTERATIONS, NON-NEOPLASTIC, ATROPHY ***************								
ATROPHY				1/1575				1/1743
*************** GROWTH ALTERATIONS, NEOPLASTIC, MALIGNANT, METASTATIC ***************								
ADENOCARCINOMA			1/327	1/1575				5/1743
ADENOCARCINOMA, MUCINOUS							1/437	
CARCINOMA, C CELL				2/1575			1/437	
CARCINOMA, SQUAMOUS CELL								1/1743
CARCINOMA, UNDIFFERENTIATED			1/327					
HISTIOCYTOMA, MALIGNANT							1/437	
INTERSTITIAL CELL TUMOR								1/1743

DIAGNOSIS	FEMALE 0 < 12	FEMALE 12 < 18	FEMALE 18 < 24	FEMALE 24+	MALE 0 < 12	MALE 12 < 18	MALE 18 < 24	MALE 24+
MESOTHELIOMA							2/437	1/1743
NEUROFIBROSARCOMA							1/437	
*************** GROWTH ALTERATIONS, NEOPLASTIC, MALIGMANT, PRIMARY SITE NOT DETERMINED ***************								
ANGIOSARCOMA				1/1575				
LEUKEMIA	1/26			1/1575				8/1743
LEUKEMIA, GRANULOCYTIC				1/1575				2/1743
LEUKEMIA, LYMPHOBLASTIC				4/1575				1/1743
LEUKEMIA, LYMPHOCYTIC				3/1575			1/437	
LEUKEMIA, STEM CELL							1/437	
LYMPHOMA				2/1575				2/1743
LYMPHOMA, HISTIOCYTIC				1/1575				

SEX:	FEMALE				MALE			
DIAGNOSIS / AGE IN MONTHS:	0 < 12	12 < 18	18 < 24	24+	0 < 12	12 < 18	18 < 24	24+
*************** MECHANICAL ABNORMALITIES AND CYST FORMATION ***************								
CYST								1/243
HEMORRHAGE				1/336				
*************** INFLAMMATION AND REPAIR ***************								
INFLAMMATION				4/336			3/104	5/243
*************** GROWTH ALTERATIONS, NON-NEOPLASTIC, HYPERPLASIA ***************								
HYPERPLASIA				1/336				1/243
HYPERPLASIA, LYMPHOID				3/336				2/243
HYPERPLASIA, PLASMA CELL				1/336				
*************** GROWTH ALTERATIONS, NEOPLASTIC, BENIGN ***************								
ANGIOMA								2/243
*************** GROWTH ALTERATIONS, NEOPLASTIC, MALIGNANT, METASTATIC ***************								
NEUROFIBROSARCOMA								1/243
*************** GROWTH ALTERATIONS, NEOPLASTIC, MALIGNANT, PRIMARY SITE NOT DETERMINED ***************								
LYMPHOMA								1/243
LYMPHOMA, LYMPHOCYTIC				1/336				

DIAGNOSIS	FEMALE 0 < 12	FEMALE 12 < 18	FEMALE 18 < 24	FEMALE 24+	MALE 0 < 12	MALE 12 < 18	MALE 18 < 24	MALE 24+
MECHANICAL ABNORMALITIES AND CYST FORMATION								
CONGESTION	1/181							
EDEMA			1/420					
ERYTHROPHAGOCYTOSIS		1/60		1/240				2/216
HEMORRHAGE						1/97		
LYMPHANGIECTASIS			1/420	4/240			1/392	
INFLAMMATION AND REPAIR								
FIBROSIS								1/216
INFLAMMATION, NONSUPPURATIVE				1/240		1/97		
INFLAMMATION, SUPPURATIVE							2/392	
PERIVASCULITIS/VASCULITIS						2/97	1/392	1/216
DEGENERATION, NECROSIS, AND DEPOSITION								
HEMOSIDERIN DEPOSITION		1/60	4/420	6/240			1/392	7/216
MINERALIZATION							1/392	1/216
GROWTH ALTERATIONS, NON-NEOPLASTIC, HYPERTROPHY								
HYPERTROPHY		3/60	1/420	28/240 *		5/97	3/392	30/216*

*VARIABILITY AMONG DATA SOURCES SIGNIFICANT AT THE 1% SIGNIFICANCE LEVEL

SEX:	FEMALE				MALE			
DIAGNOSIS / AGE IN MONTHS:	0 < 12	12 < 18	18 < 24	24+	0 < 12	12 < 18	18 < 24	24+
*************** GROWTH ALTERATIONS, NON-NEOPLASTIC, HYPERPLASIA ***************								
HYPERPLASIA	1/181		1/420					2/216
HYPERPLASIA, LYMPHOID							1/392	
HYPERPLASIA, PLASMA CELL		1/60					2/392	2/216
HYPERPLASIA, RETICULOENDOTHELIAL			1/420					
HYPERPLASIA, RETICULUM CELL					1/170			1/216
*************** GROWTH ALTERATIONS, NEOPLASTIC, BENIGN ***************								
ANGIOMA							1/392	1/216
*************** GROWTH ALTERATIONS, NEOPLASTIC, MALIGNANT, PRIMARY SITE NOT DETERMINED ***************								
LEUKEMIA, LYMPHOCYTIC			1/420			1/97		
LYMPHOMA	1/181		1/420	1/240	1/170			2/216
LYMPHOMA, RETICULUM CELL				2/240			2/392	1/216
SARCOMA, HISTIOCYTIC								3/216

SEX:	FEMALE				MALE			
DIAGNOSIS / AGE IN MONTHS:	C < 12	12 < 18	18 < 24	24+	O < 12	12 < 18	18 < 24	24+
*************** MECHANICAL ABNORMALITIES AND CYST FORMATION ***************								
ANGIECTASIS								3/94
CONGESTION				1/71				2/94
THROMBUS						1/314	2/147	2/94
*************** INFLAMMATION AND REPAIR ***************								
FIBROSIS						1/314		
*************** DEGENERATION, NECROSIS, AND DEPOSITION ***************								
INFARCT		1/369	1/193	1/71	1/171	1/314	2/147	
MINERALIZATION					1/171			
NECROSIS			2/193			1/314		
PIGMENT DEPOSITION	1/212			2/71			1/147	1/94
*************** GROWTH ALTERATIONS, NON-NEOPLASTIC, HYPERPLASIA ***************								
HEMATOPOIESIS	4/212	8/369	22/193	7/71	3/171	12/314	11/147	3/94
HYPERPLASIA, LYMPHOID	4/212	12/369	14/193	10/71	3/171	4/314	8/147	16/94
HYPERPLASIA, MEGAKARYOCYTIC							1/147	
HYPERPLASIA, RETICULOENDOTHELIAL			5/193	8/71		1/314	2/147	5/94
MASTOCYTOSIS								1/94

SEX:	FEMALE				MALE			
DIAGNOSIS / AGE IN MONTHS:	0 < 12	12 < 18	18 < 24	24+	0 < 12	12 < 18	18 < 24	24+
*************** GROWTH ALTERATIONS, NON-NEOPLASTIC, ATROPHY ***************								
ATROPHY, LYMPHOID	1/212		1/193		4/171	1/314	1/147	
*************** GROWTH ALTERATIONS, NEOPLASTIC, BENIGN ***************								
ANGIOMA		1/369	5/193			1/314	2/147	
*************** GROWTH ALTERATIONS, NEOPLASTIC, MALIGNANT, PRIMARY SITE NOT DETERMINED ***************								
ANGIOSARCOMA			1/193	1/71	1/171		1/147	1/94
LYMPHOMA	3/212	7/369	9/193	5/71	5/171	1/314	4/147	3/94
LYMPHOMA, HISTIOCYTIC								4/94
LYMPHOMA, RETICULUM CELL		5/369	24/193	32/71		3/314	6/147	14/94*

*VARIABILITY AMONG DATA SOURCES SIGNIFICANT AT THE 1% SIGNIFICANCE LEVEL

SEX:	FEMALE				MALE			
AGE IN MONTHS: / DIAGNOSIS	0 < 12	12 < 18	18 < 24	24+	0 < 12	12 < 18	18 < 24	24+
*************** MECHANICAL ABNORMALITIES AND CYST FORMATION ***************								
ANGIECTASIS			1/1356					2/958
CONGESTION	1/67		2/1356				1/1136	1/958
CYST, HEMORRHAGIC								1/958
HEMORRHAGE							1/1136	
*************** INFLAMMATION AND REPAIR ***************								
FIBROSIS							1/1136	1/958
INFLAMMATION			1/1356				2/1136	
INFLAMMATION, SUPPURATIVE		1/103						
*************** DEGENERATION, NECROSIS, AND DEPOSITION ***************								
AMYLOID DEPOSITION			2/1356		19/124*	56/199*	55/1136	
DEGENERATION				1/1015				2/958
HEMOSIDERIN DEPOSITION			1/1356	4/1015			1/1136	1/958
MINERALIZATION							1/1136	
NECROSIS			2/1356					1/958
*************** GROWTH ALTERATIONS, NON-NEOPLASTIC, HYPERPLASIA ***************								
HEMATOPOIESIS	1/67	5/103	83/1356*	123/1015*	5/124	9/199	67/1136*	74/958*
HYPERPLASIA		1/103	12/1356	2/1015			11/1136	

*VARIABILITY AMONG DATA SOURCES SIGNIFICANT AT THE 1% SIGNIFICANCE LEVEL

DIAGNOSIS / SEX:	FEMALE				MALE			
AGE IN MONTHS:	0 < 12	12 < 18	18 < 24	24+	0 < 12	12 < 18	18 < 24	24+
HYPERPLASIA, ERYTHROID			6/1356				3/1136	
HYPERPLASIA, LYMPHOID	1/67	3/103	44/1356	60/1015*	1/124		18/1136	31/958
HYPERPLASIA, MYELOID			2/1356	1/1015			1/1136	
HYPERPLASIA, RETICULOENDOTHELIAL			2/1356	4/1015			5/1136	1/958
*************** GROWTH ALTERATIONS, NON-NEOPLASTIC, ATROPHY ***************								
ATROPHY			1/1356		1/124		3/1136	
ATROPHY, LYMPHOID								1/958
*************** GROWTH ALTERATIONS, NEOPLASTIC, BENIGN ***************								
ANGIOMA			3/1356	1/1015			5/1136	5/958
*************** GROWTH ALTERATIONS, NEOPLASTIC, MALIGNANT, METASTATIC ***************								
CARCINOMA, ACINAR CELL								1/958
*************** GROWTH ALTERATIONS, NEOPLASTIC, MALIGNANT, PRIMARY SITE NOT DETERMINED ***************								
ANGIOSARCOMA			7/1356	9/1015		1/199	9/1136	14/958
LEIOMYOSARCOMA		1/103						
LYMPHOMA			6/1356	8/1015			1/1136	3/958
LYMPHOMA, HISTIOCYTIC			10/1356	2/1015			5/1136	1/958
LYMPHOMA, LYMPHOCYTIC			1/1356	1/1015		1/199	1/1136	1/958
SARCOMA			1/1356					

SEX:	FEMALE				MALE			
DIAGNOSIS — AGE IN MONTHS:	0 < 12	12 < 18	18 < 24	24+	0 < 12	12 < 18	18 < 24	24+
*************** MECHANICAL ABNORMALITIES AND CYST FORMATION ***************								
CONGESTION								1/102
THROMBUS							1/161	
*************** INFLAMMATION AND REPAIR ***************								
FIBROSIS		2/367						
INFLAMMATION		2/367	1/192					
*************** DEGENERATION, NECROSIS, AND DEPOSITION ***************								
AMYLOID DEPOSITION							1/161	5/102
INFARCT		1/367						
PIGMENT DEPOSITION	1/176	1/367	2/192	5/75		1/300		
*************** GROWTH ALTERATIONS, NON-NEOPLASTIC, HYPERPLASIA ***************								
HEMATOPOIESIS	6/176	21/367	25/192	10/75	2/131	3/300	14/161	5/102
HYPERPLASIA, LYMPHOID	4/176	11/367	7/192	6/75		7/300	6/161	9/102
HYPERPLASIA, RETICULOENDOTHELIAL		5/367	7/192	7/75		3/300	2/161	2/102
*************** GROWTH ALTERATIONS, NON-NEOPLASTIC, ATROPHY ***************								
ATROPHY, LYMPHOID	1/176	3/367			1/131	1/300	2/161	

SEX:	FEMALE				MALE			
DIAGNOSIS / AGE IN MONTHS:	0 < 12	12 < 18	18 < 24	24+	0 < 12	12 < 18	18 < 24	24+
*************** GROWTH ALTERATIONS, NON-NEOPLASTIC, METAPLASIA ***************								
METAPLASIA, OSSEOUS				1/75				
*************** GROWTH ALTERATIONS, NEOPLASTIC, BENIGN ***************								
ANGIOMA			1/192					1/102
*************** GROWTH ALTERATIONS, NEOPLASTIC, MALIGNANT, PRIMARY SITE NOT DETERMINED ***************								
ANGIOSARCOMA				2/75		1/300	2/161	
LYMPHOMA	2/176	3/367	4/192	3/75		1/300	1/161	3/102
LYMPHOMA, RETICULUM CELL	1/176	11/367	18/192	23/75		3/300	5/161	12/102

SEX:	FEMALE				MALE			
DIAGNOSIS / AGE IN MONTHS:	0 < 12	12 < 18	18 < 24	24+	0 < 12	12 < 18	18 < 24	24+
*************** DEVELOPMENTAL ANOMALIES ***************								
ACCESSORY SPLEEN						1/101		
*************** MECHANICAL ABNORMALITIES AND CYST FORMATION ***************								
ANGIECTASIS								1/1743
CONGESTION				3/1575		2/101	2/437	43/1743
CYST, HEMORRHAGIC				1/1575				
HEMATOMA				1/1575				
HEMATOMA, ORGANIZED				1/1575				
HEMORRHAGE								1/1743
THROMBUS							1/437	5/1743
*************** INFLAMMATION AND REPAIR ***************								
ABSCESS							1/437	1/1743
FIBROSIS				6/1575		1/101	2/437	15/1743
GRANULOMA				1/1575				
INFLAMMATION							1/437	1/1743
INFLAMMATION, GRANULOMATOUS				1/1575				
INFLAMMATION, NONSUPPURATIVE								1/1743

SEX:	FEMALE				MALE			
DIAGNOSIS / AGE IN MONTHS:	0 < 12	12 < 18	18 < 24	24+	0 < 12	12 < 18	18 < 24	24+
*************** DEGENERATION, NECROSIS, AND DEPOSITION ***************								
HEMOSIDERIN DEPOSITION			24/327	120/1575*	1/101		8/437	72/1743
INFARCT			2/327	1/1575			2/437	9/1743
INFARCT, HEALED								15/1743
INFARCT, HEMORRHAGIC				1/1575				2/1743
LIPID DEPOSITION								1/1743
NECROSIS				2/1575				
PIGMENT DEPOSITION			3/327					
*************** GROWTH ALTERATIONS, NON-NEOPLASTIC, HYPERPLASIA ***************								
HEMATOPOIESIS	2/26	1/24	30/327	187/1575*	1/101	2/101	14/437	117/1743*
HYPERPLASIA				1/1575			1/437	
HYPERPLASIA, ERYTHROID			5/327	35/1575			4/437	87/1743
HYPERPLASIA, LYMPHOID			2/327	1/1575				7/1743
HYPERPLASIA, MYELOID				1/1575				1/1743
HYPERPLASIA, PLASMA CELL				1/1575				
HYPERPLASIA, RETICULOENDOTHELIAL				15/1575	3/101	1/101		45/1743
HYPERPLASIA, VASCULAR				1/1575				
LEUKOCYTOSIS								1/1743

*VARIABILITY AMONG DATA SOURCES SIGNIFICANT AT THE 1% SIGNIFICANCE LEVEL

SEX:	FEMALE				MALE			
DIAGNOSIS / AGE IN MONTHS:	0 < 12	12 < 18	18 < 24	24+	0 < 12	12 < 18	18 < 24	24+
*************** GROWTH ALTERATIONS, NON-NEOPLASTIC, ATROPHY ***************								
ATROPHY			1/327				1/437	
ATROPHY, LYMPHOID							1/437	1/1743
*************** GROWTH ALTERATIONS, NEOPLASTIC, BENIGN ***************								
MESENCHYMOMA, BENIGN			1/327					
*************** GROWTH ALTERATIONS, NEOPLASTIC, MALIGNANT, METASTATIC ***************								
ADENOCARCINOMA				1/1575				
ADENOCARCINOMA, MUCINOUS							1/437	
MESOTHELIOMA							2/437	4/1743
OSTEOSARCOMA								1/1743
*************** GROWTH ALTERATIONS, NEOPLASTIC, MALIGNANT, PRIMARY SITE NOT DETERMINED ***************								
ANGIOSARCOMA				1/1575		1/101	1/437	2/1743
HISTIOCYTOMA, MALIGNANT							1/437	
LEUKEMIA			2/327	4/1575			2/437	27/1743
LEUKEMIA, GRANULOCYTIC			1/327	1/1575			1/437	3/1743
LEUKEMIA, LYMPHOBLASTIC			1/327	4/1575				2/1743
LEUKEMIA, LYMPHOCYTIC				3/1575			1/437	3/1743
LEUKEMIA, MYELOMONOCYTIC				7/1575			1/437	3/1743

SEX:	FEMALE				MALE			
DIAGNOSIS / AGE IN MONTHS:	0 < 12	12 < 18	18 < 24	24+	0 < 12	12 < 18	18 < 24	24+
LEUKEMIA, STEM CELL							1/437	
LYMPHOMA			3/327	3/1575				6/1743
LYMPHOMA, HISTIOCYTIC		1/24						1/1743
LYMPHOMA, LYMPHOCYTIC								1/1743
LYMPHOMA, RETICULUM CELL								1/1743
NEUROFIBROSARCOMA				1/1575			1/437	
SARCOMA				1/1575				

SEX:	FEMALE				MALE			
DIAGNOSIS / AGE IN MONTHS:	0 < 12	12 < 18	18 < 24	24+	0 < 12	12 < 18	18 < 24	24+
*************** MECHANICAL ABNORMALITIES AND CYST FORMATION ***************								
CONGESTION				1/336				
CYST		1/39						
HEMORRHAGE				1/336				
*************** INFLAMMATION AND REPAIR ***************								
FIBROSIS								1/243
PERIVASCULITIS/VASCULITIS								2/243
*************** DEGENERATION, NECROSIS, AND DEPOSITION ***************								
HEMOSIDERIN DEPOSITION				5/336			1/104	
PIGMENT DEPOSITION				5/336				
*************** GROWTH ALTERATIONS, NON-NEOPLASTIC, HYPERPLASIA ***************								
HEMATOPOIESIS	1/7	3/39	8/44	28/336	1/11	6/64	9/104	9/243
HYPERPLASIA, RETICULOENDOTHELIAL			1/44	1/336				
*************** GROWTH ALTERATIONS, NEOPLASTIC, BENIGN ***************								
ANGIOMA								1/243

SEX:	FEMALE				MALE			
DIAGNOSIS / AGE IN MONTHS:	0 < 12	12 < 18	18 < 24	24+	0 < 12	12 < 18	18 < 24	24+
*************** GROWTH ALTERATIONS, NEOPLASTIC, MALIGNANT, PRIMARY SITE NOT DETERMINED ***************								
ANGIOSARCOMA			1/44	2/336			6/104	8/243
FIBROSARCOMA				1/336				
LYMPHOMA								1/243

SEX:	FEMALE				MALE			
DIAGNOSIS / AGE IN MONTHS:	0 < 12	12 < 18	18 < 24	24+	0 < 12	12 < 18	18 < 24	24+
*************** MECHANICAL ABNORMALITIES AND CYST FORMATION ***************								
HEMATOMA			1/420					
HEMORRHAGE							1/392	1/216
*************** INFLAMMATION AND REPAIR ***************								
FIBROSIS				1/240			1/392	1/216
INFLAMMATION			1/420					
INFLAMMATION, SUPPURATIVE					1/170			
PERIVASCULITIS/VASCULITIS						1/97	1/392	1/216
*************** DEGENERATION, NECROSIS, AND DEPOSITION ***************								
HEMOSIDERIN DEPOSITION		5/60	14/420	29/240*		1/97	4/392	28/216*
MINERALIZATION			1/420				1/392	
MINERALIZATION, VASCULAR							1/392	1/216
PIGMENT DEPOSITION		6/60				5/97		
*************** GROWTH ALTERATIONS, NON-NEOPLASTIC, HYPERPLASIA ***************								
HEMATOPOIESIS	1/181	1/60	21/420	41/240*	1/170	3/97	2/392	33/216*
HYPERPLASIA						1/97		3/216

*VARIABILITY AMONG DATA SOURCES SIGNIFICANT AT THE 1% SIGNIFICANCE LEVEL

SEX:	FEMALE				MALE			
DIAGNOSIS / AGE IN MONTHS:	0 < 12	12 < 18	18 < 24	24+	0 < 12	12 < 18	18 < 24	24+
*************** GROWTH ALTERATIONS, NON-NEOPLASTIC, ATROPHY ***************								
ATROPHY, FIBROUS							1/392	1/216
*************** GROWTH ALTERATIONS, NON-NEOPLASTIC, HYPOPLASIA ***************								
HYPOPLASIA						1/97	1/392	
*************** GROWTH ALTERATIONS, NEOPLASTIC, BENIGN ***************								
ANGIOMA, CAPILLARY				1/240				
*************** GROWTH ALTERATIONS, NEOPLASTIC, MALIGNANT, METASTATIC ***************								
CARCINOMA, SQUAMOUS CELL							1/392	
*************** GROWTH ALTERATIONS, NEOPLASTIC, MALIGNANT, PRIMARY SITE NOT DETERMINED ***************								
ANGIOSARCOMA								1/216
LEIOMYOSARCOMA							1/392	
LEUKEMIA			1/420					
LEUKEMIA, LYMPHOCYTIC			1/420			1/97	1/392	
LYMPHOMA	1/181		1/420	1/240	1/170	1/97		
LYMPHOMA, RETICULUM CELL							1/392	
SARCOMA, HISTIOCYTIC				1/240				2/216

SEX:	FEMALE				MALE			
DIAGNOSIS / AGE IN MONTHS:	0 < 12	12 < 18	18 < 24	24+	0 < 12	12 < 18	18 < 24	24+
*************** MECHANICAL ABNORMALITIES AND CYST FORMATION ***************								
CYST	1/212	7/369	5/193	2/71	2/171	2/314	6/147	3/94
*************** GROWTH ALTERATIONS, NON-NEOPLASTIC. HYPERPLASIA ***************								
HYPERPLASIA	1/212	1/369	2/193					
HYPERPLASIA, RETICULOENDOTHELIAL				1/71				
MASTOCYTOSIS				1/71				
*************** GROWTH ALTERATIONS, NON-NEOPLASTIC, ATROPHY ***************								
ATROPHY	1/212	2/369	5/193	7/71	7/171	6/314	1/147	3/94
*************** GROWTH ALTERATIONS, NEOPLASTIC, MALIGNANT, PRIMARY SITE NOT DETERMINED ***************								
LYMPHOMA	3/212	7/369	3/193		5/171		1/147	1/94
LYMPHOMA, RETICULUM CELL		2/369	9/193	7/71		1/314	1/147	4/94

SEX:	FEMALE				MALE			
DIAGNOSIS / AGE IN MONTHS:	0 < 12	12 < 18	18 < 24	24+	0 < 12	12 < 18	18 < 24	24+
*************** MECHANICAL ABNORMALITIES AND CYST FORMATION ***************								
CONGESTION			1/1356					
CYST			1/1356					1/958
HEMORRHAGE			1/1356					
*************** INFLAMMATION AND REPAIR ***************								
INFLAMMATION, SUPPURATIVE			1/1356					
*************** DEGENERATION, NECROSIS, AND DEPOSITION ***************								
LIPID DEPOSITION				1/1015				
NECROSIS							1/1136	
*************** GROWTH ALTERATIONS, NON-NEOPLASTIC, HYPERPLASIA ***************								
HYPERPLASIA			1/1356					
HYPERPLASIA, LYMPHOID			1/1356	5/1015			1/1136	
*************** GROWTH ALTERATIONS, NON-NEOPLASTIC, ATROPHY ***************								
ATROPHY			8/1356	1/1015	1/124		10/1136	1/958
*************** GROWTH ALTERATIONS, NEOPLASTIC, MALIGNANT, PRIMARY ***************								
THYMOMA				1/1015				

SEX:	FEMALE				MALE			
DIAGNOSIS / AGE IN MONTHS:	0 < 12	12 < 18	18 < 24	24+	0 < 12	12 < 18	18 < 24	24+
*************** GROWTH ALTERATIONS, NEOPLASTIC, MALIGNANT, PRIMARY SITE NOT DETERMINED ***************								
FIBROSARCOMA							1/1136	
LYMPHOMA			3/1356	1/1015				
LYMPHOMA, HISTIOCYTIC				1/1015				
LYMPHOMA, LYMPHOCYTIC			1/1356	1/1015				1/958

SEX:	FEMALE				MALE			
DIAGNOSIS / AGE IN MONTHS:	0 < 12	12 < 18	18 < 24	24+	0 < 12	12 < 18	18 < 24	24+
*************** MECHANICAL ABNORMALITIES AND CYST FORMATION ***************								
CYST	7/176	11/367	9/192	1/75	5/131	11/300	4/161	4/102
THROMBUS			1/192					
*************** DEGENERATION, NECROSIS, AND DEPOSITION ***************								
PIGMENT DEPOSITION								1/102
*************** GROWTH ALTERATIONS, NON-NEOPLASTIC, HYPERPLASIA ***************								
HYPERPLASIA		1/367					1/161	
HYPERPLASIA, RETICULOENDOTHELIAL			1/192	1/75				
*************** GROWTH ALTERATIONS, NON-NEOPLASTIC, ATROPHY ***************								
ATROPHY			1/192	1/75			2/161	2/102
*************** GROWTH ALTERATIONS, NEOPLASTIC, MALIGNANT, PRIMARY SITE NOT DETERMINED ***************								
LYMPHOMA	1/176	3/367	2/192			1/300	2/161	1/102
LYMPHOMA, RETICULUM CELL		5/367	7/192	8/75		1/300	1/161	3/102

SEX:	FEMALE				MALE			
DIAGNOSIS / AGE IN MONTHS:	0 < 12	12 < 18	18 < 24	24+	0 < 12	12 < 18	18 < 24	24+
*************** MECHANICAL ABNORMALITIES AND CYST FORMATION ***************								
HEMORRHAGE		1/24					1/437	1/1743
*************** DEGENERATION, NECROSIS, AND DEPOSITION ***************								
HEMOSIDERIN DEPOSITION				1/1575				
*************** GROWTH ALTERATIONS, NON-NEOPLASTIC, HYPERPLASIA ***************								
HYPERPLASIA					1/101			
HYPERPLASIA, EPITHELIAL							1/437	
*************** GROWTH ALTERATIONS, NON-NEOPLASTIC, ATROPHY ***************								
ATROPHY			1/327	33/1575			2/437	12/1743
*************** GROWTH ALTERATIONS, NEOPLASTIC, BENIGN ***************								
ADENOMA				1/1575				
*************** GROWTH ALTERATIONS, NEOPLASTIC, MALIGNANT, PRIMARY ***************								
THYMOMA				1/1575				3/1743
*************** GROWTH ALTERATIONS, NEOPLASTIC, MALIGNANT, PRIMARY SITE NOT DETERMINED ***************								
CARCINOMA								1/1743

SEX:	FEMALE				MALE			
DIAGNOSIS / AGE IN MONTHS:	0 < 12	12 < 18	18 < 24	24+	0 < 12	12 < 18	18 < 24	24+
LEUKEMIA				1/1575				2/1743
LEUKEMIA, GRANULOCYTIC								1/1743
LEUKEMIA, LYMPHOBLASTIC				1/1575				1/1743
LEUKEMIA, LYMPHOCYTIC				1/1575			1/437	
LEUKEMIA, STEM CELL							1/437	
LYMPHOMA				1/1575				

SEX:	FEMALE				MALE			
AGE IN MONTHS:	C < 12	12 < 18	18 < 24	24+	0 < 12	12 < 18	18 < 24	24+
DIAGNOSIS								
*************** MECHANICAL ABNORMALITIES AND CYST FORMATION ***************								
CYST				3/336				3/243
*************** INFLAMMATION AND REPAIR ***************								
INFLAMMATION								1/243
*************** GROWTH ALTERATIONS, NON-NEOPLASTIC, ATROPHY ***************								
ATROPHY								2/243

SEX:	FEMALE				MALE			
DIAGNOSIS / AGE IN MONTHS:	0 < 12	12 < 18	18 < 24	24+	0 < 12	12 < 18	18 < 24	24+
*************** INFLAMMATION AND REPAIR ***************								
FIBROSIS				1/240				
INFLAMMATION, SUPPURATIVE					1/170			
PERIVASCULITIS/VASCULITIS			1/420					
*************** DEGENERATION, NECROSIS, AND DEPOSITION ***************								
HEMOSIDERIN DEPOSITION			2/420	1/240				1/216
MINERALIZATION								1/216
*************** GROWTH ALTERATIONS, NON-NEOPLASTIC, HYPERPLASIA ***************								
HYPERPLASIA				3/240				
*************** GROWTH ALTERATIONS, NEOPLASTIC, MALIGNANT, METASTATIC ***************								
OSTEOSARCOMA			1/420					
*************** GROWTH ALTERATIONS, NEOPLASTIC, MALIGNANT, PRIMARY SITE NOT DETERMINED ***************								
LEUKEMIA, HISTIOCYTIC							1/392	
LEUKEMIA, LYMPHOCYTIC						1/97		
LYMPHOMA			1/420					
LYMPHOMA, RETICULUM CELL			1/420				1/392	
SARCOMA, HISTIOCYTIC								1/216

SEX:	FEMALE				MALE			
DIAGNOSIS / AGE IN MONTHS:	0 < 12	12 < 18	18 < 24	24+	0 < 12	12 < 18	18 < 24	24+
*************** INFLAMMATION AND REPAIR ***************								
FIBROSIS								1/958
INFLAMMATION			4/1356	5/1015	1/124		1/1136	5/958
INFLAMMATION, GRANULOMATOUS			1/1356					
INFLAMMATION, SUPPURATIVE								1/958
*************** DEGENERATION, NECROSIS, AND DEPOSITION ***************								
INFARCT								1/958
NECROSIS, FAT				1/1015				

SEX:	FEMALE				MALE			
DIAGNOSIS / AGE IN MONTHS:	0 < 12	12 < 18	18 < 24	24+	0 < 12	12 < 18	18 < 24	24+
*************** INFLAMMATION AND REPAIR ***************								
INFLAMMATION			2/327	10/1575		1/101	2/437	6/1743
INFLAMMATION, GRANULOMATOUS				1/1575				1/1743
INFLAMMATION, NONSUPPURATIVE						1/101		
*************** DEGENERATION, NECROSIS, AND DEPOSITION ***************								
NECROSIS, FAT				1/1575			2/437	1/1743
*************** GROWTH ALTERATIONS, NON-NEOPLASTIC, ATROPHY ***************								
ATROPHY, SEROUS								1/1743
*************** GROWTH ALTERATIONS, NEOPLASTIC, BENIGN ***************								
LIPOMA			1/327					1/1743
*************** GROWTH ALTERATIONS, NEOPLASTIC, MALIGNANT, METASTATIC ***************								
MESOTHELIOMA								1/1743
*************** GROWTH ALTERATIONS, NEOPLASTIC, MALIGNANT, PRIMARY SITE NOT DETERMINED ***************								
LEUKEMIA				2/1575				
LEUKEMIA, LYMPHOCYTIC				1/1575				
LYMPHOMA								1/1743

ORGAN SYSTEM: MISCELLANEOUS ORGAN: ADIPOSE TISSUE STRAIN: FISCHER 344

SEX:	FEMALE				MALE			
DIAGNOSIS / AGE IN MONTHS:	0 < 12	12 < 18	18 < 24	24+	0 < 12	12 < 18	18 < 24	24+
NEUROFIBROSARCOMA							1/437	

SEX:	FEMALE				MALE			
AGE IN MONTHS:	0 < 12	12 < 18	18 < 24	24+	0 < 12	12 < 18	18 < 24	24+
DIAGNOSIS								
*************** INFLAMMATION AND REPAIR ***************								
INFLAMMATION								1/243

SEX:	FEMALE				MALE			
DIAGNOSIS / AGE IN MONTHS:	0 < 12	12 < 18	18 < 24	24+	0 < 12	12 < 18	18 < 24	24+
*************** INFLAMMATION AND REPAIR ***************								
INFLAMMATION	1/181		2/420				3/392	
INFLAMMATION, GRANULOMATOUS			1/420					
*************** DEGENERATION, NECROSIS, AND DEPOSITION ***************								
INFARCT								1/216

SEX:	FEMALE				MALE			
AGE IN MONTHS: / DIAGNOSIS	0 < 12	12 < 18	18 < 24	24+	0 < 12	12 < 18	18 < 24	24+
*************** MECHANICAL ABNORMALITIES AND CYST FORMATION ***************								
CONGESTION							1/1136	1/958
*************** INFLAMMATION AND REPAIR ***************								
INFLAMMATION, NONSUPPURATIVE				2/1015				
PERIVASCULITIS/VASCULITIS		1/103	2/1356					
*************** DEGENERATION, NECROSIS, AND DEPOSITION ***************								
AMYLOID DEPOSITION		1/103	1/1356	1/1015	2/124	9/199	2/1136	1/958
*************** GROWTH ALTERATIONS, NON-NEOPLASTIC, HYPERTROPHY ***************								
CYTOMEGALY							1/1136	
*************** GROWTH ALTERATIONS, NON-NEOPLASTIC, HYPERPLASIA ***************								
HEMATOPOIESIS				1/1015				1/958
HYPERPLASIA				1/1015				
HYPERPLASIA, GRANULOCYTIC							1/1136	
HYPERPLASIA, LYMPHOID			1/1356	10/1015			2/1136	4/958
HYPERPLASIA, PLASMA CELL				1/1015				
HYPERPLASIA, RETICULUM CELL			1/1356				1/1136	

SEX:	FEMALE				MALE			
DIAGNOSIS / AGE IN MONTHS:	0 < 12	12 < 18	18 < 24	24+	0 < 12	12 < 18	18 < 24	24+
*************** GROWTH ALTERATIONS, NEOPLASTIC, BENIGN ***************								
ANGIOMA								1/958
*************** GROWTH ALTERATIONS, NEOPLASTIC, MALIGNANT, METASTATIC ***************								
CARCINOMA, ISLET CELL			1/1356					
ERYTHROLEUKEMIA			1/1356					
FIBROSARCOMA		1/103	1/1356	1/1015				
LEUKEMIA			1/1356	1/1015		1/199	1/1136	
LEUKEMIA, MONONUCLEAR			1/1356	1/1015			1/1136	
MASTOCYTOMA				1/1015				
MESOTHELIOMA								2/958
NEUROFIBROSARCOMA							1/1136	
OSTEOSARCOMA							1/1136	
SARCOMA, GRANULOCYTIC				3/1015			1/1136	1/958
*************** GROWTH ALTERATIONS, NEOPLASTIC, MALIGNANT, PRIMARY SITE NOT DETERMINED ***************								
LEIOMYOSARCOMA			1/1356					
LEUKEMIA			3/1356	3/1015			3/1136	2/958
LEUKEMIA, GRANULOCYTIC		1/103	4/1356	1/1015		1/199	4/1136	2/958
LEUKEMIA, LYMPHOCYTIC		1/103	10/1356	6/1015		2/199	3/1136	
LYMPHOMA	15/67	6/103	42/1356	52/1015*	6/124	2/199	27/1136	28/958

*VARIABILITY AMONG DATA SOURCES SIGNIFICANT AT THE 1% SIGNIFICANCE LEVEL

SEX:	FEMALE				MALE			
AGE IN MONTHS: DIAGNOSIS	0 < 12	12 < 18	18 < 24	24+	0 < 12	12 < 18	18 < 24	24+
LYMPHOMA, HISTIOCYTIC	1/67	2/103	57/1356	95/1015 *	1/124	2/199	23/1136	26/958
LYMPHOMA, LYMPHOCYTIC	2/67	4/103	36/1356	37/1015	3/124	4/199	12/1136	17/958
MESOTHELIOMA				1/1015				
SARCOMA			3/1356	3/1015			1/1136	2/958

*VARIABILITY AMONG DATA SOURCES SIGNIFICANT AT THE 1% SIGNIFICANCE LEVEL

SEX:	FEMALE				MALE			
AGE IN MONTHS: / DIAGNOSIS	0 < 12	12 < 18	18 < 24	24+	0 < 12	12 < 18	18 < 24	24+
*************** MECHANICAL ABNORMALITIES AND CYST FORMATION ***************								
CONGESTION			1/327	4/1575				
*************** GROWTH ALTERATIONS, NON-NEOPLASTIC, HYPERPLASIA ***************								
HYPERPLASIA, LYMPHOID				3/1575				1/1743
*************** GROWTH ALTERATIONS, NON-NEOPLASTIC, ATROPHY ***************								
ATROPHY								14/1743
*************** GROWTH ALTERATIONS, NEOPLASTIC, MALIGNANT, METASTATIC ***************								
ADENOCARCINOMA				1/1575				
ADENOCARCINOMA, MUCINOUS						1/101		
ANGIOSARCOMA						1/101		
CARCINOMA							1/437	
CARCINOMA, C CELL				1/1575				
FIBROSARCOMA						1/101	1/437	
HISTIOCYTOMA, FIBROUS, MALIGNANT							1/437	
LEUKEMIA		2/24	14/327	23/1575			25/437 *	36/1743
LEUKEMIA, MONONUCLEAR		1/24	8/327	29/1575			11/437	38/1743
LEUKEMIA, MYELOMONOCYTIC			1/327	22/1575			11/437	21/1743
MESOTHELIOMA							1/437	4/1743

*VARIABILITY AMONG DATA SOURCES SIGNIFICANT AT THE 1% SIGNIFICANCE LEVEL

SEX:	FEMALE				MALE			
DIAGNOSIS / AGE IN MONTHS:	0 < 12	12 < 18	18 < 24	24+	0 < 12	12 < 18	18 < 24	24+
OSTEOSARCOMA							1/437	
SARCOMA, OSTEOGENIC								1/1743
*************** GROWTH ALTERATIONS, NEOPLASTIC, MALIGNANT, PRIMARY SITE NOT DETERMINED ***************								
LEUKEMIA	1/26		16/327	11/1575		3/101	17/437	27/1743
LEUKEMIA, GRANULOCYTIC			12/327	8/1575			9/437	13/1743
LEUKEMIA, LYMPHOCYTIC			1/327	4/1575			2/437	12/1743
LYMPHOMA		1/24	10/327	43/1575			17/437	60/1743
LYMPHOMA, HISTIOCYTIC			1/327	1/1575			1/437	1/1743
LYMPHOMA, LYMPHOCYTIC			2/327	4/1575	1/101		3/437	4/1743
MESOTHELIOMA							1/437	6/1743
SARCOMA							1/437	1/1743

SEX:	FEMALE				MALE			
DIAGNOSIS / AGE IN MONTHS:	0 < 12	12 < 18	18 < 24	24+	0 < 12	12 < 18	18 < 24	24+
*************** DEGENERATION, NECROSIS, AND DEPOSITION ***************								
ARTERIOSCLEROSIS							1/104	
MINERALIZATION			1/44			1/64	3/104	
*************** GROWTH ALTERATIONS, NEOPLASTIC, MALIGNANT, METASTATIC ***************								
ANGIOSARCOMA								1/243
HISTIOCYTOMA, FIBROUS, MALIGNANT		1/39	1/44	1/336				1/243
*************** GROWTH ALTERATIONS, NEOPLASTIC, MALIGNANT, PRIMARY SITE NOT DETERMINED ***************								
LEUKEMIA, LYMPHOCYTIC							1/104	1/243
LYMPHOMA					1/11		1/104	
LYMPHOMA, HISTIOCYTIC			1/44	2/336		1/64	1/104	1/243
LYMPHOMA, LYMPHOCYTIC				3/336				1/243

SEX:	FEMALE				MALE			
DIAGNOSIS / AGE IN MONTHS:	0 < 12	12 < 18	18 < 24	24+	0 < 12	12 < 18	18 < 24	24+
*************** INFLAMMATION AND REPAIR ***************								
INFLAMMATION, GRANULOMATOUS						1/97		
*************** GROWTH ALTERATIONS, NEOPLASTIC, MALIGNANT, METASTATIC ***************								
ADENOCARCINOMA		1/60						
LEUKEMIA							1/392	
LEUKEMIA, LYMPHOBLASTIC							1/392	
*************** GROWTH ALTERATIONS, NEOPLASTIC, MALIGNANT, PRIMARY SITE NOT DETERMINED ***************								
LEUKEMIA, GRANULOCYTIC						1/97	1/392	
LEUKEMIA, LYMPHOCYTIC			1/420					
LYMPHOMA, HISTIOCYTIC							1/392	
SARCOMA				1/240				

SEX:	FEMALE				MALE			
DIAGNOSIS / AGE IN MONTHS:	0 < 12	12 < 18	18 < 24	24+	0 < 12	12 < 18	18 < 24	24+
*************** INFLAMMATION AND REPAIR ***************								
FIBROSIS								1/94
*************** DEGENERATION, NECROSIS, AND DEPOSITION ***************								
MINERALIZATION								2/94
NECROSIS		2/369						5/94
*************** GROWTH ALTERATIONS, NON-NEOPLASTIC, HYPERPLASIA ***************								
HYPERPLASIA	2/212	1/369		4/71				2/94
*************** GROWTH ALTERATIONS, NEOPLASTIC, MALIGNANT, METASTATIC ***************								
ANGIOSARCOMA			1/193					
CARCINOSARCOMA, MAMMARY		1/369						
*************** GROWTH ALTERATIONS, NEOPLASTIC, MALIGNANT, PRIMARY SITE NOT DETERMINED ***************								
LYMPHOMA	1/212		1/193					

SEX:	FEMALE				MALE			
DIAGNOSIS / AGE IN MONTHS:	0 < 12	12 < 18	18 < 24	24+	0 < 12	12 < 18	18 < 24	24+
*************** MECHANICAL ABNORMALITIES AND CYST FORMATION ***************								
HEMORRHAGE			1/1356					
*************** INFLAMMATION AND REPAIR ***************								
MYELOFIBROSIS			15/1356	45/1015				
*************** DEGENERATION, NECROSIS, AND DEPOSITION ***************								
NECROSIS							1/1136	
OSTEODYSTROPHY, FIBROUS			3/1356	41/1015				
OSTEOPOROSIS				1/1015				
OSTEOSCLEROSIS			1/1356	2/1015				
*************** GROWTH ALTERATIONS, NON-NEOPLASTIC, HYPERPLASIA ***************								
HYPERPLASIA							1/1136	
HYPERPLASIA, ERYTHROID							1/1136	
HYPERPLASIA, GRANULOCYTIC			1/1356	1/1015			2/1136	2/958
HYPERPLASIA, HEMATOPOIETIC			9/1356	36/1015	2/124	1/199	9/1136	30/958
OSTEOPHYTE							1/1136	
*************** GROWTH ALTERATIONS, NON-NEOPLASTIC, ATROPHY ***************								
ATROPHY			13/1356				2/1136	2/958

SEX:	FEMALE				MALE			
DIAGNOSIS / AGE IN MONTHS:	0 < 12	12 < 18	18 < 24	24+	0 < 12	12 < 18	18 < 24	24+
RESORPTION			3/1356					
*************** GROWTH ALTERATIONS, NON-NEOPLASTIC, DYSPLASIA ***************								
DYSPLASIA, FIBROUS			2/1356					
*************** GROWTH ALTERATIONS, NEOPLASTIC, BENIGN ***************								
ANGIOMA			1/1356	1/1015				
OSTEOCHONDROMA								1/958
*************** GROWTH ALTERATIONS, NEOPLASTIC, MALIGNANT, PRIMARY ***************								
OSTEOSARCOMA			4/1356	1/1015			1/1136	
*************** GROWTH ALTERATIONS, NEOPLASTIC, MALIGNANT, METASTATIC ***************								
ANGIOSARCOMA				1/1015				
RHABDOMYOSARCOMA				1/1015				
*************** GROWTH ALTERATIONS, NEOPLASTIC, MALIGNANT, PRIMARY SITE NOT DETERMINED ***************								
FIBROSARCOMA							1/1136	

DIAGNOSIS	FEMALE 0 < 12	FEMALE 12 < 18	FEMALE 18 < 24	FEMALE 24+	MALE 0 < 12	MALE 12 < 18	MALE 18 < 24	MALE 24+
*************** INFLAMMATION AND REPAIR ***************								
FIBROSIS							2/161	
*************** DEGENERATION, NECROSIS, AND DEPOSITION ***************								
MINERALIZATION								1/102
NECROSIS				2/75			2/161	1/102
*************** GROWTH ALTERATIONS, NON-NEOPLASTIC, HYPERPLASIA ***************								
HYPERPLASIA				9/75			2/161	11/102
*************** GROWTH ALTERATIONS, NON-NEOPLASTIC, DYSPLASIA ***************								
DYSPLASIA, EPITHELIAL			1/192					

SEX:	FEMALE				MALE			
DIAGNOSIS / AGE IN MONTHS:	0 < 12	12 < 18	18 < 24	24+	0 < 12	12 < 18	18 < 24	24+
*************** MECHANICAL ABNORMALITIES AND CYST FORMATION ***************								
FRACTURE, HEALED							1/437	1/1743
HEMORRHAGE							1/437	
*************** INFLAMMATION AND REPAIR ***************								
FIBROSIS				1/1575			1/437	
INFLAMMATION, FAT				1/1575				
INFLAMMATION, SUPPURATIVE								1/1743
MYELOFIBROSIS							1/437	1/1743
*************** DEGENERATION, NECROSIS, AND DEPOSITION ***************								
OSTEODYSTROPHY, FIBROUS				1/1575				1/1743
OSTEOSCLEROSIS				1/1575				
*************** GROWTH ALTERATIONS, NON-NEOPLASTIC, HYPERPLASIA ***************								
HYPERPLASIA			2/327	5/1575		1/101	5/437	6/1743
HYPERPLASIA, ERYTHROID							1/437	1/1743
HYPERPLASIA, GRANULOCYTIC			3/327	2/1575			1/437	2/1743
HYPERPLASIA, HEMATOPOIETIC			3/327	15/1575			2/437	23/1743
HYPERPLASIA, MEGAKARYOCYTIC				1/1575				
OSTEOPETROSIS				2/1575				

SEX:	FEMALE				MALE			
DIAGNOSIS / AGE IN MONTHS:	0 < 12	12 < 18	18 < 24	24+	0 < 12	12 < 18	18 < 24	24+
*************** GROWTH ALTERATIONS, NON-NEOPLASTIC, ATROPHY ***************								
ATROPHY		1/24	4/327	25/1575	1/101	6/101*	2/437	10/1743
*************** GROWTH ALTERATIONS, NON-NEOPLASTIC, HYPOPLASIA ***************								
HYPOPLASIA, HEMATOPOIETIC			1/327					
*************** GROWTH ALTERATIONS, NEOPLASTIC, MALIGNANT, PRIMARY ***************								
OSTEOSARCOMA			1/327				1/437	4/1743
*************** GROWTH ALTERATIONS, NEOPLASTIC, MALIGNANT, METASTATIC ***************								
ADENOCARCINOMA, MUCINOUS							1/437	
*************** GROWTH ALTERATIONS, NEOPLASTIC, MALIGNANT, PRIMARY SITE NOT DETERMINED ***************								
FIBROSARCOMA							1/437	
LEUKEMIA				1/1575				1/1743
SARCOMA, GRANULOCYTIC				1/1575				

*VARIABILITY AMONG DATA SOURCES SIGNIFICANT AT THE 1% SIGNIFICANCE LEVEL

ORGAN SYSTEM: MUSCULOSKELETAL ORGAN: BONE STRAIN: OSBORNE MENDEL

SEX:	FEMALE				MALE			
AGE IN MONTHS: / DIAGNOSIS	0 < 12	12 < 18	18 < 24	24+	0 < 12	12 < 18	18 < 24	24+
*************** INFLAMMATION AND REPAIR ***************								
GRANULOMA		1/39						
*************** DEGENERATION, NECROSIS, AND DEPOSITION ***************								
FATTY CHANGE				8/336				3/243
OSTEODYSTROPHY, FIBROUS			2/44	1/336		1/64	3/104	3/243
*************** GROWTH ALTERATIONS, NON-NEOPLASTIC, HYPERPLASIA ***************								
HYPERPLASIA, HEMATOPOIETIC			2/44	3/336				1/243
*************** GROWTH ALTERATIONS, NEOPLASTIC, MALIGNANT. PRIMARY ***************								
OSTEOSARCOMA							1/104	2/243

SEX:	FEMALE				MALE			
DIAGNOSIS / AGE IN MONTHS:	0 < 12	12 < 18	18 < 24	24+	0 < 12	12 < 18	18 < 24	24+
*************** DEVELOPMENTAL ANOMALIES ***************								
DUAL STERNUM							1/392	
*************** MECHANICAL ABNORMALITIES AND CYST FORMATION ***************								
FRACTURE	1/181							
*************** DEGENERATION, NECROSIS, AND DEPOSITION ***************								
OSTEODYSTROPHY, FIBROUS			1/420			1/97	3/392	4/216
*************** GROWTH ALTERATIONS, NON-NEOPLASTIC, HYPERPLASIA ***************								
HYPERPLASIA	1/181				1/170	1/97		
OSTEOPETROSIS								1/216
*************** GROWTH ALTERATIONS, NON-NEOPLASTIC, ATROPHY ***************								
ATROPHY	1/181	2/60	84/420*			5/97*	88/392 *	
*************** GROWTH ALTERATIONS, NEOPLASTIC, BENIGN ***************								
FIBROADENOMA				1/240				
OSTEOMA							1/392	

*VARIABILITY AMONG DATA SOURCES SIGNIFICANT AT THE 1% SIGNIFICANCE LEVEL

SEX:	FEMALE				MALE			
DIAGNOSIS / AGE IN MONTHS:	0 < 12	12 < 18	18 < 24	24+	0 < 12	12 < 18	18 < 24	24+
*************** GROWTH ALTERATIONS, NEOPLASTIC, MALIGNANT, PRIMARY ***************								
OSTEOSARCOMA			1/420					
*************** GROWTH ALTERATIONS, NEOPLASTIC, MALIGNANT, PRIMARY SITE NOT DETERMINED ***************								
LEUKEMIA			1/420					
LEUKEMIA, LYMPHOCYTIC						1/97		
LYMPHOMA	1/181		1/420					
SARCOMA, HISTIOCYTIC								1/216

SEX:	FEMALE				MALE			
DIAGNOSIS / AGE IN MONTHS:	0 < 12	12 < 18	18 < 24	24+	0 < 12	12 < 18	18 < 24	24+
*************** MECHANICAL ABNORMALITIES AND CYST FORMATION ***************								
CYST							5/437	

SEX:	FEMALE				MALE			
DIAGNOSIS / AGE IN MONTHS:	0 < 12	12 < 18	18 < 24	24+	0 < 12	12 < 18	18 < 24	24+
*************** DEGENERATION, NECROSIS, AND DEPOSITION ***************								
ARTHROPATHY, DEGENERATIVE								3/94
MINERALIZATION								2/94
MYODEGENERATION							1/147	
*************** GROWTH ALTERATIONS, NON-NEOPLASTIC, HYPERPLASIA ***************								
OSSIFICATION, NODULAR								1/94
OSTEOPHYTE								1/94

 ORGAN SYSTEM: MUSCULOSKELETAL ORGAN: JOINT STRAIN: B6C3F1

SEX:	FEMALE				MALE			
DIAGNOSIS / AGE IN MONTHS:	0 < 12	12 < 18	18 < 24	24+	0 < 12	12 < 18	18 < 24	24+
***************								
GROWTH ALTERATIONS, NON-NEOPLASTIC, HYPERPLASIA								
***************								
EXOSTOSIS								1/958

SEX:	FEMALE				MALE			
DIAGNOSIS / AGE IN MONTHS:	0 < 12	12 < 18	18 < 24	24+	0 < 12	12 < 18	18 < 24	24+
*************** DEGENERATION, NECROSIS, AND DEPOSITION ***************								
MINERALIZATION								1/102

 ORGAN SYSTEM: MUSCULOSKELETAL ORGAN: JOINT STRAIN: SPRAGUE DAWLEY

SEX:	FEMALE				MALE			
AGE IN MONTHS: / DIAGNOSIS	0 < 12	12 < 18	18 < 24	24+	0 < 12	12 < 18	18 < 24	24+
*************** INFLAMMATION AND REPAIR ***************								
INFLAMMATION, SUPPURATIVE							1/392	

SEX:	FEMALE				MALE			
DIAGNOSIS / AGE IN MONTHS:	0 < 12	12 < 18	18 < 24	24+	0 < 12	12 < 18	18 < 24	24+
*************** INFLAMMATION AND REPAIR ***************								
FIBROSIS		1/369						
INFLAMMATION					1/171	2/314		
INFLAMMATION, GRANULOMATOUS					1/171			
REGENERATION					1/171			
*************** DEGENERATION, NECROSIS, AND DEPOSITION ***************								
DEGENERATION					1/171	1/314		
MINERALIZATION		1/369			1/171			
NECROSIS					1/171			
*************** GROWTH ALTERATIONS, NEOPLASTIC, MALIGNANT, METASTATIC ***************								
CARCINOSARCOMA, MAMMARY		1/369						
*************** GROWTH ALTERATIONS, NEOPLASTIC, MALIGNANT, PRIMARY SITE NOT DETERMINED ***************								
ANGIOSARCOMA			2/193					
LYMPHOMA	4/212	4/369	5/193		2/171			2/94
LYMPHOMA, RETICULUM CELL			5/193	3/71				2/94

 ORGAN SYSTEM: MUSCULOSKELETAL ORGAN: SKELETAL MUSCLE STRAIN: B6C3F1

SEX:	FEMALE				MALE			
DIAGNOSIS / AGE IN MONTHS:	0 < 12	12 < 18	18 < 24	24+	0 < 12	12 < 18	18 < 24	24+
*************** INFLAMMATION AND REPAIR ***************								
ABSCESS				1/1015				
INFLAMMATION			1/1356	2/1015			1/1136	
*************** GROWTH ALTERATIONS, NEOPLASTIC, MALIGNANT, PRIMARY ***************								
FIBROSARCOMA			2/1356				1/1136	
*************** PARASITISM ***************								
PARASITISM			1/1356	1/1015			2/1136	

SEX:	FEMALE				MALE			
DIAGNOSIS / AGE IN MONTHS:	0 < 12	12 < 18	18 < 24	24+	0 < 12	12 < 18	18 < 24	24+
*************** INFLAMMATION AND REPAIR ***************								
INFLAMMATION		2/367	1/192	1/75				1/102
INFLAMMATION, NONSUPPURATIVE							1/161	
REGENERATION							1/161	
*************** DEGENERATION, NECROSIS, AND DEPOSITION ***************								
DEGENERATION							1/161	
MINERALIZATION		1/367						
*************** GROWTH ALTERATIONS, NON-NEOPLASTIC, HYPERPLASIA ***************								
HYPERPLASIA, LYMPHOID							1/161	
*************** GROWTH ALTERATIONS, NEOPLASTIC, MALIGNANT, PRIMARY SITE NOT DETERMINED ***************								
LYMPHOMA	1/176	1/367				1/300		
LYMPHOMA, RETICULUM CELL		1/367	1/192	1/75				
SARCOMA, UNDIFFERENTIATED							1/161	

SEX:	FEMALE				MALE			
DIAGNOSIS / AGE IN MONTHS:	0 < 12	12 < 18	18 < 24	24+	0 < 12	12 < 18	18 < 24	24+
*************** MECHANICAL ABNORMALITIES AND CYST FORMATION ***************								
HEMORRHAGE							1/437	
HERNIA			1/327	3/1575				3/1743
*************** INFLAMMATION AND REPAIR ***************								
INFLAMMATION				1/1575			1/437	
REGENERATION				1/1575				
*************** DEGENERATION, NECROSIS, AND DEPOSITION ***************								
DEGENERATION				1/1575		1/101		4/1743
DYSTROPHY					1/101			
MINERALIZATION				2/1575			2/437	3/1743
*************** GROWTH ALTERATIONS, NON-NEOPLASTIC, ATROPHY ***************								
ATROPHY			1/327					9/1743
*************** GROWTH ALTERATIONS, NEOPLASTIC, BENIGN ***************								
LIPOMA							1/437	
*************** GROWTH ALTERATIONS, NEOPLASTIC, MALIGNANT, METASTATIC ***************								
MESOTHELIOMA							1/437	

SEX:	FEMALE				MALE			
DIAGNOSIS / AGE IN MONTHS:	0 < 12	12 < 18	18 < 24	24+	0 < 12	12 < 18	18 < 24	24+
OSTEOSARCOMA								1/1743
*************** GROWTH ALTERATIONS, NEOPLASTIC, MALIGNANT, PRIMARY SITE NOT DETERMINED ***************								
LEUKEMIA								1/1743
LEUKEMIA, LYMPHOBLASTIC				1/1575				
LYMPHOMA								1/1743

SEX:	FEMALE				MALE			
DIAGNOSIS / AGE IN MONTHS:	0 < 12	12 < 18	18 < 24	24+	0 < 12	12 < 18	18 < 24	24+
*************** INFLAMMATION AND REPAIR ***************								
INFLAMMATION					1/11			2/243
PERIVASCULITIS/VASCULITIS				1/336				
*************** DEGENERATION, NECROSIS, AND DEPOSITION ***************								
DEGENERATION								1/243
MINERALIZATION					1/11			
*************** GROWTH ALTERATIONS, NEOPLASTIC, MALIGNANT, PRIMARY ***************								
FIBROSARCOMA								2/243
*************** GROWTH ALTERATIONS, NEOPLASTIC, MALIGNANT, PRIMARY SITE NOT DETERMINED ***************								
ANGIOSARCOMA								1/243
HISTIOCYTOMA, FIBROUS, MALIGNANT				1/336				
SARCOMA								1/243

SEX:	FEMALE				MALE			
DIAGNOSIS / AGE IN MONTHS:	0 < 12	12 < 18	18 < 24	24+	0 < 12	12 < 18	18 < 24	24+
*************** MECHANICAL ABNORMALITIES AND CYST FORMATION ***************								
HEMORRHAGE	1/181							
*************** INFLAMMATION AND REPAIR ***************								
FATTY CHANGE				1/240				
FIBROSIS							1/392	
INFLAMMATION			2/420				1/392	
INFLAMMATION, NONSUPPURATIVE	2/181							
PERIVASCULITIS/VASCULITIS								2/216
*************** DEGENERATION, NECROSIS, AND DEPOSITION ***************								
ARTERIOLOSCLEROSIS								1/216
DEGENERATION				2/240			1/392	3/216
MINERALIZATION			2/420			1/97	1/392	2/216
MINERALIZATION, VASCULAR							1/392	1/216
*************** GROWTH ALTERATIONS, NON-NEOPLASTIC, ATROPHY ***************								
ATROPHY				3/240		1/97	2/392	4/216
ATROPHY, SENILE			1/420					2/216

SEX:	FEMALE				MALE			
AGE IN MONTHS: / DIAGNOSIS	0 < 12	12 < 18	18 < 24	24+	0 < 12	12 < 18	18 < 24	24+
*************** GROWTH ALTERATIONS, NEOPLASTIC, MALIGNANT, PRIMARY SITE NOT DETERMINED ***************								
SARCOMA							1/392	
SARCOMA, HISTIOCYTIC			1/420					1/216

SEX:	FEMALE				MALE			
DIAGNOSIS / AGE IN MONTHS:	0 < 12	12 < 18	18 < 24	24+	0 < 12	12 < 18	18 < 24	24+
*************** MECHANICAL ABNORMALITIES AND CYST FORMATION ***************								
CYST	1/212	1/369						
CYST, EPIDERMAL INCLUSION	1/212							
HEMORRHAGE	1/212	2/369	1/193	1/71	3/171			
THROMBUS				1/71				
*************** INFLAMMATION AND REPAIR ***************								
ABSCESS			1/193					
FIBROSIS								2/94
INFLAMMATION				1/71				
INFLAMMATION, NONSUPPURATIVE				1/71				
*************** DEGENERATION, NECROSIS, AND DEPOSITION ***************								
CELLULAR CHANGE, VACUOLAR								11/94
MINERALIZATION		2/369	3/193	2/71		9/314	8/147	14/94
NECROSIS		1/369	2/193	1/71				
STATUS SPONGIOSIS								1/94
*************** GROWTH ALTERATIONS, NEOPLASTIC, MALIGNANT, PRIMARY ***************								
ASTROCYTOMA				1/71				

SEX:	FEMALE				MALE			
DIAGNOSIS / AGE IN MONTHS:	0 < 12	12 < 18	18 < 24	24+	0 < 12	12 < 18	18 < 24	24+
*************** GROWTH ALTERATIONS, NEOPLASTIC, MALIGNANT, PRIMARY SITE NOT DETERMINED ***************								
LYMPHOMA	2/212	3/369	2/193		2/171		1/147	2/94
LYMPHOMA, RETICULUM CELL		1/369	4/193	2/71				1/94

SEX:	FEMALE				MALE			
DIAGNOSIS / AGE IN MONTHS:	0 < 12	12 < 18	18 < 24	24+	0 < 12	12 < 18	18 < 24	24+
*************** MECHANICAL ABNORMALITIES AND CYST FORMATION ***************								
CYST			2/1356					
CYST, EPIDERMAL INCLUSION			1/1356					
HEMORRHAGE				1/1015			1/1136	
HYDROCEPHALUS			2/1356	5/1015				
PSAMMOMA BODY				2/1015				
*************** INFLAMMATION AND REPAIR ***************								
INFLAMMATION			4/1356	4/1015	1/124			
INFLAMMATION, NONSUPPURATIVE				1/1015				
INFLAMMATION, SUPPURATIVE				1/1015				
PERIVASCULITIS/VASCULITIS				3/1015				
*************** DEGENERATION, NECROSIS, AND DEPOSITION ***************								
CELLULAR CHANGE, VACUOLAR				3/1015				
CORPORA AMYLACEA			3/1356	20/1015			5/1136	18/958
MINERALIZATION			11/1356	77/1015*			26/1136	78/958*
*************** GROWTH ALTERATIONS, NEOPLASTIC, BENIGN ***************								
MENINGIOMA								1/958

*VARIABILITY AMONG DATA SOURCES SIGNIFICANT AT THE 1% SIGNIFICANCE LEVEL

SEX:	FEMALE				MALE			
DIAGNOSIS / AGE IN MONTHS:	0 < 12	12 < 18	18 < 24	24+	0 < 12	12 < 18	18 < 24	24+
*************** GROWTH ALTERATIONS, NEOPLASTIC, MALIGNANT, PRIMARY ***************								
ASTROCYTOMA			1/1356					
GLIOMA		1/103						
OLIGODENDROGLIOMA			1/1356					
*************** GROWTH ALTERATIONS, NEOPLASTIC, MALIGNANT, PRIMARY SITE NOT DETERMINED ***************								
LYMPHOMA			1/1356					
LYMPHOMA, HISTIOCYTIC					1/124		1/1136	

SEX:	FEMALE				MALE			
DIAGNOSIS / AGE IN MONTHS:	0 < 12	12 < 18	18 < 24	24+	0 < 12	12 < 18	18 < 24	24+
*************** MECHANICAL ABNORMALITIES AND CYST FORMATION ***************								
HEMORRHAGE	1/176			1/75				
HYDROCEPHALUS		1/367					3/161	
*************** INFLAMMATION AND REPAIR ***************								
FIBROSIS		1/367						
GLIOSIS			1/192					
INFLAMMATION, GRANULOMATOUS, MENINGES								1/102
INFLAMMATION, NONSUPPURATIVE								1/102
*************** DEGENERATION, NECROSIS, AND DEPOSITION ***************								
CELLULAR CHANGE, VACUOLAR								10/102
MELANOSIS					1/131	2/300	2/161	5/102
MINERALIZATION			4/192	9/75		3/300	10/161	16/102
*************** GROWTH ALTERATIONS, NEOPLASTIC, BENIGN ***************								
MENINGIOMA			1/192					
*************** GROWTH ALTERATIONS, NEOPLASTIC, MALIGNANT, PRIMARY SITE NOT DETERMINED ***************								
LYMPHOMA						1/300		
LYMPHOMA, RETICULUM CELL			1/192	1/75				

SEX:	FEMALE				MALE			
DIAGNOSIS / AGE IN MONTHS:	0 < 12	12 < 18	18 < 24	24+	0 < 12	12 < 18	18 < 24	24+
*************** DEVELOPMENTAL ANOMALIES ***************								
VASCULAR ANOMALY							1/437	
*************** MECHANICAL ABNORMALITIES AND CYST FORMATION ***************								
CONGESTION		1/24						
HEMORRHAGE	1/26	1/24	3/327	3/1575		1/101	13/437	13/1743
HYDROCEPHALUS		1/24	4/327	6/1575	1/101	1/101	3/437	2/1743
INFARCT, HEMORRHAGIC								1/1743
MENINGOVENTRICULAR COMMUNICATION						1/101		
THROMBUS				1/1575				1/1743
*************** INFLAMMATION AND REPAIR ***************								
ABSCESS							1/437	1/1743
GLIOSIS				1/1575				2/1743
INFLAMMATION			1/327	2/1575				
INFLAMMATION, MENINGES								1/1743
INFLAMMATION, SUPPURATIVE			1/327					1/1743
*************** DEGENERATION, NECROSIS, AND DEPOSITION ***************								
CELLULAR CHANGE, VACUOLAR						3/101	9/437	55/1743
CHOLESTEROL DEPOSITION								1/1743

SEX:	FEMALE				MALE			
DIAGNOSIS / AGE IN MONTHS:	0 < 12	12 < 18	18 < 24	24+	0 < 12	12 < 18	18 < 24	24+
DEGENERATION								3/1743
DEMYELINATION			1/327					2/1743
MALACIA				1/1575			1/437	5/1743
MINERALIZATION				1/1575		1/101	2/437	1/1743
NECROSIS								2/1743
NECROSIS, HEMORRHAGIC			3/327			2/101	2/437	2/1743
RADICULONEUROPATHY								12/1743
STATUS SPONGIOSIS					1/101	1/101	4/437	29/1743
*************** GROWTH ALTERATIONS, NON-NEOPLASTIC, ATROPHY ***************								
ATROPHY				1/1575				2/1743
ATROPHY, PRESSURE			1/327	3/1575			3/437	1/1743
*************** GROWTH ALTERATIONS, NEOPLASTIC, BENIGN ***************								
EPENDYMOMA				2/1575		1/101		2/1743
MENINGIOMA								1/1743
*************** GROWTH ALTERATIONS, NEOPLASTIC, MALIGNANT, PRIMARY ***************								
ASTROCYTOMA			2/327	2/1575	1/101	7/101*	4/437	2/1743
ASTROCYTOMA, FIBROUS								1/1743
GLIOMA	1/26						1/437	4/1743
HISTIOCYTOMA, MALIGNANT							1/437	

*VARIABILITY AMONG DATA SOURCES SIGNIFICANT AT THE 1% SIGNIFICANCE LEVEL

SEX:	FEMALE				MALE			
DIAGNOSIS / AGE IN MONTHS:	0 < 12	12 < 18	18 < 24	24+	0 < 12	12 < 18	18 < 24	24+
OLIGODENDROGLIOMA			2/327	1/1575			1/437	
*************** GROWTH ALTERATIONS, NEOPLASTIC, MALIGNANT, METASTATIC ***************								
ADENOCARCINOMA, PITUITARY								1/1743
CARCINOMA			2/327	1/1575				
CARCINOMA, CERUMINOUS							1/437	
CARCINOMA, CHROMOPHOBE				2/1575				
MESOTHELIOMA						1/101		
OSTEOSARCOMA							1/437	
THYMOMA								1/1743
*************** GROWTH ALTERATIONS, NEOPLASTIC, MALIGNANT, PRIMARY SITE NOT DETERMINED ***************								
LEUKEMIA	1/26							2/1743
LEUKEMIA, GRANULOCYTIC				1/1575				4/1743
LEUKEMIA, LYMPHOBLASTIC				1/1575				1/1743
LEUKEMIA, LYMPHOCYTIC							1/437	1/1743
SARCOMA							2/437	
*************** GROWTH ALTERATIONS, NEOPLASTIC, NOT DETERMINED ***************								
MIXED GLIAL TUMOR						1/101		1/1743

SEX:	FEMALE				MALE			
DIAGNOSIS / AGE IN MONTHS:	0 < 12	12 < 18	18 < 24	24+	0 < 12	12 < 18	18 < 24	24+
*************** INFLAMMATION AND REPAIR ***************								
FIBROSIS				1/336				
INFLAMMATION						1/64	1/104	
*************** GROWTH ALTERATIONS, NEOPLASTIC, BENIGN ***************								
MENINGIOMA				1/336				
*************** GROWTH ALTERATIONS, NEOPLASTIC, MALIGNANT, PRIMARY ***************								
ASTROCYTOMA					1/11			
GLIOMA						1/64		1/243
OLIGODENDROGLIOMA				1/336				
*************** GROWTH ALTERATIONS, NEOPLASTIC, MALIGNANT, METASTATIC ***************								
CARCINOMA, CHROMOPHOBE							2/104	

SEX:	FEMALE				MALE			
DIAGNOSIS / AGE IN MONTHS:	0 < 12	12 < 18	18 < 24	24+	0 < 12	12 < 18	18 < 24	24+
*************** MECHANICAL ABNORMALITIES AND CYST FORMATION ***************								
HEMORRHAGE			2/420					
HYDROCEPHALUS	1/181	1/60	5/420	8/240	1/170		3/392	3/216
*************** INFLAMMATION AND REPAIR ***************								
FIBROSIS				1/240				
GLIOSIS				1/240				
INFLAMMATION				1/240				
INFLAMMATION, MENINGES	2/181	1/60						
INFLAMMATION, NONSUPPURATIVE	1/181							
*************** DEGENERATION, NECROSIS, AND DEPOSITION ***************								
CELLULAR CHANGE, VACUOLAR	25/181*				7/170			
CORPORA AMYLACEA							1/392	1/216
DEGENERATION								1/216
MALACIA								1/216
MINERALIZATION				1/240			1/392	
STATUS SPONGIOSIS	10/181*				11/170*			
*************** GROWTH ALTERATIONS, NEOPLASTIC, BENIGN ***************								
EPENDYMOMA							1/392	

*VARIABILITY AMONG DATA SOURCES SIGNIFICANT AT THE 1% SIGNIFICANCE LEVEL

SEX:	FEMALE				MALE			
DIAGNOSIS / AGE IN MONTHS:	0 < 12	12 < 18	18 < 24	24+	0 < 12	12 < 18	18 < 24	24+
EPENDYMOMA, NEUROHYPOPHYSEAL			1/420					
MENINGIOMA	2/181							
NEUROFIBROMA, ACOUSTIC								1/216
*************** GROWTH ALTERATIONS, NEOPLASTIC, MALIGNANT, PRIMARY ***************								
ASTROCYTOMA			3/420	1/240			2/392	1/216
GLIOBLASTOMA MULTIFORMI				1/240				
MEDULLOBLASTOMA			1/420					
*************** GROWTH ALTERATIONS, NEOPLASTIC, MALIGNANT, METASTATIC ***************								
CARCINOMA, SQUAMOUS CELL			1/420				1/392	
*************** GROWTH ALTERATIONS, NEOPLASTIC, MALIGNANT, PRIMARY SITE NOT DETERMINED ***************								
LEUKEMIA, HISTIOCYTIC							1/392	
LYMPHOMA	1/181		1/420	1/240				
LYMPHOMA, HISTIOCYTIC								1/216
SARCOMA			1/420	1/240				

SEX:	FEMALE				MALE			
DIAGNOSIS / AGE IN MONTHS:	0 < 12	12 < 18	18 < 24	24+	0 < 12	12 < 18	18 < 24	24+
***************								
DEGENERATION, NECROSIS, AND DEPOSITION								
***************								
CELLULAR CHANGE, VACUOLAR								1/94

SEX:	FEMALE				MALE			
AGE IN MONTHS:	0 < 12	12 < 18	18 < 24	24+	0 < 12	12 < 18	18 < 24	24+
DIAGNOSIS								
*************** INFLAMMATION AND REPAIR ***************								
INFLAMMATION							1/1136	2/958
*************** GROWTH ALTERATIONS, NEOPLASTIC, MALIGNANT, PRIMARY SITE NOT DETERMINED ***************								
LYMPHOMA, LYMPHOCYTIC			1/1356					

SEX:	FEMALE				MALE			
DIAGNOSIS / AGE IN MONTHS:	0 < 12	12 < 18	18 < 24	24+	0 < 12	12 < 18	18 < 24	24+
*************** DEGENERATION, NECROSIS, AND DEPOSITION ***************								
SWELLING					1/131		1/161	1/102

SEX:	FEMALE				MALE			
DIAGNOSIS / AGE IN MONTHS:	0 < 12	12 < 18	18 < 24	24+	0 < 12	12 < 18	18 < 24	24+
*************** GROWTH ALTERATIONS, NEOPLASTIC, MALIGNANT, PRIMARY ***************								
NEUROFIBROSARCOMA							1/437	

SEX:	FEMALE				MALE			
DIAGNOSIS / AGE IN MONTHS:	O < 12	12 < 18	18 < 24	24+	O < 12	12 < 18	18 < 24	24+
*************** INFLAMMATION AND REPAIR ***************								
INFLAMMATION								1/216
*************** DEGENERATION, NECROSIS, AND DEPOSITION ***************								
DEGENERATION						1/97	1/392	1/216
MINERALIZATION							1/392	1/216
*************** GROWTH ALTERATIONS, NEOPLASTIC, MALIGNANT, PRIMARY SITE NOT DETERMINED ***************								
SARCOMA, HISTIOCYTIC			1/420					1/216

SEX:	FEMALE				MALE			
AGE IN MONTHS: / DIAGNOSIS	0 < 12	12 < 18	18 < 24	24+	0 < 12	12 < 18	18 < 24	24+
*************** INFLAMMATION AND REPAIR ***************								
INFLAMMATION, SUPPURATIVE					1/171			

SEX:	FEMALE				MALE			
DIAGNOSIS / AGE IN MONTHS:	0 < 12	12 < 18	18 < 24	24+	0 < 12	12 < 18	18 < 24	24+
*************** MECHANICAL ABNORMALITIES AND CYST FORMATION ***************								
HEMORRHAGE		1/103						
THROMBUS		1/103						
*************** INFLAMMATION AND REPAIR ***************								
ABSCESS	1/67							
INFLAMMATION			1/1356					
INFLAMMATION, SUPPURATIVE	1/67		4/1356	2/1015	1/124	2/199	5/1136	2/958
*************** GROWTH ALTERATIONS, NEOPLASTIC, BENIGN ***************								
HISTIOCYTOMA, FIBROUS								1/958
*************** GROWTH ALTERATIONS, NEOPLASTIC, MALIGNANT, PRIMARY ***************								
CARCINOMA, SQUAMOUS CELL							1/1136	
MELANOMA, MALIGNANT								1/958

SEX:	FEMALE				MALE			
DIAGNOSIS / AGE IN MONTHS:	0 < 12	12 < 18	18 < 24	24+	0 < 12	12 < 18	18 < 24	24+
*************** INFLAMMATION AND REPAIR ***************								
INFLAMMATION			1/327					1/1743
INFLAMMATION, SUPPURATIVE				1/1575				1/1743
*************** GROWTH ALTERATIONS, NEOPLASTIC, BENIGN ***************								
ADENOMA, SEBACEOUS				1/1575				
ADENOMA, SEBACEOUS SQUAMOUS								1/1743
FIBROMA				2/1575				
KERATOACANTHOMA				1/1575			1/437	
PAPILLOMA, SQUAMOUS CELL								2/1743
*************** GROWTH ALTERATIONS, NEOPLASTIC, MALIGNANT, PRIMARY ***************								
CARCINOMA								4/1743
CARCINOMA, CERUMINOUS			1/327	1/1575			2/437	
CARCINOMA, SQUAMOUS CELL			1/327	2/1575		1/101	1/437	6/1743
FIBROSARCOMA				1/1575				
*************** GROWTH ALTERATIONS, NEOPLASTIC, MALIGNANT, METASTATIC ***************								
RHABDOMYOSARCOMA			1/327					

SEX:	FEMALE				MALE			
DIAGNOSIS / AGE IN MONTHS:	0 < 12	12 < 18	18 < 24	24+	0 < 12	12 < 18	18 < 24	24+
*************** MECHANICAL ABNORMALITIES AND CYST FORMATION ***************								
CONGESTION	1/181							
EDEMA	1/181							
*************** INFLAMMATION AND REPAIR ***************								
INFLAMMATION, GRANULOMATOUS	2/181							
INFLAMMATION, SUPPURATIVE			1/420					
*************** GROWTH ALTERATIONS, NON-NEOPLASTIC, HYPERPLASIA ***************								
HYPERPLASIA	2/181							
*************** GROWTH ALTERATIONS, NON-NEOPLASTIC, DYSPLASIA ***************								
DYSPLASIA, CHONDROECTODERMAL				3/240				1/216
*************** GROWTH ALTERATIONS, NEOPLASTIC, BENIGN ***************								
FIBROMA								1/216
FIBROMA, ULCERATIVE			1/420					
*************** GROWTH ALTERATIONS, NEOPLASTIC, MALIGNANT, PRIMARY ***************								
CARCINOMA, SQUAMOUS CELL							2/392	

SEX:	FEMALE				MALE			
AGE IN MONTHS:	0 < 12	12 < 18	18 < 24	24+	0 < 12	12 < 18	18 < 24	24+
DIAGNOSIS								
*************** GROWTH ALTERATIONS, NEOPLASTIC, MALIGNANT, METASTATIC ***************								
RHABDOMYOSARCOMA						1/97		

SEX:	FEMALE				MALE			
DIAGNOSIS / AGE IN MONTHS:	0 < 12	12 < 18	18 < 24	24+	0 < 12	12 < 18	18 < 24	24+
*************** INFLAMMATION AND REPAIR ***************								
INFLAMMATION, SUPPURATIVE								1/94

SEX:	FEMALE				MALE			
AGE IN MONTHS:	0 < 12	12 < 18	18 < 24	24+	0 < 12	12 < 18	18 < 24	24+
DIAGNOSIS								
*************** INFLAMMATION AND REPAIR ***************								
INFLAMMATION								1/102
INFLAMMATION, SUPPURATIVE					2/131	1/300		2/102

SEX:	FEMALE				MALE			
DIAGNOSIS / AGE IN MONTHS:	0 < 12	12 < 18	18 < 24	24+	0 < 12	12 < 18	18 < 24	24+
*************** INFLAMMATION AND REPAIR ***************								
INFLAMMATION			1/327					

SEX:	FEMALE				MALE			
DIAGNOSIS / AGE IN MONTHS:	0 < 12	12 < 18	18 < 24	24+	0 < 12	12 < 18	18 < 24	24+
*************** INFLAMMATION AND REPAIR ***************								
INFLAMMATION				1/240			3/392	

DIAGNOSIS	FEMALE 0 < 12	FEMALE 12 < 18	FEMALE 18 < 24	FEMALE 24+	MALE 0 < 12	MALE 12 < 18	MALE 18 < 24	MALE 24+
*************** INFLAMMATION AND REPAIR ***************								
SYNECHIA				1/71				
*************** DEGENERATION, NECROSIS, AND DEPOSITION ***************								
CATARACT				1/71				
MINERALIZATION								5/94
*************** GROWTH ALTERATIONS, NEOPLASTIC, MALIGNANT, PRIMARY SITE NOT DETERMINED ***************								
LYMPHOMA		2/369	1/193					
LYMPHOMA, RETICULUM CELL			1/193					

SEX:	FEMALE				MALE			
DIAGNOSIS / AGE IN MONTHS:	0 < 12	12 < 18	18 < 24	24+	0 < 12	12 < 18	18 < 24	24+
*************** INFLAMMATION AND REPAIR ***************								
ABSCESS				1/1015				
INFLAMMATION, INTERSTITIAL				1/1015				
VASCULARIZATION								2/958
*************** DEGENERATION, NECROSIS, AND DEPOSITION ***************								
CATARACT			1/1356					1/958
*************** GROWTH ALTERATIONS, NEOPLASTIC, MALIGNANT, PRIMARY ***************								
MELANOMA, MALIGNANT						1/199		

 ORGAN SYSTEM: NERVOUS SYSTEM; SPECIAL ORGANS ORGAN: EYE STRAIN: C57BL/6

SEX:	FEMALE				MALE			
DIAGNOSIS / AGE IN MONTHS:	0 < 12	12 < 18	18 < 24	24+	0 < 12	12 < 18	18 < 24	24+
*************** DEGENERATION, NECROSIS, AND DEPOSITION ***************								
CATARACT		1/367						2/102
*************** GROWTH ALTERATIONS, NON-NEOPLASTIC, HYPERPLASIA ***************								
HYPERKERATOSIS			1/192					

SEX:	FEMALE				MALE			
DIAGNOSIS / AGE IN MONTHS:	0 < 12	12 < 18	18 < 24	24+	0 < 12	12 < 18	18 < 24	24+
*************** MECHANICAL ABNORMALITIES AND CYST FORMATION ***************								
HEMORRHAGE				2/1575				3/1743
*************** INFLAMMATION AND REPAIR ***************								
INFLAMMATION, INTERSTITIAL								1/1743
INFLAMMATION, SUPPURATIVE				5/1575	2/101		2/437	6/1743
INFLAMMATION, ULCERATIVE								2/1743
SYNECHIA				1/1575				
SYNECHIA, ANTERIOR				1/1575				
SYNECHIA, POSTERIOR				1/1575			1/437	2/1743
VASCULARIZATION				2/1575				1/1743
*************** DEGENERATION, NECROSIS, AND DEPOSITION ***************								
CATARACT			2/327	34/1575		1/101	5/437	91/1743 *
MINERALIZATION				2/1575	1/101	1/101	1/437	11/1743
*************** GROWTH ALTERATIONS, NON-NEOPLASTIC, HYPERPLASIA ***************								
HYPERPLASIA, EPITHELIAL				1/1575				
*************** GROWTH ALTERATIONS, NEOPLASTIC, MALIGNANT, PRIMARY ***************								
CARCINOMA, SQUAMOUS CELL			1/327	1/1575			2/437	

*VARIABILITY AMONG DATA SOURCES SIGNIFICANT AT THE 1% SIGNIFICANCE LEVEL

SEX:	FEMALE				MALE			
DIAGNOSIS / AGE IN MONTHS:	0 < 12	12 < 18	18 < 24	24+	0 < 12	12 < 18	18 < 24	24+
*************** GROWTH ALTERATIONS, NEOPLASTIC, MALIGNANT, PRIMARY SITE NOT DETERMINED ***************								
LEUKEMIA, LYMPHOBLASTIC				1/1575				
LEUKEMIA, STEM CELL							1/437	

SEX:	FEMALE				MALE			
DIAGNOSIS / AGE IN MONTHS:	0 < 12	12 < 18	18 < 24	24+	0 < 12	12 < 18	18 < 24	24+
*************** INFLAMMATION AND REPAIR ***************								
SYNECHIA, ANTERIOR							1/104	1/243
*************** DEGENERATION, NECROSIS, AND DEPOSITION ***************								
CATARACT								2/243
*************** GROWTH ALTERATIONS, NON-NEOPLASTIC, ATROPHY ***************								
PHTHISIS BULBI								1/243

 ORGAN SYSTEM: NERVOUS SYSTEM; SPECIAL ORGANS ORGAN: EYE STRAIN: SPRAGUE DAWLEY

SEX:	FEMALE				MALE			
DIAGNOSIS / AGE IN MONTHS:	0 < 12	12 < 18	18 < 24	24+	0 < 12	12 < 18	18 < 24	24+
*************** MECHANICAL ABNORMALITIES AND CYST FORMATION ***************								
HERNIATION						1/97		
*************** INFLAMMATION AND REPAIR ***************								
ABSCESS							1/392	
INFLAMMATION			1/420					
INFLAMMATION, NONSUPPURATIVE							1/392	1/216
INFLAMMATION, SUPPURATIVE				1/240		1/97		5/216
INFLAMMATION, ULCERATIVE					1/170	1/97	1/392	
SYNECHIA, POSTERIOR	11/181 *				19/170*	1/97		
VASCULARIZATION								1/216
*************** DEGENERATION, NECROSIS, AND DEPOSITION ***************								
CATARACT			2/420	9/240			2/392	3/216
MALAKOPLAKIA							2/392	
MINERALIZATION			1/420			1/97		
*************** GROWTH ALTERATIONS, NEOPLASTIC, MALIGNANT, PRIMARY SITE NOT DETERMINED ***************								
LEUKEMIA, LYMPHOCYTIC						1/97		
LYMPHOMA			1/420					

*VARIABILITY AMONG DATA SOURCES SIGNIFICANT AT THE 1% SIGNIFICANCE LEVEL

SEX:	FEMALE				MALE			
DIAGNOSIS / AGE IN MONTHS:	0 < 12	12 < 18	18 < 24	24+	0 < 12	12 < 18	18 < 24	24+
*************** MECHANICAL ABNORMALITIES AND CYST FORMATION ***************								
CYST		1/369	2/193	2/71				
*************** INFLAMMATION AND REPAIR ***************								
FIBROSIS							1/147	
INFLAMMATION	1/212		2/193	3/71		2/314		1/94
INFLAMMATION, NONSUPPURATIVE		1/369		1/71				
*************** GROWTH ALTERATIONS, NEOPLASTIC, BENIGN ***************								
ADENOMA	6/212	6/369	12/193	4/71	1/171	17/314	10/147	11/94
*************** GROWTH ALTERATIONS, NEOPLASTIC, MALIGNANT, PRIMARY SITE NOT DETERMINED ***************								
LYMPHOMA		1/369	3/193					
LYMPHOMA, RETICULUM CELL			5/193	4/71				1/94

SEX:	FEMALE				MALE			
DIAGNOSIS / AGE IN MONTHS:	0 < 12	12 < 18	18 < 24	24+	0 < 12	12 < 18	18 < 24	24+
*************** GROWTH ALTERATIONS, NON-NEOPLASTIC, HYPERPLASIA ***************								
HYPERPLASIA			1/1356					
HYPERPLASIA, EPITHELIAL				1/1015				
*************** GROWTH ALTERATIONS, NEOPLASTIC, BENIGN ***************								
ADENOMA			1/1356	2/1015			1/1136	8/958
CYSTADENOMA				3/1015			4/1136	2/958

ORGAN SYSTEM NERVOUS SYSTEM; SPECIAL ORGANS ORGAN: EYE; ACCESSORY GLANDS STRAIN: C57BL/6

SEX:	FEMALE				MALE			
DIAGNOSIS / AGE IN MONTHS:	0 < 12	12 < 18	18 < 24	24+	0 < 12	12 < 18	18 < 24	24+
*************** MECHANICAL ABNORMALITIES AND CYST FORMATION ***************								
CYST			1/192			1/300		1/102
*************** INFLAMMATION AND REPAIR ***************								
INFLAMMATION	4/176	5/367	5/192	6/75	3/131	5/300	7/161	13/102
INFLAMMATION, NONSUPPURATIVE			3/192	6/75		4/300	9/161	17/102 *
*************** DEGENERATION, NECROSIS, AND DEPOSITION ***************								
NECROSIS					1/131			
PIGMENT DEPOSITION			1/192	4/75		2/300	1/161	7/102
*************** GROWTH ALTERATIONS, NON-NEOPLASTIC, HYPERPLASIA ***************								
HYPERPLASIA		1/367	1/192					
*************** GROWTH ALTERATIONS, NEOPLASTIC, BENIGN ***************								
ADENOMA	3/176	8/367	7/192	6/75	5/131	11/300	17/161	11/102
*************** GROWTH ALTERATIONS, NEOPLASTIC, MALIGNANT, PRIMARY SITE NOT DETERMINED ***************								
LYMPHOMA				1/75				
LYMPHOMA, RETICULUM CELL				2/75				3/102

*VARIABILITY AMONG DATA SOURCES SIGNIFICANT AT THE 1% SIGNIFICANCE LEVEL

SEX:	FEMALE				MALE			
DIAGNOSIS / AGE IN MONTHS:	0 < 12	12 < 18	18 < 24	24+	0 < 12	12 < 18	18 < 24	24+
*************** INFLAMMATION AND REPAIR ***************								
INFLAMMATION								1/1743

SEX:	FEMALE				MALE			
DIAGNOSIS / AGE IN MONTHS:	0 < 12	12 < 18	18 < 24	24+	0 < 12	12 < 18	18 < 24	24+
*************** INFLAMMATION AND REPAIR ***************								
INFLAMMATION						1/64		

SEX:	FEMALE				MALE			
DIAGNOSIS / AGE IN MONTHS:	0 < 12	12 < 18	18 < 24	24+	0 < 12	12 < 18	18 < 24	24+
*************** INFLAMMATION AND REPAIR ***************								
INFLAMMATION, ULCERATIVE			1/193					

SEX:	FEMALE				MALE			
DIAGNOSIS / AGE IN MONTHS:	0 < 12	12 < 18	18 < 24	24+	0 < 12	12 < 18	18 < 24	24+
*************** INFLAMMATION AND REPAIR ***************								
INFLAMMATION				2/1575				10/1743
INFLAMMATION, SUPPURATIVE				8/1575				12/1743

SEX:	FEMALE				MALE			
DIAGNOSIS / AGE IN MONTHS:	0 < 12	12 < 18	18 < 24	24+	0 < 12	12 < 18	18 < 24	24+
*************** INFLAMMATION AND REPAIR ***************								
VASCULARIZATION							2/104	3/243

SEX:	FEMALE				MALE			
DIAGNOSIS / AGE IN MONTHS:	0 < 12	12 < 18	18 < 24	24+	0 < 12	12 < 18	18 < 24	24+
*************** INFLAMMATION AND REPAIR ***************								
INFLAMMATION			1/420	3/240		1/97	4/392	3/216
INFLAMMATION, SUPPURATIVE			2/420	2/240		2/97	1/392	7/216
INFLAMMATION, ULCERATIVE							1/392	1/216
*************** DEGENERATION, NECROSIS, AND DEPOSITION ***************								
KERATOPATHY, BAND							1/392	3/216

SEX:	FEMALE				MALE			
DIAGNOSIS — AGE IN MONTHS:	0 < 12	12 < 18	18 < 24	24+	0 < 12	12 < 18	18 < 24	24+
*************** INFLAMMATION AND REPAIR ***************								
INFLAMMATION			3/193	1/71				1/94
*************** DEGENERATION, NECROSIS, AND DEPOSITION ***************								
DEGENERATION						1/314	2/147	
*************** GROWTH ALTERATIONS, NON-NEOPLASTIC, ATROPHY ***************								
ATROPHY	1/212	7/369	6/193	1/71		3/314	2/147	3/94

 ORGAN SYSTEM: NERVOJS SYSTEM; SPECIAL ORGANS ORGAN: EYE; RETINA STRAIN: B6C3F1

SEX:	FEMALE				MALE			
DIAGNOSIS / AGE IN MONTHS:	0 < 12	12 < 18	18 < 24	24+	0 < 12	12 < 18	18 < 24	24+
***************								
INFLAMMATION AND REPAIR								
***************								
INFLAMMATION			1/1356	1/1015				3/958

SEX:	FEMALE				MALE			
DIAGNOSIS / AGE IN MONTHS:	0 < 12	12 < 18	18 < 24	24+	0 < 12	12 < 18	18 < 24	24+
*************** INFLAMMATION AND REPAIR ***************								
INFLAMMATION			1/192	2/75				
*************** GROWTH ALTERATIONS, NON-NEOPLASTIC, ATROPHY ***************								
ATROPHY		1/367						

SEX:	FEMALE				MALE			
DIAGNOSIS / AGE IN MONTHS:	0 < 12	12 < 18	18 < 24	24+	0 < 12	12 < 18	18 < 24	24+
*************** INFLAMMATION AND REPAIR ***************								
INFLAMMATION				5/1575				10/1743
*************** DEGENERATION, NECROSIS, AND DEPOSITION ***************								
DEGENERATION				10/1575		3/101	6/437	76/1743
*************** GROWTH ALTERATIONS, NON-NEOPLASTIC, ATROPHY ***************								
ATROPHY				2/1575			1/437	17/1743

ORGAN SYSTEM: NERVOUS SYSTEM; SPECIAL ORGANS   ORGAN: EYE; RETINA   STRAIN: OSBORNE MENDEL

SEX:	FEMALE				MALE			
DIAGNOSIS / AGE IN MONTHS:	0 < 12	12 < 18	18 < 24	24+	0 < 12	12 < 18	18 < 24	24+
*************** INFLAMMATION AND REPAIR ***************								
INFLAMMATION				2/336			1/104	

SEX:	FEMALE				MALE			
DIAGNOSIS / AGE IN MONTHS:	0 < 12	12 < 18	18 < 24	24+	0 < 12	12 < 18	18 < 24	24+
*************** DEVELOPMENTAL ANOMALIES ***************								
INFOLDING OF RETINA	1/181							
*************** INFLAMMATION AND REPAIR ***************								
INFLAMMATION							1/392	
*************** GROWTH ALTERATIONS, NON-NEOPLASTIC ATROPHY ***************								
ATROPHY	8/181				7/170			
*************** GROWTH ALTERATIONS, NON-NEOPLASTIC, DYSPLASIA ***************								
RETINAL ROSETTE					1/170			

SEX:	FEMALE				MALE			
DIAGNOSIS / AGE IN MONTHS:	0 < 12	12 < 18	18 < 24	24+	0 < 12	12 < 18	18 < 24	24+
*************** INFLAMMATION AND REPAIR ***************								
INFLAMMATION			1/327	3/1575				8/1743

 ORGAN SYSTEM: RESPIRATORY SYSTEM ORGAN: LARYNX STRAIN: SPRAGUE DAWLEY

DIAGNOSIS	FEMALE				MALE			
SEX: / AGE IN MONTHS:	0 < 12	12 < 18	18 < 24	24+	0 < 12	12 < 18	18 < 24	24+
*************** INFLAMMATION AND REPAIR ***************								
INFLAMMATION			1/420					

SEX:	FEMALE				MALE			
DIAGNOSIS / AGE IN MONTHS:	0 < 12	12 < 18	18 < 24	24+	0 < 12	12 < 18	18 < 24	24+
*************** MECHANICAL ABNORMALITIES AND CYST FORMATION ***************								
ATELECTASIS				2/71		1/314		2/94
CONGESTION			1/193		1/171	3/314	1/147	
EDEMA		2/369	3/193	1/71	4/171	10/314	5/147	10/94
EMPHYSEMA							1/147	1/94
HEMORRHAGE	4/212		1/193		11/171	6/314	1/147	2/94
*************** INFLAMMATION AND REPAIR ***************								
INFLAMMATION	1/212			1/71		1/314		
INFLAMMATION, HISTIOCYTIC	1/212			5/71		1/314	1/147	7/94
INFLAMMATION, NONSUPPURATIVE		2/369	2/193	19/71		3/314	7/147	16/94*
PERIVASCULITIS/VASCULITIS					1/171	1/314		9/94
PNEUMONIA, CHRONIC MURINE		2/369	2/193			3/314	1/147	
*************** DEGENERATION, NECROSIS, AND DEPOSITION ***************								
CALCULI				1/71				
MINERALIZATION	1/212		1/193					
NECROSIS							1/147	
PIGMENT DEPOSITION		1/369			1/171		1/147	

*VARIABILITY AMONG DATA SOURCES SIGNIFICANT AT THE 1% SIGNIFICANCE LEVEL

SEX:	FEMALE				MALE			
AGE IN MONTHS:	0 < 12	12 < 18	18 < 24	24+	0 < 12	12 < 18	18 < 24	24+
DIAGNOSIS								
*************** GROWTH ALTERATIONS, NON-NEOPLASTIC, HYPERPLASIA ***************								
HYPERPLASIA, EPITHELIAL				1/71				1/94
*************** GROWTH ALTERATIONS, NEOPLASTIC, BENIGN ***************								
ADENOMA, ALVEOLAR BRONCHIOLAR								2/94
*************** GROWTH ALTERATIONS, NEOPLASTIC, MALIGNANT, PRIMARY ***************								
ADENOCARCINOMA, ALVEOLAR BRONCHIOLAR						2/314		3/94
*************** GROWTH ALTERATIONS, NEOPLASTIC, MALIGNANT, METASTATIC ***************								
ADENOCARCINOMA, MAMMARY			2/193					
CARCINOSARCOMA, MAMMARY		1/369						
*************** GROWTH ALTERATIONS, NEOPLASTIC, MALIGNANT, PRIMARY SITE NOT DETERMINED ***************								
ANGIOSARCOMA					1/171			
LYMPHOMA	3/212	9/369	6/193	3/71	3/171	1/314	1/147	3/94
LYMPHOMA, RETICULUM CELL		1/369	12/193	18/71			3/147	6/94
*************** GROWTH ALTERATIONS, NEOPLASTIC, NOT DETERMINED ***************								
ALVEOLAR CELL TUMOR	9/212	45/369	55/193	25/71	7/171	62/314	57/147	33/94*

*VARIABILITY AMONG DATA SOURCES SIGNIFICANT AT THE 1% SIGNIFICANCE LEVEL

SEX:	FEMALE				MALE			
DIAGNOSIS / AGE IN MONTHS:	0 < 12	12 < 18	18 < 24	24+	0 < 12	12 < 18	18 < 24	24+
*************** MECHANICAL ABNORMALITIES AND CYST FORMATION ***************								
ATELECTASIS				2/1015				
CONGESTION		7/103	5/1356		2/124	6/199	2/1136	3/958
EDEMA		2/103	4/1356			2/199	1/1136	
EMPHYSEMA			1/1356				1/1136	
HEMORRHAGE	1/67	1/103		3/1015			4/1136	2/958
PROTEINOSIS, ALVEOLAR				1/1015		1/199	1/1136	
*************** INFLAMMATION AND REPAIR ***************								
ABSCESS							2/1136	
FIBROSIS							1/1136	
GRANULOMA								1/958
INFLAMMATION		2/103	15/1356	26/1015	1/124	4/199	21/1136	29/958
INFLAMMATION, GRANULOMATOUS							1/1136	
INFLAMMATION, HISTIOCYTIC				7/1015			3/1136	5/958
INFLAMMATION, NONSUPPURATIVE				10/1015				8/958
INFLAMMATION, SUPPURATIVE	11/67	18/103*	44/1356	14/1015	3/124	29/199*	39/1136	5/958
PERIVASCULITIS/VASCULITIS		1/103	1/1356	15/1015			1/1136	8/958
PNEUMONIA, ASPIRATION				2/1015				1/958
PNEUMONIA, CHRONIC MURINE		2/103	23/1356	43/1015		1/199	13/1136	17/958

*VARIABILITY AMONG DATA SOURCES SIGNIFICANT AT THE 1% SIGNIFICANCE LEVEL

SEX:	FEMALE				MALE			
AGE IN MONTHS: / DIAGNOSIS	0 < 12	12 < 18	18 < 24	24+	0 < 12	12 < 18	18 < 24	24+
*************** DEGENERATION, NECROSIS, AND DEPOSITION ***************								
ARTERIOSCLEROSIS							1/1136	
CALCULI				1/1015				1/958
*************** GROWTH ALTERATIONS, NON-NEOPLASTIC, HYPERPLASIA ***************								
HYPERPLASIA			1/1356	1/1015				
HYPERPLASIA, ALVEOLAR			5/1356	7/1015			5/1136	13/958
HYPERPLASIA, EPITHELIAL			3/1356				3/1136	4/958
HYPERPLASIA, HISTIOCYTIC								2/958
HYPERPLASIA, LYMPHOID		1/103	29/1356	61/1015*			17/1136	38/958
HYPERPLASIA, PAPILLARY								1/958
HYPERPLASIA, PLASMA CELL	4/67		8/1356	5/1015		4/199	1/1136	1/958
*************** GROWTH ALTERATIONS, NON-NEOPLASTIC, METAPLASIA ***************								
METAPLASIA, OSSEOUS			1/1356					
METAPLASIA, SQUAMOUS							1/1136	
*************** GROWTH ALTERATIONS, NEOPLASTIC, BENIGN ***************								
ADENOMA, ALVEOLAR BRONCHIOLAR			34/1356	34/1015		3/199	72/1136	97/958

*VARIABILITY AMONG DATA SOURCES SIGNIFICANT AT THE 1% SIGNIFICANCE LEVEL

SEX:	FEMALE				MALE			
DIAGNOSIS / AGE IN MONTHS:	0 < 12	12 < 18	18 < 24	24+	0 < 12	12 < 18	18 < 24	24+
*************** GROWTH ALTERATIONS, NEOPLASTIC, MALIGNANT, PRIMARY ***************								
CARCINOMA, ALVEOLAR BRONCHIOLAR		1/103	21/1356	19/1015		3/199	39/1136	77/958*
*************** GROWTH ALTERATIONS, NEOPLASTIC, MALIGNANT, METASTATIC ***************								
ADENOCARCINOMA, PAPILLARY								1/958
CARCINOMA, ACINAR CELL								1/958
CARCINOMA, CORTICAL							1/1136	
CARCINOMA, HEPATOCELLULAR			3/1356	2/1015			18/1136	22/958
CARCINOMA, TRANSITIONAL CELL				1/1015				
FIBROSARCOMA			1/1356				4/1136	2/958
LEIOMYOSARCOMA			1/1356	2/1015				
OSTEOMA	1/67							
OSTEOSARCOMA		1/103	3/1356	2/1015			2/1136	1/958
*************** GROWTH ALTERATIONS, NEOPLASTIC, MALIGNANT, PRIMARY SITE NOT DETERMINED ***************								
ANGIOSARCOMA							1/1136	
HYPERPLASIA, MYELOID								1/958
LYMPHOMA, HISTIOCYTIC			1/1356					
LYMPHOMA, LYMPHOCYTIC				1/1015				
SARCOMA			1/1356					

*VARIABILITY AMONG DATA SOURCES SIGNIFICANT AT THE 1% SIGNIFICANCE LEVEL

SEX:	FEMALE				MALE			
DIAGNOSIS / AGE IN MONTHS:	0 < 12	12 < 18	18 < 24	24+	0 < 12	12 < 18	18 < 24	24+
*************** MECHANICAL ABNORMALITIES AND CYST FORMATION ***************								
ATELECTASIS		1/367	2/192					
CONGESTION				3/75		1/300		
EDEMA		1/367	1/192				2/161	3/102
HEMORRHAGE		2/367			1/131			1/102
*************** INFLAMMATION AND REPAIR ***************								
INFLAMMATION		2/367	3/192	1/75		2/300	1/161	1/102
INFLAMMATION, ATYPICAL/ADENOMATOID								1/102
INFLAMMATION, HISTIOCYTIC				10/75	1/131			10/102
INFLAMMATION, NONSUPPURATIVE	12/176	65/367	63/192	35/75	2/131	10/300	16/161	43/102*
PERIVASCULITIS/VASCULITIS					4/131	9/300	3/161	14/102
PNEUMONIA, CHRONIC MURINE	8/176	7/367	14/192	1/75	1/131		2/161	
*************** DEGENERATION, NECROSIS, AND DEPOSITION ***************								
CALCULI	7/176	4/367	9/192	4/75			2/161	5/102
PIGMENT DEPOSITION			1/192	1/75				
*************** GROWTH ALTERATIONS, NON-NEOPLASTIC, HYPERPLASIA ***************								
HYPERPLASIA, EPITHELIAL			1/192	1/75				1/102
HYPERPLASIA, LYMPHOID							1/161	2/102

*VARIABILITY AMONG DATA SOURCES SIGNIFICANT AT THE 1% SIGNIFICANCE LEVEL

SEX:	FEMALE				MALE			
DIAGNOSIS / AGE IN MONTHS:	0 < 12	12 < 18	18 < 24	24+	0 < 12	12 < 18	18 < 24	24+
*************** GROWTH ALTERATIONS, NEOPLASTIC, BENIGN ***************								
ADENOMA, ALVEOLAR BRONCHIOLAR								2/102
*************** GROWTH ALTERATIONS, NEOPLASTIC, MALIGNANT, METASTATIC ***************								
HEPATIC NEOPLASM						1/300		
*************** GROWTH ALTERATIONS, NEOPLASTIC, MALIGNANT, PRIMARY SITE NOT DETERMINED ***************								
LYMPHOMA	1/176	1/367	4/192	4/75		1/300	1/161	5/102
LYMPHOMA, RETICULUM CELL	1/176	2/367	10/192	20/75			5/161	13/102
*************** GROWTH ALTERATIONS, NEOPLASTIC, NOT DETERMINED ***************								
ALVEOLAR CELL TUMOR	1/176	1/367	4/192	2/75	1/131	5/300	5/161	9/102

SEX:	FEMALE				MALE			
DIAGNOSIS / AGE IN MONTHS:	0 < 12	12 < 18	18 < 24	24+	0 < 12	12 < 18	18 < 24	24+
*************** MECHANICAL ABNORMALITIES AND CYST FORMATION ***************								
ATELECTASIS			1/327	6/1575			3/437	10/1743
CONGESTION		1/24	1/327	38/1575	5/101	2/101	8/437	46/1743
EDEMA			1/327	1/1575	6/101	1/101	2/437	2/1743
EMPHYSEMA								4/1743
HEMORRHAGE	1/26			4/1575		2/101	2/437	5/1743
THROMBUS								1/1743
*************** INFLAMMATION AND REPAIR ***************								
ABSCESS				9/1575	1/101		7/437	11/1743
FIBROSIS				1/1575			1/437	4/1743
GRANULOMA				7/1575				3/1743
INFLAMMATION	9/26		15/327	29/1575	44/101*	2/101	11/437	77/1743
INFLAMMATION, HISTIOCYTIC				19/1575	1/101	4/101	3/437	16/1743
INFLAMMATION, NECROTIZING						1/101	1/437	
INFLAMMATION, NONSUPPURATIVE			12/327	30/1575	9/101*	20/101*	15/437	59/1743
INFLAMMATION, SUPPURATIVE		1/24	2/327	2/1575		3/101	1/437	3/1743
PERIVASCULITIS/VASCULITIS	6/26		9/327	12/1575	46/101	12/101*	9/437	19/1743
PNEUMONIA, ASPIRATION				7/1575	1/101		2/437	3/1743
PNEUMONIA, CHRONIC MURINE	1/26	1/24	10/327	115/1575*	6/101	5/101	13/437	122/1743*

*VARIABILITY AMONG DATA SOURCES SIGNIFICANT AT THE 1% SIGNIFICANCE LEVEL

SEX:	FEMALE				MALE			
DIAGNOSIS / AGE IN MONTHS:	0 < 12	12 < 18	18 < 24	24+	0 < 12	12 < 18	18 < 24	24+
*************** DEGENERATION, NECROSIS, AND DEPOSITION ***************								
ARTERIOSCLEROSIS				1/1575				
HEMOSIDERIN DEPOSITION								1/1743
MINERALIZATION				1/1575		2/101	1/437	8/1743
NECROSIS							6/437	
PIGMENT DEPOSITION						1/101	1/437	
*************** GROWTH ALTERATIONS, NON-NEOPLASTIC, HYPERTROPHY ***************								
CYTOMEGALY						1/101		
*************** GROWTH ALTERATIONS, NON-NEOPLASTIC, HYPERPLASIA ***************								
HYPERPLASIA								4/1743
HYPERPLASIA, ALVEOLAR				9/1575			2/437	14/1743
HYPERPLASIA, EPITHELIAL			3/327	9/1575			5/437	23/1743
HYPERPLASIA, HISTIOCYTIC				4/1575				3/1743
HYPERPLASIA, LYMPHOID			5/327	71/1575			4/437	53/1743
*************** GROWTH ALTERATIONS, NON-NEOPLASTIC, METAPLASIA ***************								
METAPLASIA, ALVEOLAR				1/1575				
METAPLASIA, SQUAMOUS						1/101		

SEX:	FEMALE				MALE			
AGE IN MONTHS: / DIAGNOSIS	0 < 12	12 < 18	18 < 24	24+	0 < 12	12 < 18	18 < 24	24+
*************** GROWTH ALTERATIONS, NEOPLASTIC, BENIGN ***************								
ADENOMA, ALVEOLAR BRONCHIOLAR		2/24	1/327	31/1575		1/101	10/437	37/1743
HEMANGIOPERICYTOMA							1/437	
*************** GROWTH ALTERATIONS, NEOPLASTIC, MALIGNANT, PRIMARY ***************								
CARCINOMA, ALVEOLAR BRONCHIOLAR				12/1575			3/437	19/1743
*************** GROWTH ALTERATIONS, NEOPLASTIC, MALIGNANT, METASTATIC ***************								
ADENOCARCINOMA, MAMMARY				1/1575				
ADENOCARCINOMA, SEBACEOUS				1/1575				
CARCINOMA, C CELL				2/1575				2/1743
CARCINOMA, CORTICAL				1/1575				
CARCINOMA, HEPATOCELLULAR				1/1575			2/437	1/1743
CARCINOMA, SEBACEOUS							1/437	
CARCINOMA, SQUAMOUS CELL			1/327	3/1575			2/437	3/1743
FIBROSARCOMA							2/437	
HISTIOCYTOMA, FIBROUS							1/437	
HISTIOCYTOMA, MALIGNANT							1/437	
INTERSTITIAL CELL TUMOR								1/1743
LIPOSARCOMA			1/327					
MESOTHELIOMA					1/101			

SEX:	FEMALE				MALE			
DIAGNOSIS / AGE IN MONTHS:	0 < 12	12 < 18	18 < 24	24+	0 < 12	12 < 18	18 < 24	24+
NEUROFIBROSARCOMA								1/1743
OSTEOSARCOMA							3/437	2/1743
PHEOCHROMOCYTOMA, MALIGNANT				2/1575				5/1743
RHABDOMYOSARCOMA			1/327					
THYMOMA								1/1743
*************** GROWTH ALTERATIONS, NEOPLASTIC, MALIGNANT, PRIMARY SITE NOT DETERMINED ***************								
ANGIOSARCOMA				1/1575			2/437	
LEUKEMIA	1/26			1/1575			2/437	10/1743
LEUKEMIA, GRANULOCYTIC				1/1575				3/1743
LEUKEMIA, LYMPHOBLASTIC				4/1575				
LEUKEMIA, LYMPHOCYTIC				3/1575				1/1743
LEUKEMIA, STEM CELL							1/437	
LYMPHOMA				1/1575				1/1743
SARCOMA								2/1743
*************** GROWTH ALTERATIONS, NEOPLASTIC, NOT DETERMINED ***************								
NEOPLASM								1/1743

ORGAN SYSTEM: RESPIRATORY SYSTEM   ORGAN: LUNG   STRAIN: OSBORNE MENDEL

SEX:	FEMALE				MALE			
DIAGNOSIS / AGE IN MONTHS:	0 < 12	12 < 18	18 < 24	24+	0 < 12	12 < 18	18 < 24	24+
*************** MECHANICAL ABNORMALITIES AND CYST FORMATION ***************								
CONGESTION	1/7		1/44	1/336		1/64	3/104	1/243
EDEMA							1/104	
HEMORRHAGE				5/336				
THROMBUS, ORGANIZED				1/336				
*************** INFLAMMATION AND REPAIR ***************								
ABSCESS			4/44	1/336				
INFLAMMATION	1/7		1/44	2/336			2/104	
INFLAMMATION, HISTIOCYTIC				2/336			1/104	2/243
PERIVASCULITIS/VASCULITIS							1/104	
PNEUMONIA, CHRONIC MURINE	3/7	23/39	21/44	170/336*	3/11	31/64	49/104*	103/243
*************** DEGENERATION, NECROSIS, AND DEPOSITION ***************								
MINERALIZATION			2/44	2/336			3/104	4/243
*************** GROWTH ALTERATIONS, NON-NEOPLASTIC, HYPERPLASIA ***************								
HYPERPLASIA, EPITHELIAL				3/336				
HYPERPLASIA, LYMPHOID						1/64		

*VARIABILITY AMONG DATA SOURCES SIGNIFICANT AT THE 1% SIGNIFICANCE LEVEL

SEX:	FEMALE				MALE			
DIAGNOSIS / AGE IN MONTHS:	0 < 12	12 < 18	18 < 24	24+	0 < 12	12 < 18	18 < 24	24+
GROWTH ALTERATIONS, NEOPLASTIC, BENIGN								
ADENOMA, ALVEOLAR BRONCHIOLAR								1/243
GROWTH ALTERATIONS, NEOPLASTIC, MALIGNANT, PRIMARY								
CARCINOMA, ALVEOLAR BRONCHIOLAR				1/336				
GROWTH ALTERATIONS, NEOPLASTIC, MALIGNANT, METASTATIC								
CARCINOMA, C CELL								1/243
CARCINOMA, HEPATOCELLULAR				1/336				
FIBROSARCOMA				1/336				1/243
HISTIOCYTOMA, FIBROUS				1/336				
MIXED TUMOR, MALIGNANT						1/64		
NEUROFIBROSARCOMA								1/243
OSTEOSARCOMA							1/104	1/243
GROWTH ALTERATIONS, NEOPLASTIC, MALIGNANT, PRIMARY SITE NOT DETERMINED								
ANGIOSARCOMA				1/336				2/243

SEX:	FEMALE				MALE			
DIAGNOSIS / AGE IN MONTHS:	0 < 12	12 < 18	18 < 24	24+	0 < 12	12 < 18	18 < 24	24+
*************** MECHANICAL ABNORMALITIES AND CYST FORMATION ***************								
ATELECTASIS	1/181					1/97		2/216
CONGESTION		6/60	12/420	5/240		5/97	9/392	14/216*
EDEMA	1/181		7/420			2/97	10/392	5/216
EMPHYSEMA	5/181	1/60			6/170	2/97	2/392	2/216
HEMORRHAGE		1/60	1/420	1/240	1/170		1/392	
THROMBUS								1/216
*************** INFLAMMATION AND REPAIR ***************								
ABSCESS			3/420			2/97		1/216
FIBROSIS			1/420					1/216
GRANULOMA								1/216
INFLAMMATION	47/181	3/60	17/420	24/240*	40/170*	12/97*	13/392	10/216
INFLAMMATION, GRANULOMATOUS		1/60		1/240	1/170			
INFLAMMATION, HISTIOCYTIC	1/181				2/170			1/216
INFLAMMATION, NONSUPPURATIVE	13/181*	2/60		1/240	14/170*	3/97		
INFLAMMATION, SUPPURATIVE	1/181	2/60	11/420		1/170	1/97	16/392	1/216
PERIVASCULITIS/VASCULITIS	16/181*		3/420		18/170*	1/97		
PNEUMONIA, ASPIRATION			1/420	2/240				3/216
PNEUMONIA, CHRONIC MURINE	3/181	4/60	44/420*	14/240*	1/170	25/97*	43/392*	9/216

*VARIABILITY AMONG DATA SOURCES SIGNIFICANT AT THE 1% SIGNIFICANCE LEVEL

SEX:	FEMALE				MALE			
DIAGNOSIS / AGE IN MONTHS:	0 < 12	12 < 18	18 < 24	24+	0 < 12	12 < 18	18 < 24	24+
*************** DEGENERATION, NECROSIS, AND DEPOSITION ***************								
HEMOSIDERIN DEPOSITION								1/216
MINERALIZATION			3/420	1/240		1/97	6/392	5/216
*************** GROWTH ALTERATIONS, NON-NEOPLASTIC, HYPERTROPHY ***************								
HYPERTROPHY		3/60	1/420	6/240		3/97	1/392	11/216*
*************** GROWTH ALTERATIONS, NON-NEOPLASTIC, HYPERPLASIA ***************								
HYPERPLASIA	3/181		1/420	2/240	5/170			4/216
HYPERPLASIA, EPITHELIAL							1/392	
HYPERPLASIA, LYMPHOID	10/181*		7/420		11/170*	3/97	15/392	
*************** GROWTH ALTERATIONS, NON-NEOPLASTIC, METAPLASIA ***************								
METAPLASIA, ALVEOLAR							1/392	
*************** GROWTH ALTERATIONS, NEOPLASTIC, BENIGN ***************								
ADENOMA, ALVEOLAR BRONCHIOLAR							1/392	1/216
*************** GROWTH ALTERATIONS, NEOPLASTIC, MALIGNANT, PRIMARY ***************								
CARCINOMA, ALVEOLAR BRONCHIOLAR			1/420					

*VARIABILITY AMONG DATA SOURCES SIGNIFICANT AT THE 1% SIGNIFICANCE LEVEL

SEX:	FEMALE				MALE			
DIAGNOSIS / AGE IN MONTHS:	0 < 12	12 < 18	18 < 24	24+	0 < 12	12 < 18	18 < 24	24+
*************** GROWTH ALTERATIONS, NEOPLASTIC, MALIGNANT, METASTATIC ***************								
ADENOCARCINOMA, MAMMARY			1/420	1/240				
ADENOFIBROSARCOMA, MAMMARY							1/392	
CARCINOMA, SQUAMOUS CELL							1/392	
LEIOMYOSARCOMA							1/392	
OSTEOSARCOMA	1/181		1/420					
RHABDOMYOSARCOMA				1/240		1/97		
*************** GROWTH ALTERATIONS, NEOPLASTIC, MALIGNANT, PRIMARY SITE NOT DETERMINED ***************								
LEUKEMIA, LYMPHOCYTIC			1/420			1/97		
LYMPHOMA	1/181		1/420				1/392	1/216
LYMPHOMA, RETICULUM CELL			1/420				1/392	
SARCOMA, HISTIOCYTIC			1/420	2/240			1/392	5/216

SEX:	FEMALE				MALE			
DIAGNOSIS / AGE IN MONTHS:	0 < 12	12 < 18	18 < 24	24+	0 < 12	12 < 18	18 < 24	24+
*************** MECHANICAL ABNORMALITIES AND CYST FORMATION ***************								
BRONCHIECTASIS	1/212							
*************** INFLAMMATION AND REPAIR ***************								
INFLAMMATION, GRANULOMATOUS							2/147	
INFLAMMATION, NONSUPPURATIVE								1/94
*************** GROWTH ALTERATIONS, NEOPLASTIC, MALIGNANT, PRIMARY ***************								
ADENOCARCINOMA			1/193					1/94
CARCINOMA						1/314		

SEX:	FEMALE				MALE			
DIAGNOSIS / AGE IN MONTHS:	O < 12	12 < 18	18 < 24	24+	O < 12	12 < 18	18 < 24	24+
***************								
MECHANICAL ABNORMALITIES AND CYST FCRMATION								
***************								
BRONCHIECTASIS			1/1356					
***************								
INFLAMMATION AND REPAIR								
***************								
INFLAMMATION, GRANULOMATOUS								1/958
***************								
GROWTH ALTERATIONS, NEOPLASTIC, MALIGNANT, PRIMARY								
***************								
ADENOCARCINOMA			2/1356	2/1015				
CARCINOMA	1/67		2/1356	1/1015				

SEX:	FEMALE				MALE			
DIAGNOSIS / AGE IN MONTHS:	0 < 12	12 < 18	18 < 24	24+	0 < 12	12 < 18	18 < 24	24+
*************** INFLAMMATION AND REPAIR ***************								
INFLAMMATION, GRANULOMATOUS				1/75				
INFLAMMATION, NONSUPPURATIVE								1/102
INFLAMMATION, SUPPURATIVE						1/300		

SEX:	FEMALE				MALE			
DIAGNOSIS / AGE IN MONTHS:	0 < 12	12 < 18	18 < 24	24+	0 < 12	12 < 18	18 < 24	24+
*************** MECHANICAL ABNORMALITIES AND CYST FORMATION ***************								
BRONCHIECTASIS			1/327	7/1575			5/437	16/1743
*************** INFLAMMATION AND REPAIR ***************								
INFLAMMATION, GRANULOMATOUS								1/1743
INFLAMMATION, NECROTIZING							3/437	
INFLAMMATION, NONSUPPURATIVE					9/101*	1/101		
*************** GROWTH ALTERATIONS, NEOPLASTIC, BENIGN ***************								
ADENOMA, BRONCHIAL				2/1575				
*************** GROWTH ALTERATIONS, NEOPLASTIC, MALIGNANT, PRIMARY ***************								
ADENOCARCINOMA			1/327	3/1575				5/1743
CARCINOMA			1/327	1/1575			1/437	2/1743

*VARIABILITY AMONG DATA SOURCES SIGNIFICANT AT THE 1% SIGNIFICANCE LEVEL

SEX:	FEMALE				MALE			
DIAGNOSIS / AGE IN MONTHS:	0 < 12	12 < 18	18 < 24	24+	0 < 12	12 < 18	18 < 24	24+
*************** INFLAMMATION AND REPAIR ***************								
INFLAMMATION, GRANULOMATOUS								1/243
*************** GROWTH ALTERATIONS, NEOPLASTIC, MALIGNANT, PRIMARY ***************								
ADENOCARCINOMA								1/243

ORGAN SYSTEM: RESPIRATORY SYSTEM ORGAN: LUNG; BRONCHUS STRAIN: SPRAGUE DAWLEY

DIAGNOSIS	FEMALE				MALE			
SEX: / AGE IN MONTHS:	0 < 12	12 < 18	18 < 24	24+	0 < 12	12 < 18	18 < 24	24+
*************** MECHANICAL ABNORMALITIES AND CYST FORMATION ***************								
BRONCHIECTASIS	10/181*	6/60*	4/420		11/170*	12/97 *	12/392	
*************** INFLAMMATION AND REPAIR ***************								
INFLAMMATION	8/181	1/60	4/420	6/240	7/170		5/392	9/216
INFLAMMATION, NONSUPPURATIVE	1/181	1/60						
INFLAMMATION, SUPPURATIVE						1/97		
*************** GROWTH ALTERATIONS, NEOPLASTIC, MALIGNANT, PRIMARY ***************								
ADENOCARCINOMA	1/181				1/170			

*VARIABILITY AMONG DATA SOURCES SIGNIFICANT AT THE 1% SIGNIFICANCE LEVEL

 ORGAN SYSTEM: RESPIRATORY SYSTEM ORGAN: LUNG; INTERSTITIUM STRAIN: BALB/C

SEX:	FEMALE				MALE			
DIAGNOSIS / AGE IN MONTHS:	0 < 12	12 < 18	18 < 24	24+	0 < 12	12 < 18	18 < 24	24+
*************** INFLAMMATION AND REPAIR ***************								
INFLAMMATION								1/94
INFLAMMATION, NONSUPPURATIVE								1/94

SEX:	FEMALE				MALE			
DIAGNOSIS / AGE IN MONTHS:	0 < 12	12 < 18	18 < 24	24+	0 < 12	12 < 18	18 < 24	24+
*************** INFLAMMATION AND REPAIR ***************								
INFLAMMATION		1/103	51/1356	6/1015	1/124	1/199	46/1136	2/958
INFLAMMATION, SUPPURATIVE		1/103	5/1356	1/1015	1/124			3/958

ORGAN SYSTEM: RESPIRATORY SYSTEM ORGAN: LUNG; INTERSTITIUM STRAIN: C57BL/6

SEX:	FEMALE				MALE			
DIAGNOSIS / AGE IN MONTHS:	0 < 12	12 < 18	18 < 24	24+	0 < 12	12 < 18	18 < 24	24+
*************** INFLAMMATION AND REPAIR ***************								
INFLAMMATION	1/176			1/75		2/300		1/102
INFLAMMATION, NONSUPPURATIVE					1/131			4/102
INFLAMMATION, SUPPURATIVE								1/102

SEX:	FEMALE				MALE			
DIAGNOSIS / AGE IN MONTHS:	0 < 12	12 < 18	18 < 24	24+	0 < 12	12 < 18	18 < 24	24+
*************** INFLAMMATION AND REPAIR ***************								
INFLAMMATION			9/327	28/1575	1/101	5/101*	12/437	26/1743
INFLAMMATION, NONSUPPURATIVE					1/101		2/437	5/1743
INFLAMMATION, SUPPURATIVE			1/327	2/1575		2/101	1/437	8/1743

*VARIABILITY AMONG DATA SOURCES SIGNIFICANT AT THE 1% SIGNIFICANCE LEVEL

SEX:	FEMALE				MALE			
DIAGNOSIS / AGE IN MONTHS:	0 < 12	12 < 18	18 < 24	24+	0 < 12	12 < 18	18 < 24	24+
*************** INFLAMMATION AND REPAIR ***************								
INFLAMMATION		6/60	6/420	2/240		5/97	2/392	2/216
INFLAMMATION, NONSUPPURATIVE		6/60	4/420	44/240 *		8/97	8/392	43/216*
INFLAMMATION, SUPPURATIVE					1/170		1/392	

*VARIABILITY AMONG DATA SOURCES SIGNIFICANT AT THE 1% SIGNIFICANCE LEVEL

SEX:	FEMALE				MALE			
AGE IN MONTHS:	C < 12	12 < 18	18 < 24	24+	O < 12	12 < 18	18 < 24	24+
DIAGNOSIS								
*************** INFLAMMATION AND REPAIR ***************								
INFLAMMATION					1/124			
INFLAMMATION, SUPPURATIVE			1/1356					

SEX:	FEMALE				MALE			
DIAGNOSIS — AGE IN MONTHS:	O < 12	12 < 18	18 < 24	24+	O < 12	12 < 18	18 < 24	24+
***************								
INFLAMMATION AND REPAIR								
***************								
INFLAMMATION, SUPPURATIVE							1/161	

SEX:	FEMALE				MALE			
AGE IN MONTHS: / DIAGNOSIS	0 < 12	12 < 18	18 < 24	24+	0 < 12	12 < 18	18 < 24	24+
*************** INFLAMMATION AND REPAIR ***************								
INFLAMMATION			3/327					1/1743
INFLAMMATION, SUPPURATIVE			1/327					1/1743

SEX:	FEMALE				MALE			
DIAGNOSIS / AGE IN MONTHS:	0 < 12	12 < 18	18 < 24	24+	0 < 12	12 < 18	18 < 24	24+
*************** MECHANICAL ABNORMALITIES AND CYST FORMATION ***************								
CYST				2/71			2/147	
*************** INFLAMMATION AND REPAIR ***************								
INFLAMMATION					1/171			
*************** GROWTH ALTERATIONS, NEOPLASTIC, MALIGNANT, PRIMARY SITE NOT DETERMINED ***************								
LYMPHOMA		3/369	1/193					
LYMPHOMA, RETICULUM CELL			4/193	3/71				2/94

 ORGAN SYSTEM: RESPIRATORY SYSTEM ORGAN: TRACHEA STRAIN: B6C3F1

SEX:	FEMALE				MALE			
DIAGNOSIS / AGE IN MONTHS:	0 < 12	12 < 18	18 < 24	24+	0 < 12	12 < 18	18 < 24	24+
*************** MECHANICAL ABNORMALITIES AND CYST FORMATION ***************								
DILATATION								1/958
*************** INFLAMMATION AND REPAIR ***************								
INFLAMMATION						1/199		
INFLAMMATION, SUPPURATIVE	2/67	5/103	9/1356	3/1015		14/199*	6/1136	2/958
*************** GROWTH ALTERATIONS, NON-NEOPLASTIC, HYPERPLASIA ***************								
HYPERPLASIA, PLASMA CELL			1/1356	1/1015				

*VARIABILITY AMONG DATA SOURCES SIGNIFICANT AT THE 1% SIGNIFICANCE LEVEL

SEX:	FEMALE				MALE			
DIAGNOSIS / AGE IN MONTHS:	0 < 12	12 < 18	18 < 24	24+	0 < 12	12 < 18	18 < 24	24+
*************** INFLAMMATION AND REPAIR ***************								
INFLAMMATION		1/367						
INFLAMMATION, NONSUPPURATIVE						1/300		
*************** GROWTH ALTERATIONS, NEOPLASTIC, MALIGNANT, PRIMARY SITE NOT DETERMINED ***************								
LYMPHOMA, RETICULUM CELL				1/75				

DIAGNOSIS	FEMALE 0 < 12	FEMALE 12 < 18	FEMALE 18 < 24	FEMALE 24+	MALE 0 < 12	MALE 12 < 18	MALE 18 < 24	MALE 24+
*************** MECHANICAL ABNORMALITIES AND CYST FORMATION ***************								
DILATATION								3/1743
*************** INFLAMMATION AND REPAIR ***************								
INFLAMMATION	10/26		14/327	76/1575	46/101*	5/101*	18/437	126/1743*
INFLAMMATION, NONSUPPURATIVE		1/24	5/327	2/1575			3/437	1/1743
INFLAMMATION, SUPPURATIVE		1/24	4/327	34/1575	1/101		6/437	47/1743
POLYP, INFLAMMATORY				2/1575				
*************** DEGENERATION, NECROSIS, AND DEPOSITION ***************								
MINERALIZATION						1/101		
*************** GROWTH ALTERATIONS, NON-NEOPLASTIC, HYPERPLASIA ***************								
HYPERPLASIA, LYMPHOID				1/1575			1/437	2/1743
*************** GROWTH ALTERATIONS, NON-NEOPLASTIC, METAPLASIA ***************								
METAPLASIA, SQUAMOUS				1/1575				
*************** GROWTH ALTERATIONS, NEOPLASTIC, BENIGN ***************								
FIBROMA								1/1743

*VARIABILITY AMONG DATA SOURCES SIGNIFICANT AT THE 1% SIGNIFICANCE LEVEL

SEX:	FEMALE				MALE			
DIAGNOSIS / AGE IN MONTHS:	0 < 12	12 < 18	18 < 24	24+	0 < 12	12 < 18	18 < 24	24+
*************** GROWTH ALTERATIONS, NEOPLASTIC, MALIGNANT, PRIMARY ***************								
ADENOCARCINOMA			1/327	1/1575				5/1743
CARCINOMA IN SITU			1/327					
CARCINOMA, SQUAMOUS CELL							1/437	

SEX:	FEMALE				MALE			
DIAGNOSIS / AGE IN MONTHS:	0 < 12	12 < 18	18 < 24	24+	0 < 12	12 < 18	18 < 24	24+
*************** INFLAMMATION AND REPAIR ***************								
INFLAMMATION			2/44	6/336	1/11	1/64	1/104	6/243
*************** GROWTH ALTERATIONS, NEOPLASTIC, MALIGNANT, PRIMARY SITE NOT DETERMINED ***************								
SARCOMA								1/243

SEX:	FEMALE				MALE			
DIAGNOSIS / AGE IN MONTHS:	0 < 12	12 < 18	18 < 24	24+	0 < 12	12 < 18	18 < 24	24+
*************** MECHANICAL ABNORMALITIES AND CYST FORMATION ***************								
DILATATION				1/240		1/97	1/392	3/216
*************** INFLAMMATION AND REPAIR ***************								
INFLAMMATION			23/420*				31/392	
INFLAMMATION, NONSUPPURATIVE	16/181*		1/420		32/170*		1/392	1/216
INFLAMMATION, SUPPURATIVE			2/420	1/240			4/392	
*************** DEGENERATION, NECROSIS, AND DEPOSITION ***************								
MINERALIZATION							1/392	
*************** GROWTH ALTERATIONS, NEOPLASTIC, MALIGNANT, PRIMARY ***************								
CARCINOMA			1/420					
*************** GROWTH ALTERATIONS, NEOPLASTIC, MALIGNANT, METASTATIC ***************								
RHABDOMYOSARCOMA								1/216

SEX:	FEMALE				MALE			
DIAGNOSIS / AGE IN MONTHS:	0 < 12	12 < 18	18 < 24	24+	0 < 12	12 < 18	18 < 24	24+
*************** MECHANICAL ABNORMALITIES AND CYST FORMATION ***************								
CYST		1/369					1/147	
*************** INFLAMMATION AND REPAIR ***************								
FIBROSIS		1/369						
INFLAMMATION				1/71	2/171	1/314		
INFLAMMATION, NONSUPPURATIVE			1/193					5/94
*************** GROWTH ALTERATIONS, NON-NEOPLASTIC, ATROPHY ***************								
ATROPHY								1/94
*************** GROWTH ALTERATIONS, NEOPLASTIC, BENIGN ***************								
ADENOMA	5/212	18/369	23/193	16/71	5/171	19/314	12/147	7/94
CYSTADENOMA, PAPILLARY								1/94
*************** GROWTH ALTERATIONS, NEOPLASTIC, MALIGNANT, PRIMARY SITE NOT DETERMINED ***************								
LYMPHOMA	1/212	4/369	5/193					
LYMPHOMA, RETICULUM CELL		1/369	3/193	4/71		1/314		1/94

SEX:	FEMALE				MALE			
AGE IN MONTHS: / DIAGNOSIS	0 < 12	12 < 18	18 < 24	24+	0 < 12	12 < 18	18 < 24	24+
*************** GROWTH ALTERATIONS, NON-NEOPLASTIC, HYPERPLASIA ***************								
HYPERPLASIA			1/1356	2/1015				
*************** GROWTH ALTERATIONS, NEOPLASTIC, BENIGN ***************								
ADENOMA		1/103	1/1356	1/1015			2/1136	4/958
ADENOMA, PAPILLARY			1/1356					1/958
CYSTADENOMA							1/1136	1/958
CYSTADENOMA, PAPILLARY				1/1015			1/1136	1/958

SEX:	FEMALE				MALE			
DIAGNOSIS / AGE IN MONTHS:	0 < 12	12 < 18	18 < 24	24+	0 < 12	12 < 18	18 < 24	24+
*************** INFLAMMATION AND REPAIR ***************								
INFLAMMATION		1/367		7/75				1/102
INFLAMMATION, NONSUPPURATIVE		2/367		5/75		1/300	1/161	7/102
*************** GROWTH ALTERATIONS, NON-NEOPLASTIC, HYPERPLASIA ***************								
HYPERPLASIA		1/367						
*************** GROWTH ALTERATIONS, NON-NEOPLASTIC, ATROPHY ***************								
ATROPHY								1/102
*************** GROWTH ALTERATIONS, NEOPLASTIC, BENIGN ***************								
ADENOMA		4/367	4/192	6/75		6/300	6/161	2/102
ADENOMA, PAPILLARY								1/102
*************** GROWTH ALTERATIONS, NEOPLASTIC, MALIGNANT, PRIMARY SITE NOT DETERMINED ***************								
LYMPHOMA				2/75				1/102
LYMPHOMA, RETICULUM CELL		1/367	1/192	1/75				1/102

SEX:	FEMALE				MALE			
AGE IN MONTHS:	0 < 12	12 < 18	18 < 24	24+	0 < 12	12 < 18	18 < 24	24+
DIAGNOSIS								
*************** INFLAMMATION AND REPAIR ***************								
INFLAMMATION					1/101	3/101	3/437	3/1743
INFLAMMATION, NONSUPPURATIVE					3/101	1/101		8/1743
*************** GROWTH ALTERATIONS, NON-NEOPLASTIC, HYPERPLASIA ***************								
HYPERPLASIA				2/1575				
*************** GROWTH ALTERATIONS, NON-NEOPLASTIC, ATROPHY ***************								
ATROPHY							4/437	6/1743
*************** GROWTH ALTERATIONS, NEOPLASTIC, MALIGNANT, PRIMARY ***************								
CARCINOMA				1/1575				
CARCINOMA, SQUAMOUS CELL			1/327					

SEX:	FEMALE				MALE			
AGE IN MONTHS:	0 < 12	12 < 18	18 < 24	24+	0 < 12	12 < 18	18 < 24	24+
DIAGNOSIS								
*************** INFLAMMATION AND REPAIR ***************								
INFLAMMATION				1/336		1/64	1/104	

ORGAN SYSTEM: SKIN AND APPENDAGES ORGAN: MAMMARY GLAND STRAIN: BALB/C

SEX:	FEMALE				MALE			
DIAGNOSIS / AGE IN MONTHS:	0 < 12	12 < 18	18 < 24	24+	0 < 12	12 < 18	18 < 24	24+
*************** INFLAMMATION AND REPAIR ***************								
INFLAMMATION		1/369						
*************** DEGENERATION, NECROSIS, AND DEPOSITION ***************								
FIBROCYSTIC DISEASE			1/193					
MINERALIZATION		3/369	1/193	1/71				
*************** GROWTH ALTERATIONS, NON-NEOPLASTIC, HYPERPLASIA ***************								
HYPERPLASIA	6/212	2/369	1/193	2/71				
*************** GROWTH ALTERATIONS, NEOPLASTIC, BENIGN ***************								
ANGIOMA		1/369						
*************** GROWTH ALTERATIONS, NEOPLASTIC, MALIGNANT, PRIMARY ***************								
ADENOCARCINOMA	1/212	1/369	3/193	2/71				
CARCINOSARCOMA		3/369						
*************** GROWTH ALTERATIONS, NEOPLASTIC, MALIGNANT, PRIMARY SITE NOT DETERMINED ***************								
LYMPHOMA	1/212	3/369	3/193					
LYMPHOMA, RETICULUM CELL		1/369	8/193	6/71				

SEX:	FEMALE				MALE			
DIAGNOSIS / AGE IN MONTHS:	0 < 12	12 < 18	18 < 24	24+	0 < 12	12 < 18	18 < 24	24+
*************** MECHANICAL ABNORMALITIES AND CYST FORMATION ***************								
CYST							1/1136	
GALACTOCELE			2/1356					
*************** GROWTH ALTERATIONS, NON-NEOPLASTIC, HYPERPLASIA ***************								
HYPERPLASIA			6/1356	1/1015				
HYPERPLASIA, CYSTIC			1/1356					
*************** GROWTH ALTERATIONS, NON-NEOPLASTIC, METAPLASIA ***************								
METAPLASIA, SQUAMOUS			1/1356					
*************** GROWTH ALTERATIONS, NEOPLASTIC, BENIGN ***************								
ADENOMA				5/1015				
FIBROADENOMA		1/103	2/1356	2/1015				
MYOEPITHELIOMA				1/1015				
*************** GROWTH ALTERATIONS, NEOPLASTIC, MALIGNANT, PRIMARY ***************								
ADENOCARCINOMA		1/103	11/1356	5/1015				
CARCINOMA			1/1356	1/1015				
CARCINOMA, ACINAR CELL				1/1015				

DIAGNOSIS	SEX:	FEMALE				MALE			
	AGE IN MONTHS:	0 < 12	12 < 18	18 < 24	24+	0 < 12	12 < 18	18 < 24	24+
*************** GROWTH ALTERATIONS, NEOPLASTIC, MALIGNANT, PRIMARY SITE NOT DETERMINED ***************									
LEIOMYOSARCOMA				1/1356					

SEX:	FEMALE				MALE			
DIAGNOSIS / AGE IN MONTHS:	0 < 12	12 < 18	18 < 24	24+	0 < 12	12 < 18	18 < 24	24+
*************** MECHANICAL ABNORMALITIES AND CYST FORMATION ***************								
CYST				1/75				
*************** INFLAMMATION AND REPAIR ***************								
INFLAMMATION			1/192	2/75				
*************** GROWTH ALTERATIONS, NON-NEOPLASTIC, HYPERPLASIA ***************								
HYPERPLASIA		2/367	2/192	4/75				1/102
HYPERPLASIA, NODULAR		1/367						
*************** GROWTH ALTERATIONS, NEOPLASTIC, MALIGNANT, PRIMARY ***************								
ADENOCARCINOMA							1/161	
*************** GROWTH ALTERATIONS, NEOPLASTIC, MALIGNANT, PRIMARY SITE NOT DETERMINED ***************								
LYMPHOMA				1/75				

SEX:	FEMALE				MALE			
DIAGNOSIS / AGE IN MONTHS:	0 < 12	12 < 18	18 < 24	24+	0 < 12	12 < 18	18 < 24	24+
*************** MECHANICAL ABNORMALITIES AND CYST FORMATION ***************								
CYST			2/327	13/1575			1/437	3/1743
CYST, HEMORRHAGIC								1/1743
ECTASIA, DUCT			9/327	46/1575			1/437	7/1743
GALACTOCELE			6/327	91/1575 *			6/437	6/1743
HEMORRHAGE								1/1743
*************** INFLAMMATION AND REPAIR ***************								
FIBROSIS				2/1575				
INFLAMMATION			1/327	4/1575				1/1743
INFLAMMATION, GRANULOMATOUS								1/1743
*************** GROWTH ALTERATIONS, NON-NEOPLASTIC, HYPERTROPHY ***************								
HYPERTROPHY								1/1743
*************** GROWTH ALTERATIONS, NON-NEOPLASTIC, HYPERPLASIA ***************								
ADENOSIS				1/1575				
GYNECOMASTIA								3/1743
HYPERPLASIA		1/24	13/327	57/1575			12/437	38/1743
HYPERPLASIA, CYSTIC	1/26		4/327	36/1575			2/437	1/1743
HYPERPLASIA, EPITHELIAL				4/1575				

*VARIABILITY AMONG DATA SOURCES SIGNIFICANT AT THE 1% SIGNIFICANCE LEVEL

SEX:	FEMALE				MALE			
AGE IN MONTHS: / DIAGNOSIS	0 < 12	12 < 18	18 < 24	24+	0 < 12	12 < 18	18 < 24	24+
HYPERPLASIA, NODULAR			1/327					
*************** GROWTH ALTERATIONS, NON-NEOPLASTIC, DYSPLASIA ***************								
DYSPLASIA			6/327	2/1575				
*************** GROWTH ALTERATIONS, NEOPLASTIC, BENIGN ***************								
ADENOMA			3/327	30/1575				6/1743
ADENOMA, PAPILLARY				1/1575				
FIBROADENOMA	2/26		34/327	315/1575			1/437	26/1743
FIBROMA			3/327	4/1575				1/1743
PAPILLOMA, INTRADUCTAL								1/1743
*************** GROWTH ALTERATIONS, NEOPLASTIC, MALIGNANT, PRIMARY ***************								
ADENOCARCINOMA			8/327	22/1575				4/1743
CARCINOMA	1/26	1/24						1/1743
CARCINOMA, UNDIFFERENTIATED				1/1575				
*************** GROWTH ALTERATIONS, NEOPLASTIC, MALIGNANT, PRIMARY SITE NOT DETERMINED ***************								
HISTIOCYTOMA, FIBROUS, MALIGNANT			1/327					
LEUKEMIA								1/1743
LEUKEMIA, LYMPHOCYTIC				1/1575				

SEX:	FEMALE				MALE			
DIAGNOSIS / AGE IN MONTHS:	0 < 12	12 < 18	18 < 24	24+	0 < 12	12 < 18	18 < 24	24+
*************** GROWTH ALTERATIONS, NEOPLASTIC, NOT DETERMINED ***************								
BASAL CELL TUMOR			1/327					

SEX:	FEMALE				MALE			
DIAGNOSIS / AGE IN MONTHS:	0 < 12	12 < 18	18 < 24	24+	0 < 12	12 < 18	18 < 24	24+
*************** MECHANICAL ABNORMALITIES AND CYST FORMATION ***************								
CYST								1/243
GALACTOCELE				2/336				2/243
*************** DEGENERATION, NECROSIS, AND DEPOSITION ***************								
NECROSIS, FAT				2/336				
*************** GROWTH ALTERATIONS, NON-NEOPLASTIC, HYPERPLASIA ***************								
HYPERPLASIA				2/336				
*************** GROWTH ALTERATIONS, NEOPLASTIC, BENIGN ***************								
ADENOMA				2/336				1/243
FIBROADENOMA		3/39	10/44	87/336			1/104	4/243
FIBROMA				1/336				1/243
*************** GROWTH ALTERATIONS, NEOPLASTIC, MALIGNANT, PRIMARY ***************								
ADENOCARCINOMA		1/39	1/44	12/336		1/64	2/104	5/243
*************** GROWTH ALTERATIONS, NEOPLASTIC, MALIGNANT, METASTATIC ***************								
CARCINOMA, MEDULLARY								1/243

SEX:	FEMALE				MALE			
AGE IN MONTHS: / DIAGNOSIS	0 < 12	12 < 18	18 < 24	24+	0 < 12	12 < 18	18 < 24	24+
*************** MECHANICAL ABNORMALITIES AND CYST FORMATION ***************								
CYST	1/181	4/60	35/420*				2/392	
CYST, HEMORRHAGIC			1/420					
GALACTOCELE		1/60	7/420	11/240				3/216
*************** INFLAMMATION AND REPAIR ***************								
ABSCESS				1/240		1/97		
FIBROSIS		1/60						
GRANULOMA				1/240				
INFLAMMATION			1/420	2/240				
INFLAMMATION, GRANULOMATOUS			1/420					
INFLAMMATION, NONSUPPURATIVE			1/420					
*************** DEGENERATION, NECROSIS, AND DEPOSITION ***************								
FIBROCYSTIC DISEASE			1/420	13/240*				
HEMOSIDERIN DEPOSITION							1/392	
MINERALIZATION							1/392	
NECROSIS				1/240				
*************** GROWTH ALTERATIONS, NON-NEOPLASTIC, HYPERTROPHY ***************								
HYPERTROPHY		3/60						

*VARIABILITY AMONG DATA SOURCES SIGNIFICANT AT THE 1% SIGNIFICANCE LEVEL

SEX:	FEMALE				MALE			
DIAGNOSIS / AGE IN MONTHS:	0 < 12	12 < 18	18 < 24	24+	0 < 12	12 < 18	18 < 24	24+
*************** GROWTH ALTERATIONS, NON-NEOPLASTIC, HYFERPLASIA ***************								
HYPERPLASIA		2/60	8/420	27/240*				3/216
HYPERPLASIA, CYSTIC		4/60	39/420*	28/240*		1/97	4/392	15/216*
HYPERPLASIA, EPITHELIAL				1/240				
HYPERPLASIA, HISTIOCYTIC		1/60						
*************** GROWTH ALTERATIONS, NON-NEOPLASTIC, ATROPHY ***************								
ATROPHY		1/60		1/240				
*************** GROWTH ALTERATIONS, NON-NEOPLASTIC, DYSPLASIA ***************								
DYSPLASIA, STROMAL			1/420					
*************** GROWTH ALTERATIONS, NEOPLASTIC, BENIGN ***************								
ADENOMA	1/181	2/60	16/420	37/240*			1/392	4/216
ADENOMA, PAPILLARY			1/420					
FIBROADENOMA	1/181	12/60*	92/420*	79/240*			9/392	1/216
FIBROMA			1/420					
FIBROMA, STROMAL				1/240				
PAPILLOMA, INTRADUCTAL			2/420	4/240				1/216

*VARIABILITY AMONG DATA SOURCES SIGNIFICANT AT THE 1% SIGNIFICANCE LEVEL

SEX:	FEMALE				MALE			
DIAGNOSIS — AGE IN MONTHS:	0 < 12	12 < 18	18 < 24	24+	0 < 12	12 < 18	18 < 24	24+
*************** GROWTH ALTERATIONS, NEOPLASTIC, MALIGNANT, PRIMARY ***************								
ADENOCARCINOMA	1/181	3/60	12/420	4/240		1/97	1/392	1/216
*************** GROWTH ALTERATIONS, NEOPLASTIC, MALIGNANT, METASTATIC ***************								
CARCINOMA, C CELL				1/240				
*************** GROWTH ALTERATIONS, NEOPLASTIC, MALIGNANT, PRIMARY SITE NOT DETERMINED ***************								
LYMPHOMA			1/420					

SEX:	FEMALE				MALE			
DIAGNOSIS / AGE IN MONTHS:	0 < 12	12 < 18	18 < 24	24+	0 < 12	12 < 18	18 < 24	24+
*************** MECHANICAL ABNORMALITIES AND CYST FORMATION ***************								
EDEMA					1/171			
*************** INFLAMMATION AND REPAIR ***************								
FIBROSIS						2/314		1/94
INFLAMMATION	1/212		2/193		2/171	2/314		2/94
INFLAMMATION, ULCERATIVE			1/193			2/314		
*************** DEGENERATION, NECROSIS, AND DEPOSITION ***************								
MINERALIZATION			2/193					1/94
*************** GROWTH ALTERATIONS, NON-NEOPLASTIC, HYPERPLASIA ***************								
ACANTHOSIS						2/314		
HYPERKERATOSIS						3/314		
*************** GROWTH ALTERATIONS, NON-NEOPLASTIC, METAPLASIA ***************								
METAPLASIA, OSSEOUS								1/94
*************** GROWTH ALTERATIONS, NEOPLASTIC, BENIGN ***************								
MYOEPITHELIOMA			1/193					

SEX:	FEMALE				MALE			
DIAGNOSIS / AGE IN MONTHS:	0 < 12	12 < 18	18 < 24	24+	0 < 12	12 < 18	18 < 24	24+
*************** GROWTH ALTERATIONS, NEOPLASTIC, MALIGNANT, PRIMARY ***************								
ANGIOSARCOMA		1/369	1/193					
CARCINOMA, SQUAMOUS CELL								1/94
FIBROSARCOMA							1/147	
*************** GROWTH ALTERATIONS, NEOPLASTIC, MALIGNANT, PRIMARY SITE NOT DETERMINED ***************								
LYMPHOMA		2/369						

SEX:	FEMALE				MALE			
DIAGNOSIS / AGE IN MONTHS:	0 < 12	12 < 18	18 < 24	24+	0 < 12	12 < 18	18 < 24	24+
*************** MECHANICAL ABNORMALITIES AND CYST FORMATION ***************								
CYST					1/124		1/1136	3/958
CYST, EPIDERMAL INCLUSION						1/199	6/1136	4/958
CYST, SEBACEOUS							1/1136	
EDEMA							1/1136	
HEMORRHAGE				2/1015				
THROMBUS				1/1015				
*************** INFLAMMATION AND REPAIR ***************								
ABSCESS			2/1356		1/124	5/199	13/1136	2/958
FIBROSIS			3/1356				4/1136	2/958
GRANULOMA							1/1136	3/958
INFLAMMATION			4/1356	2/1015	3/124	5/199	20/1136	8/958
INFLAMMATION, GRANULOMATOUS				1/1015			3/1136	1/958
INFLAMMATION, NECROTIZING	1/67						1/1136	1/958
INFLAMMATION, NONSUPPURATIVE				1/1015				
INFLAMMATION, SUPPURATIVE		1/103					3/1136	
INFLAMMATION, ULCERATIVE			3/1356	1/1015			5/1136	4/958
POLYP, INFLAMMATORY							1/1136	

SEX:	FEMALE				MALE			
DIAGNOSIS / AGE IN MONTHS:	0 < 12	12 < 18	18 < 24	24+	0 < 12	12 < 18	18 < 24	24+
*************** DEGENERATION, NECROSIS, AND DEPOSITION ***************								
MINERALIZATION			1/1356					
NECROSIS							1/1136	
NECROSIS, FAT							1/1136	3/958
*************** GROWTH ALTERATIONS, NON-NEOPLASTIC, HYPERPLASIA ***************								
ACANTHOSIS			8/1356				2/1136	4/958
HYPERKERATOSIS				1/1015			1/1136	4/958
HYPERPLASIA, CYSTIC								1/958
*************** GROWTH ALTERATIONS, NEOPLASTIC, BENIGN ***************								
ADENOMA, SEBACEOUS	1/67						1/1136	1/958
ANGIOMA			2/1356	6/1015				2/958
FIBROMA				1/1015		1/199	6/1136	15/958
KERATOACANTHOMA			1/1356					1/958
LIPOMA				2/1015				2/958
PAPILLOMA							1/1136	
PAPILLOMA, SQUAMOUS CELL								1/958
*************** GROWTH ALTERATIONS, NEOPLASTIC, MALIGNANT, PRIMARY ***************								
ADENOCARCINOMA			1/1356					

SEX:	FEMALE				MALE			
DIAGNOSIS / AGE IN MONTHS:	0 < 12	12 < 18	18 < 24	24+	0 < 12	12 < 18	18 < 24	24+
ANGIOSARCOMA			2/1356	2/1015		1/199	4/1136	
CARCINOMA, BASAL CELL							1/1136	1/958
FIBROSARCOMA			9/1356	1/1015		2/199	27/1136	22/958
HISTIOCYTOMA, FIBROUS, MALIGNANT								1/958
MELANOMA, MALIGNANT				1/1015				
NEURILEMOMA, MALIGNANT			2/1356					1/958
NEUROFIBROSARCOMA				2/1015		1/199		
SARCOMA			2/1356				3/1136	
*************** GROWTH ALTERATIONS, NEOPLASTIC, MALIGNANT, METASTATIC ***************								
OSTEOSARCOMA			2/1356				2/1136	
RHABDOMYOSARCOMA				1/1015				
*************** GROWTH ALTERATIONS, NEOPLASTIC, MALIGNANT, PRIMARY SITE NOT DETERMINED ***************								
LEIOMYOSARCOMA		1/103	2/1356				2/1136	
*************** GROWTH ALTERATIONS, NEOPLASTIC, NOT DETERMINED ***************								
HISTIOCYTOMA, FIBROUS								2/958
MAST CELL TUMOR								1/958

SEX:	FEMALE				MALE			
DIAGNOSIS / AGE IN MONTHS:	0 < 12	12 < 18	18 < 24	24+	0 < 12	12 < 18	18 < 24	24+
*************** MECHANICAL ABNORMALITIES AND CYST FORMATION ***************								
CYST						1/300		
*************** INFLAMMATION AND REPAIR ***************								
FIBROSIS	2/176	5/367				11/300	1/161	
INFLAMMATION	4/176	2/367	1/192	2/75	1/131		1/161	
INFLAMMATION, ULCERATIVE	4/176	4/367	1/192	1/75			1/161	1/102
*************** DEGENERATION, NECROSIS, AND DEPOSITION ***************								
NECROSIS, FAT				1/75				
*************** GROWTH ALTERATIONS, NON-NEOPLASTIC, HYPERPLASIA ***************								
ACANTHOSIS	3/176		1/192					
HYPERKERATOSIS		1/367						
*************** GROWTH ALTERATIONS, NON-NEOPLASTIC, ATROPHY ***************								
ATROPHY			1/192					
*************** GROWTH ALTERATIONS, NEOPLASTIC, BENIGN ***************								
ANGIOMA				1/75				

SEX:	FEMALE				MALE			
DIAGNOSIS / AGE IN MONTHS:	0 < 12	12 < 18	18 < 24	24+	0 < 12	12 < 18	18 < 24	24+
*************** GROWTH ALTERATIONS, NEOPLASTIC, MALIGNANT, PRIMARY ***************								
ANGIOSARCOMA				1/75				
CARCINOMA, SQUAMOUS CELL				1/75				
FIBROSARCOMA			1/192					

SEX:	FEMALE				MALE			
DIAGNOSIS / AGE IN MONTHS:	0 < 12	12 < 18	18 < 24	24+	0 < 12	12 < 18	18 < 24	24+
*************** MECHANICAL ABNORMALITIES AND CYST FORMATION ***************								
CYST								5/1743
CYST, EPIDERMAL INCLUSION		1/24	1/327	4/1575			2/437	14/1743
CYST, HEMORRHAGIC							1/437	
CYST, SEBACEOUS				1/1575				
HEMATOMA						1/101	1/437	
*************** INFLAMMATION AND REPAIR ***************								
ABSCESS				4/1575		1/101	2/437	6/1743
CELLULITIS						1/101		
FIBROSIS								1/1743
GRANULOMA				1/1575				
INFLAMMATION		1/24		4/1575			2/437	4/1743
INFLAMMATION, GRANULOMATOUS							1/437	
INFLAMMATION, NECROTIZING							1/437	
INFLAMMATION, SUPPURATIVE			1/327	1/1575				2/1743
INFLAMMATION, ULCERATIVE				1/1575				
KELOID							1/437	
*************** DEGENERATION, NECROSIS, AND DEPOSITION ***************								
MINERALIZATION				1/1575				

SEX:	FEMALE				MALE			
DIAGNOSIS / AGE IN MONTHS:	0 < 12	12 < 18	18 < 24	24+	0 < 12	12 < 18	18 < 24	24+
NECROSIS		1/24	1/327					1/1743
NECROSIS, FAT			2/327	2/1575				16/1743
*************** GROWTH ALTERATIONS, NON-NEOPLASTIC, HYPERPLASIA ***************								
HYPERKERATOSIS							1/437	2/1743
HYPERPLASIA, BASAL CELL								1/1743
*************** GROWTH ALTERATIONS, NON-NEOPLASTIC, METAPLASIA ***************								
METAPLASIA, OSSEOUS								3/1743
METAPLASIA, SQUAMOUS								1/1743
*************** GROWTH ALTERATIONS, NON-NEOPLASTIC, DYSPLASIA ***************								
DYSPLASIA, FIBROUS								1/1743
*************** GROWTH ALTERATIONS, NEOPLASTIC, BENIGN ***************								
ADENOMA, APOCRINE GLAND		1/24					1/437	
ADENOMA, SEBACEOUS				2/1575			2/437	1/1743
FIBROADENOMA			4/327	2/1575				1/1743
FIBROMA			4/327	10/1575		1/101	11/437	74/1743
KERATOACANTHOMA				1/1575			3/437	5/1743
LEIOMYOMA				1/1575				3/1743
LIPOMA				5/1575				4/1743

SEX:	FEMALE				MALE			
DIAGNOSIS / AGE IN MONTHS:	0 < 12	12 < 18	18 < 24	24+	0 < 12	12 < 18	18 < 24	24+
NEURILEMMOMA								1/1743
NEUROFIBROMA							1/437	1/1743
PAPILLOMA								5/1743
PAPILLOMA, BASOSQUAMOUS								1/1743
PAPILLOMA, SQUAMOUS CELL				4/1575				11/1743
TRICHOEPITHELIOMA			1/327	2/1575			1/437	4/1743
*************** GROWTH ALTERATIONS, NEOPLASTIC, MALIGNANT, PRIMARY ***************								
ADENOCARCINOMA				1/1575				
ADENOCARCINOMA, SEBACEOUS				2/1575				
ANGIOSARCOMA							2/437	1/1743
CARCINOMA		1/24		1/1575				
CARCINOMA, BASAL CELL			2/327	4/1575			1/437	3/1743
CARCINOMA, SEBACEOUS							1/437	
CARCINOMA, SQUAMOUS CELL			2/327	6/1575		1/101	8/437	14/1743
CARCINOMA, UNDIFFERENTIATED								1/1743
CARCINOSARCOMA						1/101	1/437	
FIBROSARCOMA			3/327				4/437	9/1743
HEMANGIOPERICYTOMA, MALIGNANT							1/437	
HISTIOCYTOMA, FIBROUS, MALIGNANT							1/437	
NEUROFIBROSARCOMA								1/1743
NEUROMA, MALIGNANT				1/1575				

SEX:	FEMALE				MALE			
AGE IN MONTHS: / DIAGNOSIS	0 < 12	12 < 18	18 < 24	24+	0 < 12	12 < 18	18 < 24	24+
SARCOMA			1/327	3/1575			3/437	5/1743
*************** GROWTH ALTERATIONS, NEOPLASTIC, MALIGNANT, METASTATIC ***************								
OSTEOSARCOMA			1/327				1/437	1/1743
*************** GROWTH ALTERATIONS, NEOPLASTIC, MALIGNANT, PRIMARY SITE NOT DETERMINED ***************								
LEIOMYOSARCOMA							1/437	
LEUKEMIA								1/1743
LEUKEMIA, LYMPHOBLASTIC								1/1743
LEUKEMIA, STEM CELL							1/437	
LYMPHOMA, HISTIOCYTIC								2/1743
*************** GROWTH ALTERATIONS, NEOPLASTIC, NOT DETERMINED ***************								
NEOPLASM						1/101		

SEX:	FEMALE				MALE			
DIAGNOSIS / AGE IN MONTHS:	0 < 12	12 < 18	18 < 24	24+	0 < 12	12 < 18	18 < 24	24+
*************** MECHANICAL ABNORMALITIES AND CYST FORMATION ***************								
CYST, EPIDERMAL INCLUSION				1/336				7/243
*************** INFLAMMATION AND REPAIR ***************								
ABSCESS		1/39		1/336	1/11			6/243
FIBROSIS								1/243
INFLAMMATION			1/44	3/336				10/243
INFLAMMATION, ULCERATIVE			1/44					1/243
*************** GROWTH ALTERATIONS, NON-NEOPLASTIC, HYPERPLASIA ***************								
ACANTHOSIS								1/243
HYPERKERATOSIS								1/243
*************** GROWTH ALTERATIONS, NEOPLASTIC, BENIGN ***************								
FIBROMA		1/39	2/44	6/336		1/64	1/104	9/243
LIPOMA				2/336				
*************** GROWTH ALTERATIONS, NEOPLASTIC, MALIGNANT, PRIMARY ***************								
ANGIOSARCOMA			2/44	2/336		1/64	2/104	2/243
CARCINOMA, BASAL CELL				1/336				
CARCINOMA, SQUAMOUS CELL				1/336				

SEX:	FEMALE				MALE			
DIAGNOSIS / AGE IN MONTHS:	0 < 12	12 < 18	18 < 24	24+	0 < 12	12 < 18	18 < 24	24+
FIBROSARCOMA				7/336		1/64	3/104	5/243
HEMANGIOPERICYTOMA, MALIGNANT								2/243
HISTIOCYTOMA, FIBROUS, MALIGNANT						1/64		
LIPOSARCOMA							1/104	1/243
NEUROFIBROSARCOMA								1/243
SARCOMA								1/243
*************** GROWTH ALTERATIONS, NEOPLASTIC, MALIGNANT, PRIMARY SITE NOT DETERMINED ***************								
LYMPHOMA, HISTIOCYTIC	1/7							

SEX:	FEMALE				MALE			
DIAGNOSIS / AGE IN MONTHS:	0 < 12	12 < 18	18 < 24	24+	0 < 12	12 < 18	18 < 24	24+
*************** DEVELOPMENTAL ANOMALIES ***************								
HAMARTOMA							1/392	
*************** MECHANICAL ABNORMALITIES AND CYST FORMATION ***************								
CYST	1/181			1/240			1/392	1/216
CYST, EPIDERMAL INCLUSION			2/420	1/240		2/97	6/392	5/216
CYST, SEBACEOUS			2/420				1/392	1/216
HEMORRHAGE						1/97	1/392	
THROMBUS								1/216
*************** INFLAMMATION AND REPAIR ***************								
ABSCESS				3/240		1/97	2/392	3/216
FIBROSIS							1/392	
GRANULOMA			1/420					1/216
INFLAMMATION		1/60	2/420	1/240			9/392	2/216
INFLAMMATION, GRANULOMATOUS			1/420				1/392	
INFLAMMATION, NONSUPPURATIVE					2/170			
INFLAMMATION, SUPPURATIVE			2/420			1/97	1/392	7/216
INFLAMMATION, ULCERATIVE			1/420				1/392	1/216
KELOID								1/216

SEX:	FEMALE				MALE			
DIAGNOSIS / AGE IN MONTHS:	0 < 12	12 < 18	18 < 24	24+	0 < 12	12 < 18	18 < 24	24+
*************** DEGENERATION, NECROSIS, AND DEPOSITION ***************								
HEMOSIDERIN DEPOSITION							1/392	
MINERALIZATION							1/392	
NECROSIS			1/420					1/216
NECROSIS, FAT				1/240				
*************** GROWTH ALTERATIONS, NON-NEOPLASTIC, HYPERPLASIA ***************								
ACANTHOSIS					2/170			
HYPERKERATOSIS					2/170		1/392	
HYPERPLASIA, SEBACEOUS								1/216
PARAKERATOSIS							1/392	
*************** GROWTH ALTERATIONS, NEOPLASTIC, BENIGN ***************								
ADENOMA			1/420	3/240			1/392	
ADENOMA, APOCRINE GLAND								2/216
ADENOMA, PAPILLARY SEBACEOUS								1/216
ADENOMA, SEBACEOUS			1/420					2/216
CHONDROMA							1/392	
EPITHELIOMA, BASAL CELL								1/216
FIBROADENOMA			2/420	4/240				
FIBROMA		1/60	5/420	10/240		2/97	3/392	11/216*

*VARIABILITY AMONG DATA SOURCES SIGNIFICANT AT THE 1% SIGNIFICANCE LEVEL

SEX:	FEMALE				MALE			
DIAGNOSIS / AGE IN MONTHS:	0 < 12	12 < 18	18 < 24	24+	0 < 12	12 < 18	18 < 24	24+
HEMANGIOPERICYTOMA			1/420					
KERATOACANTHOMA							1/392	
LEIOMYOMA		1/60	1/420					1/216
LIPOMA				1/240			3/392	3/216
MYXOFIBROMA								1/216
MYXOMA							1/392	
PAPILLOMA							1/392	2/216
PAPILLOMA, SEBACEOUS								1/216
PAPILLOMA, SQUAMOUS CELL							1/392	
*************** GROWTH ALTERATIONS, NEOPLASTIC, MALIGNANT, PRIMARY ***************								
ANGIOSARCOMA			1/420					1/216
CARCINOMA, BASAL CELL								1/216
CARCINOMA, BASOSQUAMOUS					1/170			
CARCINOMA, SQUAMOUS CELL			1/420				3/392	1/216
FIBROSARCOMA						1/97	1/392	
SARCOMA				1/240			1/392	
SARCOMA, HISTIOCYTIC			1/420					
*************** GROWTH ALTERATIONS, NEOPLASTIC, MALIGNANT, METASTATIC ***************								
ADENOCARCINOMA, MAMMARY			1/420					
OSTEOSARCOMA	1/181							

SEX:	FEMALE				MALE			
DIAGNOSIS / AGE IN MONTHS:	0 < 12	12 < 18	18 < 24	24+	0 < 12	12 < 18	18 < 24	24+
RHABDOMYOSARCOMA				3/240				2/216
*************** GROWTH ALTERATIONS, NEOPLASTIC, MALIGNANT, PRIMARY SITE NOT DETERMINED ***************								
LEUKEMIA, HISTIOCYTIC							1/392	
LYMPHOMA			1/420					
LYMPHOMA, RETICULUM CELL			1/420	1/240			2/392	
SARCOMA, HISTIOCYTIC							1/392	4/216

SEX:	FEMALE				MALE			
DIAGNOSIS / AGE IN MONTHS:	0 < 12	12 < 18	18 < 24	24+	0 < 12	12 < 18	18 < 24	24+
MECHANICAL ABNORMALITIES AND CYST FORMATION								
CYST	4/212	8/369	5/193	9/71	2/171	25/314	26/147	27/94*
DILATATION, TUBULAR			2/193			1/314		
HYDRONEPHROSIS								1/94
THROMBUS						1/314		
INFLAMMATION AND REPAIR								
ABSCESS						1/314		
GLOMERULOSCLEROSIS			1/193	1/71				1/94
INFLAMMATION	1/212	1/369	1/193			2/314		
INFLAMMATION, NONSUPPURATIVE			4/193	13/71		2/314	4/147	20/94*
PERIVASCULITIS/VASCULITIS						2/314		6/94
PYELONEPHRITIS	1/212	3/369	3/193	2/71		3/314		3/94
DEGENERATION, NECROSIS, AND DEPOSITION								
AMYLOID DEPOSITION			1/193			1/314		1/94
CAST, PROTEINACEOUS		3/369	1/193	7/71		4/314	4/147	17/94*
DEGENERATION		1/369	1/193			1/314		3/94
FATTY CHANGE			1/193				1/147	
HYALINOSIS					1/171			9/94
HYDRONEPHROSIS		1/369		2/71		2/314	2/147	1/94

*VARIABILITY AMONG DATA SOURCES SIGNIFICANT AT THE 1% SIGNIFICANCE LEVEL

SEX:	FEMALE				MALE			
DIAGNOSIS / AGE IN MONTHS:	0 < 12	12 < 18	18 < 24	24+	0 < 12	12 < 18	18 < 24	24+
INFARCT			2/193			3/314		
MINERALIZATION			1/193			2/314		1/94
NEPHROPATHY								1/94
*************** GROWTH ALTERATIONS, NON-NEOPLASTIC, HYPERPLASIA ***************								
REGENERATION						1/314		1/94
*************** GROWTH ALTERATIONS, NON-NEOPLASTIC, METAPLASIA ***************								
METAPLASIA, OSSEOUS			1/193	1/71		1/314		1/94
*************** GROWTH ALTERATIONS, NEOPLASTIC, BENIGN ***************								
ADENOMA		1/369					2/147	
*************** GROWTH ALTERATIONS, NEOPLASTIC, MALIGNANT, PRIMARY ***************								
ADENOCARCINOMA							1/147	1/94
CARCINOMA, TRANSITIONAL CELL						1/314		1/94
*************** GROWTH ALTERATIONS, NEOPLASTIC, MALIGNANT, PRIMARY SITE NOT DETERMINED ***************								
LYMPHOMA	3/212	5/369	7/193		2/171			2/94
LYMPHOMA, HISTIOCYTIC								1/94
LYMPHOMA, RETICULUM CELL		2/369	12/193	23/71			1/147	6/94
SARCOMA, UNDIFFERENTIATED				1/71				

 ORGAN SYSTEM: URINARY SYSTEM ORGAN: KIDNEY STRAIN: B6C3F1

SEX:	FEMALE				MALE			
DIAGNOSIS / AGE IN MONTHS:	0 < 12	12 < 18	18 < 24	24+	0 < 12	12 < 18	18 < 24	24+
*************** MECHANICAL ABNORMALITIES AND CYST FORMATION ***************								
CYST			1/1356	1/1015			4/1136	
*************** INFLAMMATION AND REPAIR ***************								
ABSCESS						1/199		
FIBROSIS				1/1015			2/1136	2/958
GLOMERULONEPHRITIS	2/67	3/103	27/1356				11/1136	1/958
GLOMERULONEPHRITIS, MEMBRANOUS			2/1356					
GLOMERULOSCLEROSIS			2/1356	2/1015				3/958
INFLAMMATION			8/1356	5/1015	7/124*	5/199	8/1136	2/958
INFLAMMATION, INTERSTITIAL		1/103	47/1356	3/1015	3/124		66/1136*	6/958
INFLAMMATION, NONSUPPURATIVE		1/103	11/1356	6/1015	1/124	1/199	2/1136	7/958
INFLAMMATION, SUPPURATIVE				1/1015	6/124	2/199	2/1136	
PERIVASCULITIS/VASCULITIS			5/1356	14/1015			4/1136	14/958
PYELONEPHRITIS		1/103	31/1356	27/1015	6/124*	69/199*	124/1136*	45/958
*************** DEGENERATION, NECROSIS, AND DEPOSITION ***************								
AMYLOID DEPOSITION	1/67	1/103	2/1356	2/1015	6/124*	43/199*	47/1136	2/958
ARTERIOSCLEROSIS							1/1136	
CALCULI							20/1136	
CAST, TUBULE								1/958

*VARIABILITY AMONG DATA SOURCES SIGNIFICANT AT THE 1% SIGNIFICANCE LEVEL

SEX:	FEMALE				MALE			
DIAGNOSIS / AGE IN MONTHS:	0 < 12	12 < 18	18 < 24	24+	0 < 12	12 < 18	18 < 24	24+
CELLULAR CHANGE, VACUOLAR							3/1136	7/958
DEGENERATION				2/1015			4/1136	1/958
FATTY CHANGE							10/1136	
HYDRONEPHROSIS			3/1356	3/1015		5/199	17/1136	1/958
INFARCT			1/1356	2/1015				4/958
LIPID DEPOSITION								3/958
MINERALIZATION			2/1356				2/1136	
NECROSIS							2/1136	
NECROSIS, FAT			1/1356					
NEPHROPATHY			1/1356	2/1015			1/1136	1/958
PIGMENT DEPOSITION		1/103						1/958
*************** GROWTH ALTERATIONS, NON-NEOPLASTIC HYPERPLASIA ***************								
HYPERPLASIA, EPITHELIAL								1/958
HYPERPLASIA, LYMPHOID			12/1356	52/1015*			5/1136	37/958
HYPERPLASIA, RETICULUM CELL							1/1136	
HYPERPLASIA, TUBULAR CELL				2/1015			5/1136	15/958
*************** GROWTH ALTERATIONS, NON-NEOPLASTIC, ATROPHY ***************								
ATROPHY			1/1356				2/1136	

*VARIABILITY AMONG DATA SOURCES SIGNIFICANT AT THE 1% SIGNIFICANCE LEVEL

SEX:	FEMALE				MALE			
DIAGNOSIS — AGE IN MONTHS:	0 < 12	12 < 18	18 < 24	24+	0 < 12	12 < 18	18 < 24	24+
*************** GROWTH ALTERATIONS, NON-NEOPLASTIC, METAPLASIA ***************								
METAPLASIA, OSSEOUS				1/1015			1/1136	
*************** GROWTH ALTERATIONS, NEOPLASTIC, BENIGN ***************								
ADENOMA								1/958
ADENOMA, RENAL TUBULAR								1/958
PAPILLOMA, TRANSITIONAL CELL								1/958
*************** GROWTH ALTERATIONS, NEOPLASTIC, MALIGNANT, PRIMARY ***************								
ANGIOSARCOMA			1/1356					
CARCINOMA, ACINAR CELL								1/958
CARCINOMA, TRANSITIONAL CELL				1/1015				
*************** GROWTH ALTERATIONS, NEOPLASTIC, MALIGNANT, METASTATIC ***************								
CARCINOMA, HEPATOCELLULAR			1/1356				1/1136	1/958
*************** GROWTH ALTERATIONS, NEOPLASTIC, MALIGNANT, PRIMARY SITE NOT DETERMINED ***************								
LEIOMYOSARCOMA			1/1356	2/1015				
LYMPHOMA			2/1356	3/1015				
LYMPHOMA, HISTIOCYTIC			2/1356	1/1015				
LYMPHOMA, LYMPHOCYTIC			1/1356			1/199	1/1136	

ORGAN SYSTEM: URINARY SYSTEM ORGAN: KIDNEY STRAIN: C57BL/6

SEX:	FEMALE				MALE			
AGE IN MONTHS: / DIAGNOSIS	0 < 12	12 < 18	18 < 24	24+	0 < 12	12 < 18	18 < 24	24+
*************** DEVELOPMENTAL ANOMALIES ***************								
DOUBLE PELVIS						1/300		
*************** MECHANICAL ABNORMALITIES AND CYST FORMATION ***************								
CYST	1/176	19/367	14/192	17/75		8/300	9/161	28/102 *
DILATATION, TUBULAR		1/367	2/192	5/75		2/300	2/161	1/102
THROMBUS			1/192					
*************** INFLAMMATION AND REPAIR ***************								
FIBROSIS			1/192					2/102
GLOMERULOSCLEROSIS		7/367	4/192	22/75		1/300	2/161	13/102
INFLAMMATION						3/300		3/102
INFLAMMATION, NONSUPPURATIVE		3/367	1/192	9/75		5/300	4/161	31/102 *
INFLAMMATION, SUPPURATIVE							1/161	
NEPHROSCLEROSIS							1/161	1/102
PERIVASCULITIS/VASCULITIS							1/161	1/102
PYELONEPHRITIS	1/176	5/367	4/192	10/75	1/131	5/300	2/161	11/102
*************** DEGENERATION, NECROSIS, AND DEPOSITION ***************								
AMYLOID DEPOSITION		2/367	1/192			2/300	4/161	20/102
CALCULI	1/176							

*VARIABILITY AMONG DATA SOURCES SIGNIFICANT AT THE 1% SIGNIFICANCE LEVEL

SEX:	FEMALE				MALE			
DIAGNOSIS / AGE IN MONTHS:	0 < 12	12 < 18	18 < 24	24+	0 < 12	12 < 18	18 < 24	24+
CAST, PROTEINACEOUS	3/176	11/367	11/192	43/75		2/300	2/161	33/102
CAST, TUBULE								1/102
DEGENERATION		1/367		1/75		1/300		11/102
FATTY CHANGE				1/75				
HYALINOSIS						2/300		1/102
HYDRONEPHROSIS	3/176		2/192	3/75		3/300		6/102
INFARCT							1/161	1/102
MINERALIZATION								1/102
NECROSIS							1/161	
NEPHROPATHY						3/300	2/161	4/102
*************** GROWTH ALTERATIONS, NON-NEOPLASTIC, ATROPHY ***************								
ATROPHY			1/192					
*************** GROWTH ALTERATIONS, NON-NEOPLASTIC, METAPLASIA ***************								
METAPLASIA, OSSEOUS				1/75				1/102
*************** GROWTH ALTERATIONS, NEOPLASTIC, MALIGNANT, PRIMARY ***************								
CARCINOMA, TRANSITIONAL CELL			1/192					
*************** GROWTH ALTERATIONS, NEOPLASTIC, MALIGNANT, PRIMARY SITE NOT DETERMINED ***************								
LYMPHOMA		1/367	2/192	2/75		1/300		4/102

SEX:	FEMALE				MALE			
DIAGNOSIS / AGE IN MONTHS:	0 < 12	12 < 18	18 < 24	24+	0 < 12	12 < 18	18 < 24	24+
LYMPHOMA, RETICULUM CELL		2/367	8/192	12/75			1/161	7/102
SARCOMA, UNDIFFERENTIATED				1/75				

ORGAN SYSTEM: URINARY SYSTEM ORGAN: KIDNEY STRAIN: FISCHER 344

SEX:	FEMALE				MALE			
DIAGNOSIS / AGE IN MONTHS:	0 < 12	12 < 18	18 < 24	24+	0 < 12	12 < 18	18 < 24	24+
*************** DEVELOPMENTAL ANOMALIES ***************								
HAMARTOMA							1/437	1/1743
*************** MECHANICAL ABNORMALITIES AND CYST FORMATION ***************								
CONGESTION					1/101	1/101	1/437	1/1743
CYST				3/1575				6/1743
HYDRONEPHROSIS						1/101	1/437	
*************** INFLAMMATION AND REPAIR ***************								
ABSCESS								2/1743
FIBROSIS				2/1575			3/437	26/1743
GLOMERULONEPHRITIS			20/327	103/1575*			76/437*	286/1743*
GLOMERULONEPHRITIS, MEMBRANOUS			1/327					
GLOMERULOSCLEROSIS				2/1575	7/101*	19/101*	10/437	47/1743
INFLAMMATION	1/26		6/327	35/1575		12/101	11/437	64/1743
INFLAMMATION, INTERSTITIAL			5/327	10/1575		5/101*	7/437	22/1743
INFLAMMATION, MEMBRANOUS			6/327	1/1575			8/437	1/1743
INFLAMMATION, NONSUPPURATIVE					1/101	2/101	7/437	35/1743
INFLAMMATION, SUPPURATIVE			1/327			1/101		
NEPHROSCLEROSIS			1/327	8/1575			3/437	34/1743
PERIVASCULITIS/VASCULITIS						4/101	3/437	5/1743

*VARIABILITY AMONG DATA SOURCES SIGNIFICANT AT THE 1% SIGNIFICANCE LEVEL

SEX:	FEMALE				MALE			
DIAGNOSIS / AGE IN MONTHS:	0 < 12	12 < 18	18 < 24	24+	0 < 12	12 < 18	18 < 24	24+
PYELONEPHRITIS	1/26	3/24	53/327	399/1575*	1/101	14/101*	85/437*	720/1743*
*************** DEGENERATION, NECROSIS, AND DEPOSITION ***************								
AMYLOID DEPOSITION								1/1743
CALCULI				7/1575		1/101		2/1743
CAST, TUBULE			2/327	19/1575		1/101	3/437	29/1743
CELLULAR CHANGE								1/1743
DEGENERATION			6/327*	82/1575	1/101	1/101	7/437	135/1743*
FATTY CHANGE				1/1575				
HEMOSIDERIN DEPOSITION				1/1575			1/437	
HYALINOSIS			1/327					
HYDRONEPHROSIS			2/327	3/1575			4/437	
INFARCT				2/1575			1/437	3/1743
MINERALIZATION		1/24	3/327	15/1575	1/101		3/437	14/1743
NECROSIS		1/24	2/327	7/1575			3/437	2/1743
NECROSIS, ISCHEMIC	1/26							
NECROSIS, MEDULLARY			1/327				1/437	1/1743
NEPHROPATHY			6/327	93/1575*	10/101*	12/101*	21/437	158/1743*
NEPHROPATHY, TOXIC								9/1743
PIGMENT DEPOSITION			5/327	32/1575			1/437	8/1743
PIGMENT DEPOSITION, LIPOCHROME						2/101	1/437	1/1743
PIGMENT DEPOSITION, LIPOFUSCIN					2/101			

*VARIABILITY AMONG DATA SOURCES SIGNIFICANT AT THE 1% SIGNIFICANCE LEVEL

ORGAN SYSTEM: URINARY SYSTEM ORGAN: KIDNEY STRAIN: FISCHER 344

SEX:	FEMALE				MALE			
DIAGNOSIS / AGE IN MONTHS:	0 < 12	12 < 18	18 < 24	24+	0 < 12	12 < 18	18 < 24	24+
*************** GROWTH ALTERATIONS, NON-NEOPLASTIC, HYPERPLASIA ***************								
HYPERPLASIA, EPITHELIAL				4/1575				8/1743
HYPERPLASIA, LYMPHOID				1/1575				
REGENERATION				1/1575	1/101			
*************** GROWTH ALTERATIONS, NON-NEOPLASTIC, ATROPHY ***************								
ATROPHY							1/437	
*************** GROWTH ALTERATIONS, NON-NEOPLASTIC, DYSPLASIA ***************								
DYSPLASIA, NODULAR								1/1743
*************** GROWTH ALTERATIONS, NEOPLASTIC, BENIGN ***************								
ADENOMA				1/1575				1/1743
LIPOMA				1/1575				1/1743
*************** GROWTH ALTERATIONS, NEOPLASTIC, MALIGNANT, PRIMARY ***************								
ADENOCARCINOMA				2/1575				2/1743
ANGIOSARCOMA				1/1575				
CARCINOMA			1/327	1/1575				1/1743
CARCINOMA, TRANSITIONAL CELL			1/327					1/1743
FIBROSARCOMA							1/437	

SEX:	FEMALE				MALE			
DIAGNOSIS / AGE IN MONTHS:	0 < 12	12 < 18	18 < 24	24+	0 < 12	12 < 18	18 < 24	24+
HISTIOCYTOMA, MALIGNANT							1/437	
LIPOSARCOMA								1/1743
MIXED TUMOR, MALIGNANT				1/1575				1/1743
NEUROFIBROSARCOMA							1/437	
*************** GROWTH ALTERATIONS, NEOPLASTIC, MALIGNANT, METASTATIC ***************								
MESOTHELIOMA							2/437	1/1743
THYMOMA								1/1743
*************** GROWTH ALTERATIONS, NEOPLASTIC, MALIGNANT, PRIMARY SITE NOT DETERMINED ***************								
LEUKEMIA	1/26			3/1575			1/437	6/1743
LEUKEMIA, GRANULOCYTIC				1/1575				3/1743
LEUKEMIA, LYMPHOBLASTIC				3/1575				1/1743
LEUKEMIA, LYMPHOCYTIC				3/1575				2/1743
LEUKEMIA, STEM CELL							1/437	
LYMPHOMA				1/1575				1/1743
SARCOMA							1/437	

SEX:	FEMALE				MALE			
DIAGNOSIS / AGE IN MONTHS:	0 < 12	12 < 18	18 < 24	24+	0 < 12	12 < 18	18 < 24	24+
*************** DEVELOPMENTAL ANOMALIES ***************								
HAMARTOMA				12/336				
*************** MECHANICAL ABNORMALITIES AND CYST FORMATION ***************								
CONGESTION	1/7		1/44	3/336		2/64	2/104	1/243
CYST				1/336				6/243
*************** INFLAMMATION AND REPAIR ***************								
ABSCESS							1/104	
GLOMERULONEPHRITIS				1/336			1/104	1/243
INFLAMMATION	2/7		2/44	4/336		7/64	1/104	2/243
INFLAMMATION, INTERSTITIAL				1/336				
INFLAMMATION, SUPPURATIVE				1/336			1/104	
PYELONEPHRITIS	2/7	10/39	12/44	134/336	1/11	33/64	71/104	177/243
*************** DEGENERATION, NECROSIS, AND DEPOSITION ***************								
CALCULI				4/336				1/243
CAST, TUBULE			1/44	1/336				
DEGENERATION						1/64		
HYDRONEPHROSIS			2/44					
MINERALIZATION			4/44	27/336	1/11	8/64	13/104	8/243

SEX:	FEMALE				MALE			
AGE IN MONTHS: / DIAGNOSIS	0 < 12	12 < 18	18 < 24	24+	0 < 12	12 < 18	18 < 24	24+
NEPHROPATHY				1/336				
*************** GROWTH ALTERATIONS, NON-NEOPLASTIC, HYPERPLASIA ***************								
HYPERPLASIA, EPITHELIAL	1/7			9/336			4/104	8/243
*************** GROWTH ALTERATIONS, NEOPLASTIC, BEN-GN ***************								
ADENOMA, RENAL TUBULAR							1/104	1/243
LIPOMA				1/336				2/243
PAPILLOMA				1/336				
*************** GROWTH ALTERATIONS, NEOPLASTIC, MALIGNANT, PRIMARY ***************								
ANGIOSARCOMA								1/243
FIBROSARCOMA				1/336				
MIXED TUMOR, MALIGNANT			1/44	3/336			1/104	3/243
OSTEOSARCOMA							1/104	
*************** GROWTH ALTERATIONS, NEOPLASTIC, MALIGNANT, PRIMARY SITE NOT DETERMINED ***************								
LYMPHOMA								1/243
*************** GROWTH ALTERATIONS, NEOPLASTIC, NOT DETERMINED ***************								
HISTIOCYTOMA, FIBROUS								1/243

SEX:	FEMALE				MALE			
DIAGNOSIS / AGE IN MONTHS:	0 < 12	12 < 18	18 < 24	24+	0 < 12	12 < 18	18 < 24	24+
*************** MECHANICAL ABNORMALITIES AND CYST FORMATION ***************								
CONGESTION						2/97		
CYST				2/240		1/97	2/392	2/216
HYDRONEPHROSIS	15/181 *		2/420	16/240 *	10/170	1/97	2/392	24/216 *
*************** INFLAMMATION AND REPAIR ***************								
ABSCESS				2/240	1/170	1/97		
FIBROSIS	3/181		2/420	7/240	2/170	2/97	2/392	25/216 *
GLOMERULONEPHRITIS			1/420					
INFLAMMATION	25/181 *	3/60	20/420	22/240 *	25/170*	16/97 *	31/392 *	36/216 *
INFLAMMATION, HISTIOCYTIC	1/181							
INFLAMMATION, INTERSTITIAL			5/420	4/240		1/97	44/392 *	7/216
INFLAMMATION, NONSUPPURATIVE	16/181	2/60	3/420	17/240 *	14/170	5/97	2/392	26/216 *
INFLAMMATION, SUPPURATIVE						1/97	1/392	1/216
INFLAMMATION, ULCERATIVE			1/420					
NEPHROSCLEROSIS		1/60	6/420	3/240	1/170	10/97 *	31/392 *	51/216 *
NEPHROSCLEROSIS, SENILE								2/216
PERIVASCULITIS/VASCULITIS				1/240				1/216
PYELONEPHRITIS	5/181	2/60	23/420*		7/170	1/97	100/392	16/216 *

*VARIABILITY AMONG DATA SOURCES SIGNIFICANT AT THE 1% SIGNIFICANCE LEVEL

SEX:	FEMALE				MALE			
DIAGNOSIS / AGE IN MONTHS:	0 < 12	12 < 18	18 < 24	24+	0 < 12	12 < 18	18 < 24	24+
*************** DEGENERATION, NECROSIS, AND DEPOSITION ***************								
ARTERIOLONEPHROSCLEROSIS						1/97		
CALCULI	2/181	7/60*	41/420*	54/240*		4/97	11/392	7/216
CAST, TUBULE	2/181			1/240		1/97	1/392	
DEGENERATION								3/216
HEMOSIDERIN DEPOSITION			1/420					
HYDRONEPHROSIS	2/181	1/60	3/420	1/240		2/97	2/392	1/216
INFARCT				1/240	1/170			
LIPID DEPOSITION			1/420					1/216
MINERALIZATION	6/181	2/60	3/420	2/240	4/170	3/97	15/392	11/216*
MINERALIZATION, VASCULAR		1/60					3/392	2/216
NEPHROPATHY						1/97		
NEPHROPATHY, SENILE			1/420	7/240			2/392	25/216*
PIGMENT DEPOSITION, LIPOFUSCIN		1/60	6/420	8/240				
*************** GROWTH ALTERATIONS, NON-NEOPLASTIC, HYPERPLASIA ***************								
HYPERPLASIA, EPITHELIAL		2/60		5/240		1/97	2/392	6/216
HYPERPLASIA, PROLIFERATIVE			1/420			1/97		
REGENERATION					1/170			

*VARIABILITY AMONG DATA SOURCES SIGNIFICANT AT THE 1% SIGNIFICANCE LEVEL

SEX:	FEMALE				MALE			
DIAGNOSIS / AGE IN MONTHS:	0 < 12	12 < 18	18 < 24	24+	0 < 12	12 < 18	18 < 24	24+
*************** GROWTH ALTERATIONS, NON-NEOPLASTIC, ATROPHY ***************								
ATROPHY				2/240		1/97	1/392	2/216
*************** GROWTH ALTERATIONS, NEOPLASTIC, BENIGN ***************								
ADENOMA, PAPILLARY TUBULAR								1/216
ADENOMA, RENAL TUBULAR								1/216
ANGIOLIPOMA				1/240				
PAPILLOMA				1/240				
*************** GROWTH ALTERATIONS, NEOPLASTIC, MALIGNANT, PRIMARY ***************								
ADENOCARCINOMA					1/170			
CARCINOMA								1/216
CARCINOMA, TRANSITIONAL CELL				1/240				
*************** GROWTH ALTERATIONS, NEOPLASTIC, MALIGNANT, METASTATIC ***************								
RHABDOMYOSARCOMA				1/240				
*************** GROWTH ALTERATIONS, NEOPLASTIC, MALIGNANT, PRIMARY SITE NOT DETERMINED ***************								
LEUKEMIA								1/216
LEUKEMIA, LYMPHOCYTIC						1/97		
LYMPHOMA			2/420		1/170			

 ORGAN SYSTEM: URINARY SYSTEM ORGAN: KIDNEY STRAIN: SPRAGUE DAWLEY

SEX:	FEMALE				MALE			
DIAGNOSIS / AGE IN MONTHS:	0 < 12	12 < 18	18 < 24	24+	0 < 12	12 < 18	18 < 24	24+
LYMPHOMA, RETICULUM CELL			1/420				1/392	
SARCOMA, HISTIOCYTIC				1/240			1/392	2/216

SEX:	FEMALE				MALE			
DIAGNOSIS / AGE IN MONTHS:	0 < 12	12 < 18	18 < 24	24+	0 < 12	12 < 18	18 < 24	24+
*************** MECHANICAL ABNORMALITIES AND CYST FORMATION ***************								
DILATATION						1/314		
*************** INFLAMMATION AND REPAIR ***************								
INFLAMMATION			1/193	1/71				
*************** DEGENERATION, NECROSIS, AND DEPOSITION ***************								
CELLULAR CHANGE, VACUOLAR			2/193					
*************** GROWTH ALTERATIONS, NON-NEOPLASTIC, HYPERPLASIA ***************								
HYPERPLASIA			2/193			1/314		
*************** GROWTH ALTERATIONS, NEOPLASTIC, MALIGNANT, PRIMARY ***************								
CARCINOMA, TRANSITIONAL CELL						1/314		
*************** GROWTH ALTERATIONS, NEOPLASTIC, MALIGNANT, PRIMARY SITE NOT DETERMINED ***************								
LYMPHOMA		1/369	1/193		1/171			
LYMPHOMA, RETICULUM CELL		1/369	4/193	4/71				

SEX:	FEMALE				MALE			
DIAGNOSIS / AGE IN MONTHS:	0 < 12	12 < 18	18 < 24	24+	0 < 12	12 < 18	18 < 24	24+
*************** MECHANICAL ABNORMALITIES AND CYST FORMATION ***************								
DILATATION		1/367						
*************** INFLAMMATION AND REPAIR ***************								
INFLAMMATION		1/367	1/192					1/102
INFLAMMATION, NONSUPPURATIVE				2/75				
*************** DEGENERATION, NECROSIS, AND DEPOSITION ***************								
AMYLOID DEPOSITION								1/102
CELLULAR CHANGE, VACUOLAR				1/75				
*************** GROWTH ALTERATIONS, NON-NEOPLASTIC, HYPERPLASIA ***************								
HYPERPLASIA								1/102
*************** GROWTH ALTERATIONS, NEOPLASTIC, MALIGNANT, PRIMARY SITE NOT DETERMINED ***************								
LYMPHOMA				1/75				2/102
LYMPHOMA, RETICULUM CELL			1/192	2/75				1/102

SEX:	FEMALE				MALE			
DIAGNOSIS / AGE IN MONTHS:	0 < 12	12 < 18	18 < 24	24+	0 < 12	12 < 18	18 < 24	24+
*************** GROWTH ALTERATIONS, NON-NEOPLASTIC, HYPERPLASIA ***************								
HYPERPLASIA, EPITHELIAL				1/1575				

SEX:	FEMALE				MALE			
AGE IN MONTHS:	0 < 12	12 < 18	18 < 24	24+	0 < 12	12 < 18	18 < 24	24+
DIAGNOSIS								
*************** GROWTH ALTERATIONS, NON-NEOPLASTIC, HYPERPLASIA ***************								
HYPERPLASIA, EPITHELIAL							1/104	

DIAGNOSIS	SEX: FEMALE				MALE			
AGE IN MONTHS:	0 < 12	12 < 18	18 < 24	24+	0 < 12	12 < 18	18 < 24	24+
*************** INFLAMMATION AND REPAIR ***************								
INFLAMMATION	1/212					1/314	1/147	
INFLAMMATION, NONSUPPURATIVE				1/71				
*************** DEGENERATION, NECROSIS, AND DEPOSITION ***************								
CALCULI							1/147	
*************** GROWTH ALTERATIONS, NON-NEOPLASTIC, HYPERPLASIA ***************								
HYPERPLASIA			1/193				1/147	
*************** GROWTH ALTERATIONS, NEOPLASTIC, MALIGNANT, PRIMARY SITE NOT DETERMINED ***************								
LYMPHOMA			1/193		2/171			
LYMPHOMA, RETICULUM CELL			3/193	2/71				

SEX:	FEMALE				MALE			
DIAGNOSIS / AGE IN MONTHS:	0 < 12	12 < 18	18 < 24	24+	0 < 12	12 < 18	18 < 24	24+
*************** MECHANICAL ABNORMALITIES AND CYST FORMATION ***************								
OBSTRUCTION					1/124			
*************** DEGENERATION, NECROSIS, AND DEPOSITION ***************								
CALCULI					1/124			

SEX:	FEMALE				MALE			
DIAGNOSIS / AGE IN MONTHS:	0 < 12	12 < 18	18 < 24	24+	0 < 12	12 < 18	18 < 24	24+
*************** INFLAMMATION AND REPAIR ***************								
INFLAMMATION, NONSUPPURATIVE								1/102
INFLAMMATION, SUPPURATIVE							1/161	
*************** GROWTH ALTERATIONS, NON-NEOPLASTIC, HYPERPLASIA ***************								
HYPERPLASIA		1/367						
*************** GROWTH ALTERATIONS, NEOPLASTIC, MALIGNANT, PRIMARY SITE NOT DETERMINED ***************								
LYMPHOMA, RETICULUM CELL			1/192	1/75				

 ORGAN SYSTEM: URINARY SYSTEM ORGAN: URETHRA STRAIN: FISCHER 344

SEX:	FEMALE				MALE			
DIAGNOSIS / AGE IN MONTHS:	0 < 12	12 < 18	18 < 24	24+	0 < 12	12 < 18	18 < 24	24+
****************								
GROWTH ALTERATIONS, NON-NEOPLASTIC, HYPERPLASIA								
****************								
HYPERPLASIA, EPITHELIAL								1/1743

SEX:	FEMALE				MALE			
DIAGNOSIS / AGE IN MONTHS:	0 < 12	12 < 18	18 < 24	24+	0 < 12	12 < 18	18 < 24	24+
*************** INFLAMMATION AND REPAIR ***************								
INFLAMMATION						1/64		1/243
*************** GROWTH ALTERATIONS, NON-NEOPLASTIC, HYPERPLASIA ***************								
HYPERPLASIA, EPITHELIAL							1/104	

SEX:	FEMALE				MALE			
DIAGNOSIS / AGE IN MONTHS:	0 < 12	12 < 18	18 < 24	24+	0 < 12	12 < 18	18 < 24	24+
*************** INFLAMMATION AND REPAIR ***************								
INFLAMMATION		4/369	2/193	2/71	1/171	3/314	1/147	
INFLAMMATION, NONSUPPURATIVE		1/369	1/193	8/71			2/147	17/94
INFLAMMATION, ULCERATIVE			2/193			1/314		
PERIVASCULITIS/VASCULITIS								1/94
*************** DEGENERATION, NECROSIS, AND DEPOSITION ***************								
CALCULI						2/314		
CELLULAR CHANGE, VACUOLAR			4/193				2/147	
MINERALIZATION		2/369			1/171			2/94
*************** GROWTH ALTERATIONS, NON-NEOPLASTIC, HYPERPLASIA ***************								
HYPERPLASIA		3/369	4/193	2/71	1/171	3/314	2/147	1/94
HYPERPLASIA, LYMPHOID								1/94
*************** GROWTH ALTERATIONS, NEOPLASTIC, MALIGNANT, PRIMARY ***************								
CARCINOMA						1/314		
*************** GROWTH ALTERATIONS, NEOPLASTIC, MALIGNANT, PRIMARY SITE NOT DETERMINED ***************								
LYMPHOMA	2/212	4/369	4/193	3/71	1/171			2/94
LYMPHOMA, RETICULUM CELL		2/369	12/193	20/71			1/147	4/94

SEX:	FEMALE				MALE			
DIAGNOSIS / AGE IN MONTHS:	0 < 12	12 < 18	18 < 24	24+	0 < 12	12 < 18	18 < 24	24+
*************** GROWTH ALTERATIONS, NEOPLASTIC, NOT DETERMINED ***************								
NEOPLASM								1/94

SEX:	FEMALE				MALE			
DIAGNOSIS / AGE IN MONTHS:	0 < 12	12 < 18	18 < 24	24+	0 < 12	12 < 18	18 < 24	24+
*************** MECHANICAL ABNORMALITIES AND CYST FORMATION ***************								
DILATATION					2/124			
*************** INFLAMMATION AND REPAIR ***************								
FIBROSIS			1/1356					
INFLAMMATION			27/1356	5/1015	9/124	9/199	10/1136	
INFLAMMATION, NONSUPPURATIVE			3/1356	10/1015				1/958
INFLAMMATION, SUPPURATIVE	1/67				2/124		1/1136	
PERIVASCULITIS/VASCULITIS			2/1356	4/1015				
*************** DEGENERATION, NECROSIS, AND DEPOSITION ***************								
CALCULI					1/124			1/958
MINERALIZATION	1/67		1/1356					
*************** GROWTH ALTERATIONS, NON-NEOPLASTIC, HYPERPLASIA ***************								
HYPERPLASIA				1/1015				
HYPERPLASIA, EPITHELIAL		1/103	10/1356		1/124		17/1136	1/958
HYPERPLASIA, LYMPHOID			2/1356	31/1015				16/958
NODULE				1/1015				

SEX:	FEMALE				MALE			
DIAGNOSIS / AGE IN MONTHS:	0 < 12	12 < 18	18 < 24	24+	0 < 12	12 < 18	18 < 24	24+
*************** GROWTH ALTERATIONS, NON-NEOPLASTIC, METAPLASIA ***************								
METAPLASIA, SQUAMOUS					1/124	1/199	1/1136	
*************** GROWTH ALTERATIONS, NON-NEOPLASTIC, DYSPLASIA ***************								
DYSPLASIA							1/1136	
*************** GROWTH ALTERATIONS, NEOPLASTIC, BENIGN ***************								
PAPILLOMA							1/1136	
*************** GROWTH ALTERATIONS, NEOPLASTIC, MALIGNANT, PRIMARY ***************								
LEIOMYOSARCOMA				2/1015				

SEX:	FEMALE				MALE			
DIAGNOSIS / AGE IN MONTHS:	0 < 12	12 < 18	18 < 24	24+	0 < 12	12 < 18	18 < 24	24+
*************** MECHANICAL ABNORMALITIES AND CYST FORMATION ***************								
CYST		1/367						
*************** INFLAMMATION AND REPAIR ***************								
INFLAMMATION	1/176	2/367	1/192	2/75		1/300	2/161	1/102
INFLAMMATION, NONSUPPURATIVE			1/192	9/75				2/102
INFLAMMATION, SUPPURATIVE							1/161	
*************** DEGENERATION, NECROSIS, AND DEPOSITION ***************								
CALCULI						4/300	2/161	4/102
CELLULAR CHANGE, VACUOLAR		1/367		4/75				2/102
MINERALIZATION		1/367				1/300		2/102
*************** GROWTH ALTERATIONS, NON-NEOPLASTIC, HYPERPLASIA ***************								
HYPERPLASIA		1/367	1/192	2/75	2/131	7/300	4/161	2/102
*************** GROWTH ALTERATIONS, NEOPLASTIC, MALIGNANT, PRIMARY SITE NOT DETERMINED ***************								
LYMPHOMA		1/367	1/192	2/75		1/300		2/102
LYMPHOMA, RETICULUM CELL			2/192	5/75				2/102
SARCOMA, UNDIFFERENTIATED				1/75				

SEX:	FEMALE				MALE			
DIAGNOSIS / AGE IN MONTHS:	0 < 12	12 < 18	18 < 24	24+	0 < 12	12 < 18	18 < 24	24+
*************** MECHANICAL ABNORMALITIES AND CYST FORMATION ***************								
CAST, PROTEINACEOUS								3/1743
CONGESTION								1/1743
CYST						1/101	1/437	
HEMORRHAGE				1/1575			1/437	3/1743
*************** INFLAMMATION AND REPAIR ***************								
INFLAMMATION		1/24	4/327	5/1575		1/101	2/437	8/1743
INFLAMMATION, HEMORRHAGIC							1/437	
INFLAMMATION, SUPPURATIVE								2/1743
INFLAMMATION, ULCERATIVE			1/327			1/101		
POLYP, INFLAMMATORY				1/1575				
*************** DEGENERATION, NECROSIS, AND DEPOSITION ***************								
CALCULI				3/1575		1/101	2/437	4/1743
*************** GROWTH ALTERATIONS, NON-NEOPLASTIC, HYPERPLASIA ***************								
HYPERPLASIA, EPITHELIAL	1/26		3/327	3/1575			6/437	14/1743
*************** GROWTH ALTERATIONS, NON-NEOPLASTIC, METAPLASIA ***************								
METAPLASIA, SQUAMOUS			2/327	1/1575				2/1743

SEX:	FEMALE				MALE			
AGE IN MONTHS: / DIAGNOSIS	0 < 12	12 < 18	18 < 24	24+	0 < 12	12 < 18	18 < 24	24+
*************** GROWTH ALTERATIONS, NEOPLASTIC, BENIGN ***************								
PAPILLOMA			1/327	1/1575				3/1743
PAPILLOMA, TRANSITIONAL CELL			1/327	1/1575				1/1743
POLYP				1/1575				
*************** GROWTH ALTERATIONS, NEOPLASTIC, MALIGNANT, PRIMARY ***************								
CARCINOMA							1/437	
CARCINOMA, PAPILLARY				1/1575				
CARCINOMA, TRANSITIONAL CELL			3/327	3/1575				
CARCINOMA, UNDIFFERENTIATED							1/437	
LEIOMYOSARCOMA				1/1575				
NEUROFIBROSARCOMA							1/437	
*************** GROWTH ALTERATIONS, NEOPLASTIC, MALIGNANT, METASTATIC ***************								
INTERSTITIAL CELL TUMOR								1/1743
MESOTHELIOMA							3/437	3/1743
*************** GROWTH ALTERATIONS, NEOPLASTIC, MALIGNANT, PRIMARY SITE NOT DETERMINED ***************								
LEUKEMIA				1/1575				2/1743
LEUKEMIA, LYMPHOBLASTIC				1/1575				

SEX:	FEMALE				MALE			
DIAGNOSIS / AGE IN MONTHS:	0 < 12	12 < 18	18 < 24	24+	0 < 12	12 < 18	18 < 24	24+
*************** INFLAMMATION AND REPAIR ***************								
INFLAMMATION			1/44	1/336		5/64	5/104	5/243
*************** GROWTH ALTERATIONS, NON-NEOPLASTIC, HYPERTROPHY ***************								
HYPERTROPHY			1/44					
*************** GROWTH ALTERATIONS, NON-NEOPLASTIC, HYPERPLASIA ***************								
HYPERPLASIA, EPITHELIAL	1/7			4/336		1/64	4/104	4/243
*************** GROWTH ALTERATIONS, NEOPLASTIC, BENIGN ***************								
PAPILLOMA								3/243
*************** GROWTH ALTERATIONS, NEOPLASTIC, MALIGNANT, PRIMARY ***************								
CARCINOMA, TRANSITIONAL CELL								1/243

SEX:	FEMALE				MALE			
DIAGNOSIS / AGE IN MONTHS:	0 < 12	12 < 18	18 < 24	24+	0 < 12	12 < 18	18 < 24	24+
*************** MECHANICAL ABNORMALITIES AND CYST FORMATION ***************								
CYST						1/97	1/392	
*************** INFLAMMATION AND REPAIR ***************								
EROSION			1/420					
INFLAMMATION			1/420	1/240	1/170		4/392	1/216
INFLAMMATION, NONSUPPURATIVE			1/420		1/170			
INFLAMMATION, SUPPURATIVE				1/240			1/392	
PERIVASCULITIS/VASCULITIS			1/420					
*************** DEGENERATION, NECROSIS, AND DEPOSITION ***************								
CALCULI			1/420		1/170		1/392	
CAST, PROTEINACEOUS					3/170			
*************** GROWTH ALTERATIONS, NON-NEOPLASTIC, HYPERPLASIA ***************								
HYPERPLASIA			2/420			1/97	1/392	1/216
HYPERPLASIA, EPITHELIAL	1/181				1/170			2/216
HYPERPLASIA, POLYPOID								1/216
HYPERPLASIA, PSEUDOPAPILLARY							1/392	

SEX:	FEMALE				MALE			
DIAGNOSIS / AGE IN MONTHS:	0 < 12	12 < 18	18 < 24	24+	0 < 12	12 < 18	18 < 24	24+
*************** GROWTH ALTERATIONS, NON-NEOPLASTIC, DYSPLASIA ***************								
DYSPLASIA								1/216
*************** GROWTH ALTERATIONS, NEOPLASTIC, BENIGN ***************								
PAPILLOMA	1/181			1/240			1/392	1/216
PAPILLOMA, TRANSITIONAL CELL			1/420					
*************** GROWTH ALTERATIONS, NEOPLASTIC, MALIGNANT, PRIMARY ***************								
CARCINOMA, TRANSITIONAL CELL							1/392	
*************** GROWTH ALTERATIONS, NEOPLASTIC, MALIGNANT, PRIMARY SITE NOT DETERMINED ***************								
LYMPHOMA			1/420					
SARCOMA, HISTIOCYTIC							1/392	

APPENDIX

PREFERRED TERM	SNOVET CODE	LADB TERMS
ABSCESS	M-41740	ABSCESS
ADENOCARCINOMA	M-81403	ADENOCARCINOMA
ANGIOMA	M-91200	HEMANGIOMA
ANGIOSARCOMA	M-91203	HEMANGIOSARCOMA
ARTERIOSCLEROSIS	M-52000	ARTERIOSCLEROSIS
ASCITES	M-36300	ASCITES
CARCINOMA	M-80103	CARCINOMA
CARCINOMA, ALVEOLAR BRONCHIOLAR	M-82506	ALVEOLAR/BRONCHIOLAR CARCINOMA
CARCINOMA, CORTICAL	M-82703	CORTICAL CARCINOMA
CYST	M-33400	CYST
FIBROMA	M-81000	FIBROMA
FIBROSARCOMA	M-88103	FIBROSARCOMA
FIBROSIS	M-49000	ADHESION
		FIBROSIS
HEMATOMA	M-37100	HEMATOMA
		HEMATOMA, ORGANIZED
HEMORRHAGE	M-37000	HEMOPERITONEUM
		HEMORRHAGE
HISTIOCYTOMA, FIBROUS, MALIGNANT	M-88303	FIBROUS HISTIOCYTOMA, MALIGNANT
INFARCT	M-54700	INFARCT
INFLAMMATION	M-40000	INFLAMMATION
		PERITONITIS
INFLAMMATION, GRANULOMATOUS	M-44000	INFLAMMATION, GRANULOMATOUS
		INFLAMMATION, PYOGRANULOMATOUS
		LIPOGRANULOMA
INFLAMMATION, NECROTIZING	M-40700	INFLAMMATION, NECROTIZING
INFLAMMATION, SUPPURATIVE	M-40400	INFLAMMATION, SUPPURATIVE
LEIOMYOSARCOMA	M-88903	LEIOMYOSARCOMA
LEUKEMIA	M-98003	LEUKEMIC INFILTRATION
LIPOMA	M-88500	ABDOMINAL LIPOMA
		LIPOMA
		LIPOMA, INFARCTED
		LIPOMA, NODULAR
		PEDUNCULATED LIPOMA
LIPOSARCOMA	M-88503	LIPOSARCOMA
LYMPHOMA, RETICULUM CELL	M-96403	RETICULUM CELL SARCOMA
MESOTHELIOMA	M-90503	MESOTHELIOMA
		MESOTHELIOMA, MALIGNANT
		MESOTHELIOMATOSIS
METAPLASIA, OSSEOUS	M-73400	METAPLASIA, OSSEOUS
MINERALIZATION	M-55400	CALCIFICATION
		MINERALIZATION
MINERALIZATION, VASCULAR (T-40000)	M-55400	CALCIFICATION, MEDIAL
NECROSIS	M-54000	NECROSIS
NECROSIS, FAT	M-54110	NECROSIS, FAT
NEUROFIBROSARCOMA	M-95403	NEUROFIBROSARCOMA
OSTEOSARCOMA	M-91803	OSTEOSARCOMA
PERIVASCULITIS/VASCULITIS	M-47800	ANGIITIS
		PERIARTERITIS
		PERIVASCULITIS

PREFERRED TERM	SNOVET CODE	LADB TERMS
PERIVASCULITIS/VASCULITIS	M-47800	POLYARTERITIS
RHABDOMYOSARCOMA	M-89003	RHABDOMYOSARCOMA
SARCOMA	M-88003	SARCOMA
THROMBUS	M-35100	THROMBOSIS

ORGAN SYSTEM. BODY CAVITIES   ORGAN: BODY CAVITY (NOT SPECIFIED)

PREFERRED TERM	SNOVET CODE	LADB TERMS
MESOTHELIOMA	M-90503	MESOTHELIOMA
		MESOTHELIOMA, MALIGNANT

PREFERRED TERM	SNOVET CODE	LADB TERMS
ABSCESS	M-41740	ABSCESS
ANGIOSARCOMA	M-91203	HEMANGIOSARCOMA
CARCINOMA, ALVEOLAR BRONCHIOLAR	M-82506	ALVEOLAR/BRONCHIOLAR CARCINOMA
EMPYEMA (T-2Y700)	M-41400	EMPYEMA
FIBROMA	M-81000	FIBROMA
HEMORRHAGE	M-37000	HEMOTHORAX
HYDROTHORAX (T-Y2100)	M-36300	HYDROTHORAX
INFLAMMATION	M-40000	INFLAMMATION
		MEDIASTINITIS
		PLEURITIS
INFLAMMATION, FIBRINOUS	M-40500	PLEURITIS, FIBRINOUS
INFLAMMATION, NONSUPPURATIVE	M-47130	PLEURITIS, EFFUSIVE, NONSUPPURATIVE
LIPOSARCOMA	M-88503	LIPOSARCOMA
MESOTHELIOMA	M-90503	MESOTHELIOMA
MESOTHELIOMA, BENIGN	M-90500	MESOTHELIOMA, BENIGN
PERIVASCULITIS/VASCULITIS	M-47800	PERIARTERITIS
SARCOMA	M-88003	SARCOMA
SARCOMA, HISTIOCYTIC	M-96403	HISTIOCYTIC SARCOMA

PREFERRED TERM	SNOVET CODE	LADB TERMS
ANGIOSARCOMA	M-91203	HEMANGIOSARCOMA
ARTERIOSCLEROSIS	M-52000	ARTERIOSCLEROSIS
ATHEROSCLEROSIS	M-52110	ATHEROSCLEROSIS
CARCINOSARCOMA, MAMMARY (T-04000)	M-89806	MAMMARY CARCINOSARCOMA
CONGESTION	M-36100	CONGESTION
DEGENERATION	M-50000	DEGENERATION
		DEGENERATION, HYALINE
DEGENERATION, FIBRINOID	M-54090	DEGENERATION, FIBRINOID
EMBOLISM	M-35300	EMBOLISM
FIBROSIS	M-49000	FIBROSIS
HEMATOPOIESIS	F-50010	HYPERPLASIA, HEMATOPOIETIC
HEMOSIDERIN DEPOSITION	M-57500	HEMOSIDERIN
HYPERPLASIA	M-72000	HYPERPLASIA
HYPERTROPHY	M-71000	HYPERTROPHY
		HYPERTROPHY, MEDIAL
INFLAMMATION	M-40000	ARTERITIS
		INFLAMMATION
INFLAMMATION, GRANULOMATOUS	M-44000	INFLAMMATION, GRANULOMATOUS
INFLAMMATION, NONSUPPURATIVE	M-47130	AORTITIS, LYMPHOCYTIC
		INFLAMMATION, NONSUPPURATIVE
LEUKEMIA	M-98003	LEUKEMIA
LEUKEMIA, GRANULOCYTIC	M-98603	GRANULOCYTIC LEUKEMIA
LEUKEMIA, HISTIOCYTIC	M-98903	LEUKEMIC HISTIOCYTIC SARCOMA
LEUKEMIA, LYMPHOBLASTIC	M-98213	LYMPHOBLASTIC LEUKEMIA
LEUKEMIA, LYMPHOCYTIC	M-98203	LYMPHATIC LEUKEMIA
LYMPHOMA	M-95903	LYMPHOMA
LYMPHOMA, RETICULUM CELL	M-96403	RETICULUM CELL SARCOMA
METAPLASIA, CARTILAGINOUS	M-73600	METAPLASIA, CARTILAGINOUS
MINERALIZATION	M-55400	CALCIFICATION
		CALCIFICATION, DYSTROPHIC
		CALCIFICATION, MEDIAL
		MEDIAL CALCIFICATION
		MINERALIZATION
NECROSIS	M-54000	NECROSIS
PERIVASCULITIS/VASCULITIS	M-47800	PANARTERITIS, NODOSA
		PERIARTERITIS
		PERIARTERITIS NODOSA
		PERIVASCULITIS
		POLYARTERITIS
		POLYARTERITIS NODOSA
RUPTURE	M-14430	RUPTURE
SARCOMA, HISTIOCYTIC	M-96403	HISTIOCYTIC SARCOMA
SCLEROSIS	M-49020	SCLEROSIS
THROMBUS	M-35100	THROMBOANGITIS OBLITERANS
		THROMBOSIS
		THROMBUS
		THROMBUS, ORGANIZED
THROMBUS, MURAL	M-35130	THROMBUS, MURAL

PREFERRED TERM	SNOVET CODE	LADB TERMS
ABSCESS	M-41740	ABSCESS
ADENOCARCINOMA, PULMONARY (T-28000)	M-81403	PULMONARY ADENOCARCINOMA
AMYLOID DEPOSITION		AMYLOIDOSIS
ANGIOMA	M-91200	HEMANGIOMA
ANGIOSARCOMA	M-91203	HEMANGIOSARCOMA
ARTERIOSCLEROSIS	M-52000	ARTERIOSCLEROSIS
ATROPHY	M-58000	ATROPHY
CARCINOMA, HEPATOCELLULAR	M-81703	HEPATOCELLULAR CARCINOMA
CARCINOMA, HURTHLE CELL (T-96...)	M-82903	HURTHLE CELL CARCINOMA
CARCINOSARCOMA, MAMMARY (T-04000)	M-89806	MAMMARY CARCINOSARCOMA
CONGESTION	M-36100	CONGESTION
DEGENERATION, MYOCARDIAL (T-33010)	M-50000	DEGENERATION
		DEGENERATION & INCREASED CELLULARITY
		DEGENERATION, ACTIVE
		DEGENERATION, BASOPHILIC
		DEGENERATION, GRANULAR
		DEGENERATION, HYALINE
		DEGENERATION, MUCINOUS
		DEGENERATION, MYOCARDIAL
		DEGENERATION, MYOCARDIAL BASOPHILIC
		DEGENERATION, MYOFIBER
		DYSTROPHY
		DYSTROPHY, ACTIVE
		DYSTROPHY, FIBROSING
		DYSTROPHY, FIBROSING, MYOFIBER
		DYSTROPHY, MYOFIBER
		MYODEGENERATION
		MYODEGENERATION, ACTIVE
DILATATION	M-32100	DILATATION
EMBOLUS, SEPTIC	M-35390	EMBOLUS, SEPTIC
FATTY CHANGE	M-50080	DEGENERATION, FATTY
		METAMORPHOSIS, FATTY
FIBROMA	M-81000	FIBROMA
FIBROSARCOMA	M-88103	FIBROSARCOMA
FIBROSIS	M-49000	FIBROSIS
		FIBROSIS, INTERSTITIAL
HEMORRHAGE	M-37000	HEMORRHAGE
		HEMORRHAGE, RESOLVED
HEMOSIDERIN DEPOSITION	M-57500	HEMOSIDERIN
		HEMOSIDEROSIS
HYPERPLASIA	M-72000	HYPERPLASIA
		INCREASED CELLULARITY
HYPERPLASIA, MESOTHELIAL	M-72080	HYPERPLASIA, MESOTHELIAL
HYPERTROPHY	M-71000	HYPERTROPHY
		HYPERTROPHY, MEDIAL
INFLAMMATION	M-40000	INFLAMMATION, INTERSTITIAL
		INFLAMMATION, PROLIFERATIVE
		INFLAMMATION, REACTIVE
		MYOCARDITIS
		MYOCARDITIS, FIBROSING

PREFERRED TERM	SNOVET CODE	LADB TERMS
INFLAMMATION	M-40000	MYOCARDITIS, REACTIVE
INFLAMMATION, GRANULOMATOUS	M-44000	GRANULOMATOUS
		GRANULOMATOUS, EOSINOPHILIC
		INFLAMMATION, GRANULOMATOUS
		MYOCARDITIS, GRANULOMATOUS
INFLAMMATION, MIXED		MYOCARDITIS, MIXED LEUKOCYTIC
INFLAMMATION, NONSUPPURATIVE	M-47130	INFILTRATION, MONONUCLEAR CELL
		LYMPHOCYTES, FOCUS OF
		LYMPHOID COLLECTION
		MAST CELLS
		MYOCARDITIS, NONSUPPURATIVE
INFLAMMATION, SUPPURATIVE	M-40400	INFLAMMATION, SUPPURATIVE
		MYOCARDITIS, PURULENT
LEIOMYOSARCOMA	M-88903	LEIOMYOSARCOMA
LEUKEMIA	M-98003	LEUKEMIA
		LEUKEMIC INFILTRATION
LEUKEMIA, GRANULOCYTIC	M-98603	GRANULOCYTIC LEUKEMIA
LEUKEMIA, LYMPHOCYTIC	M-98203	LEUKEMIC LYMPHOSARCOMA
		LYMPHATIC LEUKEMIA
		LYMPHOCYTIC LEUKEMIA
LEUKEMIA, STEM CELL	M-98013	STEM CELL LEUKEMIA
LYMPHOMA	M-95903	LYMPHOMA
		MALIGNANT LYMPHOMA
LYMPHOMA, RETICULUM CELL	M-96403	RETICULUM CELL SARCOMA
MELANOSIS	M-57210	MELANIN
		MELANOSIS
METAPLASIA	M-73000	METAPLASIA
METAPLASIA, CARTILAGINOUS	M-73600	METAPLASIA, CARTILAGINOUS
		METAPLASIA, CHONDROID
METAPLASIA, OSSEOUS	M-73400	METAPLASIA, OSSEOUS
METAPLASIA, OSSEOUS, CHONDROID	M-73600	METAPLASIA, OSSEOUS, CHONDROID
MINERALIZATION	M-55400	CALCIFICATION
		CALCIFICATION, DYSTROPHIC
		CALCIFICATION, MYOCARDIAL
		CALCIFICATION, SECONDARY
		CALCIUM DEPOSIT
		MINERALIZATION
		MINERALIZATION, DYSTROPHIC
MIXED TUMOR	M-89400	MIXED TUMOR
NECROSIS	M-54000	NECROSIS
NEUROFIBROSARCOMA	M-95403	NEUROFIBROSARCOMA
PERIVASCULITIS/VASCULITIS	M-47800	ARTERITIS, GRANULOMATOUS
		PERIARTERITIS
		PERIVASCULITIS
		POLYARTERITIS
PIGMENT DEPOSITION	M-57000	PIGMENTATION
RHABDOMYOSARCOMA	M-89003	RHABDOMYOSARCOMA
SARCOMA	M-88003	SARCOMA
SARCOMA, ANITSCHKOW CELL		ANITSCHKOW-CELL SARCOMA
SARCOMA, HISTIOCYTIC	M-96403	HISTIOCYTIC SARCOMA

PREFERRED TERM	SNOVET CODE	LADB TERMS
THROMBUS	M-35100	THROMBOSIS THROMBUS
THROMBUS, MURAL	M-35130	THROMBUS, CARDIAC MURAL THROMBUS, MURAL
THROMBUS, ORGANIZED	M-35103	THROMBUS, ORGANIZED

PREFERRED TERM	SNOVET CODE	LADB TERMS
INFLAMMATION	M-40000	ENDOCARDITIS
		ENDOCARDITIS, BACTERIAL
		ENDOCARDITIS, MURAL

PREFERRED TERM	SNOVET CODE	LADB TERMS
INFLAMMATION	M-40000	EPICARDITIS
		INFLAMMATION
		PERICARDITIS
INFLAMMATION, NONSUPPURATIVE	M-47130	EPICARDITIS, NONSUPPURATIVE
		INFLAMMATION, NONSUPPURATIVE

ORGAN SYSTEM: CARDIOVASCULAR SYSTEM ORGAN: LYMPHATIC VESSEL (T-09010)

PREFERRED TERM	SNOVET CODE	LADB TERMS
INFLAMMATION	M-40000	INFLAMMATION

PREFERRED TERM	SNOVET CODE	LADB TERMS
CALCULI	M-30000	CRYSTAL
		GALL STONE
CELLULAR CHANGE, EOSINOPHILIC	M-50300	EOSINOPHILIC CYTOPLASMIC CHANGE
CYST	M-33400	CYST
HYPERPLASIA	M-72000	HYPERPLASIA
INFLAMMATION	M-40000	INFLAMMATION
PAPILLOMA	M-80500	PAPILLOMA

PREFERRED TERM	SNOVET CODE	LADB TERMS
ABSCESS	M-41740	ABSCESS
		MICROABSCESS, NONSPECIFIC
ADENOCARCINOMA	M-81403	ADENOCARCINOMA
ADENOMA	M-81400	ADENOMA
		HEPATOCELLULAR ADENOMA
AMYLOID DEPOSITION		AMYLOIDOSIS
ANGIECTASIS	M-76300	ANGIECTASIS
		PELIOSIS HEPATITIS
		TELANGIECTASIS
ANGIOMA	M-91200	HEMANGIOMA
ANGIOSARCOMA	M-91203	ANGIOSARCOMA
		HEMANGIOSARCOMA
ATROPHY	M-58000	ATROPHY
		ATROPHY, PRESSURE
CARCINOMA	M-80103	HEPATOCELLULAR CARCINOMA
CARCINOMA, ACINAR CELL	M-85506	ACINAR-CELL CARCINOMA
CARCINOMA, CORTICAL	M-82703	CORTICAL CARCINOMA
CELLULAR CHANGE	M-66040	ALTERATION, CYTOLOGIC
		CELLULAR CHANGE
		CYTOPLASMIC CHANGE
		CYTOPLASMIC SWELLING
CELLULAR CHANGE, BASOPHILIC	M-50310	BASOPHILIC CYTOPLASMIC CHANGE
CELLULAR CHANGE, CLEAR CELL	M-66040	CLEAR-CELL CHANGE
CELLULAR CHANGE, EOSINOPHILIC	M-50300	EOSINOPHILIC CYTOPLASMIC CHANGE
CELLULAR CHANGE, GLYCOGEN	M-69530	INFILTRATION, GLYCOGEN
CELLULAR CHANGE, VACUOLAR	M-66040	CYTOPLASMIC VACUOLATION
		CYTOPLASMIC VACUOLIZATION
		DEGENERATION, HYDROPIC
		HYDROPIC CHANGE
		SWELLING, CYTOPLASMIC
CIRRHOSIS	M-49500	CIRRHOSIS
		CIRRHOSIS, BILIARY
		CIRRHOSIS, CARDIAC
		CIRRHOSIS, PORTAL
		CIRRHOSIS, POST NECROTIC
CONGESTION	M-36100	CONGESTION
		CONGESTION, PASSIVE
		CONGESTION, SECONDARY
CYST	M-33400	CYST, BENIGN
		CYST, EPIDERMAL INCLUSION
		CYST, MULTILOCULAR
		CYST, MULTILOCULAR, FATTY
		DILATATION, CYSTIC
		MULTILOCULAR CYST
CYST, HEMORRHAGIC	M-33540	HEMATOCYST
		HEMORRHAGIC CYST
CYTOMEGALY	M-69220	HEPATOCYTOMEGALY
		MEGALOCYTOSIS
DEGENERATION	M-50000	CYTOLOGIC DEGENERATION
		DEGENERATION

PREFERRED TERM	SNOVET CODE	LADB TERMS
DEGENERATION	M-50000	DEGENERATION, CYSTIC
		DEGENERATION, EOSINOPHILIC
		DEGENERATION, GRANULAR
		DEGENERATION, HYALINE
		DEGENERATION, PARENCHYMATOUS
		DEGENERATION, VACUOLAR
FATTY CHANGE	M-50080	DEGENERATION, FATTY
		DEGENERATION, FATTY, ANOXIC
		DEGENERATION, SECONDARY, FATTY
		FATTY CHANGE
		INFILTRATION, FATTY
		METAMORPHOSIS
		METAMORPHOSIS, FATTY
FIBROSARCOMA	M-88103	FIBROSARCOMA
FIBROSIS	M-49000	FIBROSIS, SEPTAL LIVER
		SCAR
		SCLEROSIS
GRANULOMA	M-44000	GRANULOMA
		GRANULOMA, MICRO
		MICROGRANULOMA
		MICROGRANULOMA, NONSPECIFIC
		REACTION, FOREIGN BODY
HAMARTOMA	M-75500	HAMARTOMA
HEMATOPOIESIS	F-50010	ERYTHROPOIESIS
		GRANULOPOIESIS
		HEMATOPOIESIS
		HEMATOPOIESIS, EXTRAMEDULLARY
		HYPERPLASIA, HEMATOPOIETIC
		MYELOID METAPLASIA
HEMORRHAGE	M-37000	HEMORRHAGE
HEMOSIDERIN DEPOSITION	M-57500	HEMOSIDERIN
HEPATOMA	M-81703	HEPATOMA
HERNIATION	M-31600	HERNIA
HISTIOCYTOMA, FIBROUS	M-88300	FIBROUS HISTIOCYTOMA
HISTIOCYTOMA, MALIGNANT	M-88323	MALIGNANT HISTIOCYTOSIS
HISTIOCYTOSIS	M-77800	HISTIOCYTOSIS
HYALINOSIS, VASCULAR	M-50060	HYALINOSIS, VASCULAR
HYPERPLASIA	M-72000	HYPERPLASIA, BENIGN
		HYPERPLASIA, ECTATIC
		HYPERPLASIA, REGENERATIVE
HYPERPLASIA, BASOPHILIC (T-91130)	M-72000	HYPERPLASIA, BASOPHILIC
HYPERPLASIA, ERYTHROID	M-72000	HYPERPLASIA, ERYTHROID
HYPERPLASIA, MYELOID (T-06200)	M-72000	HYPERPLASIA, GRANULOCYTIC
		LEUKOCYTOSIS
	72000	LEUKEMOID REACTION
HYPERPLASIA, NODULAR	M-72030	HYPERPLASIA, NODULAR
		HYPERPLASIA, NODULAR, BENIGN
		HYPERPLASTIC NODULE
HYPERPLASIA, RETICULUM CELL	M-72180	HYPERPLASIA, RETICULUM CELL
HYPERPLASIA, SCLEROSING		HYPERPLASIA, SCLEROSING

PREFERRED TERM	SNOVET CODE	LADB TERMS
HYPERTROPHY	M-71000	HYPERTROPHY
INFARCT	M-54700	INFARCT
INFLAMMATION	M-40000	HEPATITIS
		HEPATITIS, INTERSTITIAL
		HEPATITIS, MIXED LEUKOCYTIC
		HEPATITIS, NONSPECIFIC, REACTIVE
		HEPATITIS, TOXIC
		LEUCOCYTIC FOCI
		LEUCOCYTIC INFILTRATION
		TRIADITIS
INFLAMMATION WITH FIBROSIS	M-45000	INFLAMMATION WITH FIBROSIS
INFLAMMATION, GRANULOMATOUS	M-44000	HEPATITIS, GRANULOMATOUS
		HEPATITIS, GRANULOMATOUS, NONSUPPURATIVE
		INFLAMMATION, GRANULOMATOUS
INFLAMMATION, NECROTIZING	M-40700	HEPATITIS, NECROTIC
		INFLAMMATION, NECROTIZING
		NECROSIS, WITH INFLAMMATION
INFLAMMATION, NONSUPPURATIVE	M-47130	CAPSULITIS, LYMPHOCYTIC
		CELLULAR INFILTRATION
		HEPATITIS, NODULAR LYMPHOCYTIC
		HEPATITIS, NON SUPPURATIVE
		HEPATITIS, NONSUPPURATIVE
		HYPERPLASIA, LYMPHOID
		INFILTRATION, LYMPHOID
		INFILTRATION, MONONUCLEAR CELL
		INFLAMMATION, LYMPHOCYTIC
		LYMPHOCYTIC ACCUMULATION
		LYMPHOCYTIC ACCUMULATION, NODULAR
		LYMPHOCYTIC INFLAMMATORY INFILTRATE
		LYMPHOCYTIC NODULE
		LYMPHOCYTOSIS
		LYMPHOID CELLS
		ROUND CELL FOCI
INFLAMMATION, SUPPURATIVE	M-40400	HEPATITIS, NECROPURULENT
		HEPATITIS, SUPPURATIVE
		INFLAMMATION, SUPPURATIVE
		MICROABSCESS
LEIOMYOSARCOMA	M-88903	LEIOMYOSARCOMA
LEUKEMIA	M-98003	LEUKEMIA
		LEUKEMIC INFILTRATION
LEUKEMIA, GRANULOCYTIC	M-98603	GRANULOCYTIC LEUKEMIA
LEUKEMIA, HISTIOCYTIC	M-98903	LEUKEMIC HISTIOCYTIC SARCOMA
LEUKEMIA, LYMPHOBLASTIC	M-98213	LYMPHOBLASTIC LEUKEMIA
LEUKEMIA, LYMPHOCYTIC	M-98203	LEUKEMIC LYMPHOSARCOMA
		LYMPHATIC LEUKEMIA
		LYMPHOCYTIC LEUKEMIA
LEUKEMIA, STEM CELL	M-98013	STEM CELL LEUKEMIA
LEUKEMIA, UNDIFFERENTIATED	M-98013	UNDIFFERENTIATED LEUKEMIA
LIPID DEPOSITION		LIPOIDOSIS
LIPOMA	M-88500	LIPOMA

PREFERRED TERM	SNOVET CODE	LADB TERMS
LYMPHANGIECTASIS	M-76400	LYMPHANGIECTASIS
LYMPHOMA	M-95903	LYMPHOMA
		LYMPHOSARCOMATOUS INFILTRATION
		MALIGNANT LYMPHOMA
		MALIGNANT LYMPHOMA, MIXED TYPE
LYMPHOMA, HISTIOCYTIC	M-96403	MALIGNANT LYMPHOMA, HISTIOCYTIC TYPE
LYMPHOMA, LYMPHOCYTIC	M-96203	MALIGNANT LYMPHOMA, LYMPHOCYTIC TYPE
LYMPHOMA, RETICULUM CELL	M-96403	RETICULUM CELL SARCOMA
MASTOCYTOSIS	M-97413	MASTOCYTOSIS
MINERALIZATION	M-55400	CALCIFICATION
		CALCIFICATION, DYSTROPHIC
		CALCIUM DEPOSIT
MYELOSARCOMA	M-99303	MYELOSARCOMA
NECROSIS	M-54000	NECROSIS
		NECROSIS, CASEOUS
		NECROSIS, GENERALIZED
		NECROSIS, ISCHEMIC
NECROSIS, COAGULATIVE	M-54060	NECROSIS, COAGULATION
		NECROSIS, COAGULATIVE
NECROSIS, FAT	M-54110	NECROSIS, FAT
NECROSIS, HEMORRHAGIC	M-54040	NECROSIS, HEMORRHAGIC
NEOPLASM	M-72030	NEOPLASTIC NODULE
	M-80001	NEOPLASM
NEUROFIBROSARCOMA	M-95403	NEUROFIBROSARCOMA
NODULE	M-03010	NODULE
PARASITISM	D-06500	PARASITISM
PERIVASCULITIS/VASCULITIS	M-47800	CUFFING, PERIVASCULAR
		PERIARTERITIS
		PERIVASCULAR CUFFING
		PERIVASCULITIS
		VASCULITIS, LYMPHOCYTIC
PIGMENT DEPOSITION	M-57000	PIGMENT
		PIGMENTATION
RHABDOMYOSARCOMA	M-89003	RHABDOMYOSARCOMA
SARCOMA	M-88003	SARCOMA
SARCOMA, HISTIOCYTIC	M-96403	HISTIOCYTIC SARCOMA
THROMBUS	M-35100	THROMBOSIS
		THROMBUS
THROMBUS, ORGANIZED	M-35103	THROMBUS, ORGANIZED

PREFERRED TERM	SNOVET CODE	LADB TERMS
CARCINOMA	M-80103	BILE DUCT CARCINOMA
		CARCINOMA
CHOLANGIECTASIS		CHOLANGIECTASIS
CYST	M-33400	CYST
		CYST, BILE DUCT
		CYST, BILIARY
		CYST, INTRAHEPATIC BILIARY
DILATATION	M-32100	DILATATION
		DILATATION, CYSTIC BILIARY
FIBROSIS	M-49000	FIBROSIS
HYPERPLASIA	M-72000	HYPERPLASIA
		HYPERPLASIA, BILIARY
INFLAMMATION	M-40000	CHOLANGITIS
		INFLAMMATION
		PERICHOLANGITIS
INFLAMMATION, NONSUPPURATIVE	M-47130	CHOLANGITIS, LYMPHOCYTIC
		INFLAMMATION, NONSUPPURATIVE
		PERICHOLANGITIS, LYMPHORETICULAR
		PERICHOLANGITIS, NONSUPPURATIVE

PREFERRED TERM	SNOVET CODE	LADB TERMS
ADENOCARCINOMA	M-81403	ADENOCARCINOMA
ADENOMA	M-81400	ADENOMA
AMYLOID DEPOSITION		AMYLOIDOSIS
ANGIOSARCOMA	M-91203	ANGIOSARCOMA
ATROPHY	M-58000	ATROPHY
CARCINOMA	M-80103	CARCINOMA
CARCINOMA IN SITU	M-80102	CARCINOMA-IN-SITU
CYST	M-33400	CYST
DEGENERATION	M-50000	DEGENERATION
DYSPLASIA	M-74000	DYSPLASIA
EDEMA	M-36500	EDEMA
FATTY CHANGE	M-50080	METAMORPHOSIS, FATTY
FIBROADENOCARCINOMA	M-90103	FIBROADENOCARCINOMA
FIBROMA	M-81000	FIBROMA
FIBROSIS	M-49000	FIBROSIS
HISTIOCYTOMA, FIBROUS	M-88300	FIBROUS HISTIOCYTOMA
HYPERPLASIA	M-72000	HYPERPLASIA
HYPERPLASIA, INTRADUCTAL	M-72170	HYPERPLASIA, INTRADUCTAL
INFLAMMATION	M-40000	INFLAMMATION
		INFLAMMATION, INTERSTITIAL
		PAROTITIS
INFLAMMATION, NECROTIZING	M-40700	INFLAMMATION, NECROTIZING
INFLAMMATION, NONSUPPURATIVE	M-47130	INFILTRATION, LYMPHOID
		INFLAMMATION, LYMPHOCYTIC
		LYMPHOCYTIC INFLAMMATORY INFILTRATE
		SIALITIS, NONSUPPURATIVE
INFLAMMATION, SUPPURATIVE	M-40400	INFLAMMATION, SUPPURATIVE
		SIALOADENITIS, PURULOGRANULOMATOUS
		SIALOADENITIS, SUPPURATIVE
KARYOMEGALY	M-65110	KARYOMEGALY
LEIOMYOSARCOMA	M-88903	LEIOMYOSARCOMA
LEUKEMIA, LYMPHOCYTIC	M-98203	LEUKEMIC LYMPHOSARCOMA
		LYMPHATIC LEUKEMIA
LYMPHOMA	M-95903	LYMPHOMA
		LYMPHOSARCOMATOUS INFILTRATION
		MALIGNANT LYMPHOMA
LYMPHOMA, RETICULUM CELL	M-96403	RETICULUM CELL SARCOMA
MINERALIZATION	M-55400	CALCIFICATION
MYOEPITHELIOMA	M-89820	MYOEPITHELIOMA
NECROSIS	M-54000	NECROSIS
NEUROFIBROSARCOMA	M-95403	NEUROFIBROSARCOMA
PERIVASCULITIS/VASCULITIS	M-47800	PERIVASCULAR CUFFING
		PERIVASCULITIS
SARCOMA	M-88003	SARCOMA

PREFERRED TERM	SNOVET CODE	LADB TERMS
CARCINOMA	M-80103	CARCINOMA
DILATATION	M-32100	DILATATION
DYSPLASIA	M-74000	DYSPLASIA
INFLAMMATION	M-40000	INFLAMMATION
INFLAMMATION, SUPPURATIVE	M-40400	INFLAMMATION, SUPPURATIVE
INFLAMMATION, ULCERATIVE	M-40030	ULCER
LYMPHOMA	M-95903	LYMPHOMA
LYMPHOMA, RETICULUM CELL	M-96403	RETICULUM CELL SARCOMA
NECROSIS, FAT	M-54110	NECROSIS, FAT
PAPILLOMA	M-80500	PAPILLOMA
RHABDOMYOSARCOMA	M-89003	RHABDOMYOSARCOMA
SARCOMA	M-88003	SARCOMA

PREFERRED TERM	SNOVET CODE	LADB TERMS
ADENOCARCINOMA	M-81403	ADENOCARCINOMA
		ADENOCARCINOMA IN ADENOMATOUS POLYP
ADENOCARCINOMA, MUCINOUS	M-84803	MUCINOUS ADENOCARCINOMA
ADENOMA	M-81400	ADENOMA
AMYLOID DEPOSITION		AMYLOID
		AMYLOIDOSIS
ANGIOSARCOMA	M-91203	HEMANGIOSARCOMA
ARTERIOLOSCLEROSIS	M-52200	ARTERIOLOSCLEROSIS
ARTERIOSCLEROSIS	M-52000	ARTERIOSCLEROSIS
ATROPHY	M-58000	ATROPHY
CARCINOMA, SQUAMOUS CELL	M-80703	SQUAMOUS CELL CARCINOMA
CONGESTION	M-36100	CONGESTION
CYST	M-33400	CYST
		MUCOCELE
DYSPLASIA, EPITHELIAL (T-50135)	M-74000	DYSPLASIA, EPITHELIAL
EDEMA	M-36500	EDEMA
FIBROMYXOMA	M-88110	FIBROMYXOMA
FIBROSARCOMA	M-88103	FIBROSARCOMA
FIBROSIS	M-49000	FIBROSIS
GRANULOMA	M-44000	GRANULOMA
HEMORRHAGE	M-37000	HEMORRHAGE
HYPERPLASIA	M-72000	HYPERPLASIA
		HYPERPLASIA, CYSTIC
HYPERPLASIA, LYMPHOID	M-72200	HYPERPLASIA, LYMPHOID
HYPERPLASIA, RETICULUM CELL	M-72180	HYPERPLASIA, RETICULUM CELL
IMPACTION	M-33100	IMPACTION
		IMPACTION, FECAL
INFLAMMATION	M-40000	INFLAMMATION
INFLAMMATION, GRANULOMATOUS	M-44000	INFLAMMATION, GRANULOMATOUS
INFLAMMATION, HEMORRHAGIC	M-40600	ENTERITIS, HEMORRHAGIC
INFLAMMATION, NONSUPPURATIVE	M-47130	COLITIS, LYMPHOCYTIC
		INFILTRATION, LYMPHOID
		INFLAMMATION, NONSUPPURATIVE
		TYPHLITIS, LYMPHOCYTIC
INFLAMMATION, SUPPURATIVE	M-40400	COLITIS, PURULENT
		ENTERITIS, NECROPURULENT
		INFLAMMATION, SUPPURATIVE
INFLAMMATION, ULCERATIVE	M-40030	ULCER
		ULCERATION
INTUSSUSCEPTION	M-31130	INTUSSUSCEPTION
LEIOMYOSARCOMA	M-88903	LEIOMYOSARCOMA
LEUKEMIA	M-98003	LEUKEMIC INFILTRATION
LEUKEMIA, LYMPHOCYTIC	M-98203	LYMPHATIC LEUKEMIA
LEUKEMIA, STEM CELL	M-98013	STEM CELL LEUKEMIA
LYMPHOMA	M-95903	LYMPHOMA
		MALIGNANT LYMPHOMA
		MALIGNANT LYMPHOMA, MIXED TYPE
		MALIGNANT LYMPHOMA, UNDIFFERENTIATED TYPE
LYMPHOMA, HISTIOCYTIC	M-96403	MALIGNANT LYMPHOMA, HISTIOCYTIC TYPE
LYMPHOMA, LYMPHOCYTIC	M-96203	MALIGNANT LYMPHOMA, LYMPHOCYTIC TYPE

PREFERRED TERM	SNOVET CODE	LADB TERMS
LYMPHOMA, RETICULUM CELL	M-96403	RETICULUM CELL SARCOMA
MESOTHELIOMA	M-90503	MESOTHELIOMA
MINERALIZATION	M-55400	CALCIFICATION
		CALCIFICATION, DYSTROPHIC
NECROSIS	M-54000	NECROSIS
NEUROFIBROSARCOMA	M-95403	NEUROFIBROSARCOMA
PARASITISM	D-06500	NEMATODIASIS
		NEMATODIASIS, INTESTINAL
		PARASITISM
		PARASITOSIS
PERIVASCULITIS/VASCULITIS	M-47800	PANARTERITIS
		PERIARTERITIS
		POLYARTERITIS
POLYP	M-76800	POLYP
POLYP, ADENOMATOUS	M-82100	ADENOMATOUS POLYP
PROLAPSE, RECTAL (T-68000)	M-31050	PROLAPSE
		PROLAPSE, RECTAL
RHABDOMYOSARCOMA	M-89003	RHABDOMYOSARCOMA
SARCOMA	M-88003	SARCOMA
SARCOMA, UNDIFFERENTIATED	M-88003	UNDIFFERENTIATED SARCOMA
THROMBUS	M-35100	THROMBUS

ORGAN SYSTEM: DIGESTIVE SYSTEM; ALIMENTARY TRACT ORGAN: MOUTH (T-51000)

PREFERRED TERM	SNOVET CODE	LADB TERMS
CARCINOMA, SQUAMOUS CELL	M-80703	SQUAMOUS CELL CARCINOMA
CONGESTION	M-36100	CONGESTION
CYST	M-33400	CYST
CYST, EPIDERMAL	M-33410	CYST, EPIDERMOID
FIBROSARCOMA	M-88103	FIBROSARCOMA
INFLAMMATION	M-40000	INFLAMMATION
LYMPHOMA	M-95903	LYMPHOMA
LYMPHOMA, RETICULUM CELL	M-96403	RETICULUM CELL SARCOMA
PAPILLOMA, SQUAMOUS CELL	M-80520	SQUAMOUS CELL PAPILLOMA

PREFERRED TERM	SNOVET CODE	LADB TERMS
ACANTHOSIS	M-72710	ACANTHOSIS
ADENOCARCINOMA	M-81403	ADENOCARCINOMA
		GASTRIC ADENOCARCINOMA
ADENOCARCINOMA, MUCINOUS	M-84803	MUCINOUS ADENOCARCINOMA
ADENOMA	M-81400	ADENOMA
ADENOMA, PAPILLARY	M-82600	PAPILLARY ADENOMA
AMYLOID DEPOSITION		AMYLOID
		AMYLOIDOSIS
ANGIOSARCOMA	M-91203	ANGIOSARCOMA
ATROPHY	M-58000	ATROPHY
CARCINOMA, ACINAR CELL	M-85503	ACINAR-CELL CARCINOMA
CARCINOMA, BASAL CELL	M-80903	BASAL-CELL CARCINOMA
CARCINOMA, C CELL	M-85106	C-CELL CARCINOMA
CARCINOMA, SQUAMOUS CELL	M-80703	SQUAMOUS CELL CARCINOMA
CYST	M-33400	CYST
		CYST, CONGENITAL MUCINOUS
		CYST, RETENTION
		CYSTIC CRYPTS
		DILATATION, CYSTIC
		EPIDERMAL INCLUSION CYST
DILATATION	M-32100	DILATATION
		DILATATION, GASTRIC
DIVERTICULITIS	M-46420	DIVERTICULITIS
DIVERTICULUM	M-32700	DIVERTICULUM
DYSPLASIA, EPITHELIAL	M-74000	DYSPLASIA, EPITHELIAL
EDEMA	M-36500	EDEMA
EROSION	M-14110	EROSION
FIBROSARCOMA	M-88103	FIBROSARCOMA
FIBROSIS	M-49000	FIBROSIS
GRANULOMA	M-44000	GRANULOMA
		GRANULOMA, FOREIGN BODY
HAMARTOMA	M-75500	HAMARTOMA
HEMORRHAGE	M-37000	HEMORRHAGE
HYPERKERATOSIS	M-72600	HYPERKERATOSIS
HYPERPLASIA	M-72000	HYPERPLASIA
HYPERPLASIA, ADENOMATOUS	M-72420	HYPERPLASIA, ADENOMATOUS
HYPERPLASIA, BASAL CELL	M-72120	HYPERPLASIA, BASAL CELL
HYPERPLASIA, EPITHELIAL	M-72000	HYPERPLASIA, EPITHELIAL
HYPERPLASIA, LYMPHOID	M-72200	HYPERPLASIA, LYMPHOID
HYPERPLASIA, MAST CELL (T-05140)	M-72000	MASTOCYTOSIS
HYPERPLASIA, PAPILLARY	M-72050	HYPERPLASIA, PAPILLARY
HYPERPLASIA, PLASMOCYTIC		HYPERPLASIA, PLASMOCYTIC
INFARCT	M-54700	INFARCT
INFLAMMATION	M-40000	GASTRITIS
		GASTRITIS, NONSPECIFIC
		INFLAMMATION
		INFLAMMATION, ALLERGIC
		INFLAMMATION, INTERSTITIAL
INFLAMMATION, GRANULOMATOUS	M-44000	INFLAMMATION, GRANULOMATOUS
INFLAMMATION, NECROTIZING	M-40700	INFLAMMATION, NECROTIZING

PREFERRED TERM	SNOVET CODE	LADB TERMS
INFLAMMATION, NONSUPPURATIVE	M-47130	INFLAMMATION, LYMPHOCYTIC
		INFLAMMATION, NONSUPPURATIVE
INFLAMMATION, SUPPURATIVE	M-40400	INFLAMMATION, SUPPURATIVE
INFLAMMATION, ULCERATIVE	M-40030	GASTRIC ULCER
		GASTRITIS, ULCERATIVE
		ULCER
		ULCER, GASTRIC
		ULCER, HEMORRHAGIC
		ULCERATION
LEIOMYOSARCOMA	M-88903	LEIOMYOSARCOMA
LEUKEMIA	M-98003	LEUKEMIA
		LEUKEMIC INFILTRATION
LYMPHOMA	M-95903	LYMPHOMA
		MALIGNANT LYMPHOMA
LYMPHOMA, HISTIOCYTIC	M-96403	MALIGNANT LYMPHOMA, HISTIOCYTIC TYPE
LYMPHOMA, RETICULUM CELL	M-96403	RETICULUM CELL SARCOMA
MAST CELL TUMOR	M-97401	MAST CELL TUMOR
MESOTHELIOMA	M-90503	MESOTHELIOMA
METAPLASIA	M-73000	METAPLASIA
MINERALIZATION	M-55400	CALCIFICATION
		CALCIFICATION, DYSTROPHIC
		CALCIUM DEPOSIT
		MINERALIZATION
		MINERALIZATION, DYSTROPHIC
MINERALIZATION, VASCULAR (T-40000)	M-55400	CALCIFICATION, MEDIAL
NECROSIS	M-54000	NECROSIS
PAPILLOMA	M-80500	PAPILLOMA
PAPILLOMA, SQUAMOUS CELL	M-80520	SQUAMOUS CELL PAPILLOMA
PERIVASCULITIS/VASCULITIS	M-47800	PERIARTERITIS
		POLYARTERITIS
POLYP	M-76800	POLYP
POLYP, ADENOMATOUS	M-82100	ADENOMATOUS POLYP
RHABDOMYOSARCOMA	M-89003	RHABDOMYOSARCOMA
SARCOMA	M-88003	SARCOMA
SARCOMA, HISTIOCYTIC	M-96403	HISTIOCYTIC SARCOMA

PREFERRED TERM	SNOVET CODE	LADB TERMS
ABSCESS	M-41740	ABSCESS
ACCESSORY ADRENAL TISSUE	M-22300	ACCESSORY ADRENAL TISSUE
		SUPERNUMERARY
ACCESSORY STRUCTURE	M-22300	ACCESSORY STRUCTURE
ADENOMA, CORTICAL	M-83700	ADENOMA
		CORTICAL ADENOMA
AGENESIS	M-21000	AGENESIS
AMYLOID DEPOSITION	M-55100	AMYLOIDOSIS
ANGIECTASIS	M-76300	ANGIECTASIS
		ECTASIA, VACUOLAR
		TELANGIECTASIS
ANGIOLIPOMA	M-88610	ANGIOLIPOMA
ANGIOMA	M-91200	HEMANGIOMA
ATROPHY	M-58000	ATROPHY
		ATROPHY, SECONDARY
ATROPHY, PRESSURE	M-58120	ATROPHY, PRESSURE
CARCINOMA	M-80103	CARCINOMA
CARCINOMA, ACINAR CELL	M-85503	ACINAR-CELL CARCINOMA
CARCINOMA, CORTICAL	M-82703	ADENOCARCINOMA
		CORTICAL CARCINOMA
CELLULAR CHANGE, VACUOLAR	M-66040	CYTOPLASMIC VACUOLIZATION
		DEGENERATION, HYDROPIC
		DEGENERATION, VACUOLAR
		VACUOLATION
CONGESTION	M-36100	CONGESTION
		CONGESTION, SINUSOIDAL
		ENGORGEMENT
CYST	M-33400	CYST
CYST, HEMORRHAGIC	M-33540	HEMATOCYST
		HEMORRHAGIC CYST
CYTOMEGALY	M-69220	CYTOMEGALY
DEGENERATION	M-50000	DEGENERATION
		DEGENERATION, CYTOLOGIC
		LIPOLYSIS
DEGENERATION, CYSTIC	M-50030	DEGENERATION, CYSTIC
DILATATION, SINUSOIDAL	M-32100	DILATATION, SINUSOIDAL
FATTY CHANGE	M-50080	DEGENERATION, FATTY
		INFILTRATION, FATTY
		METAMORPHOSIS, FATTY
FIBROSIS	M-49000	FIBROSIS
GANGLIONEUROBLASTOMA	M-94903	ADRENAL GANGLIONEUROBLASTOMA
GANGLIONEUROMA	M-94900	GANGLIONEUROMA
HEMATOPOIESIS	F-50010	GRANULOPOIESIS
		HEMATOPOIESIS
		HEMATOPOIESIS, EXTRAMEDULLARY
HEMORRHAGE	M-37000	HEMORRHAGE
HEMOSIDERIN DEPOSITION	M-57500	HEMOSIDEROSIS
HISTIOCYTOMA, MALIGNANT	M-88323	MALIGNANT HISTIOCYTOSIS
HYPERPLASIA	M-72000	HYPERPLASIA
		HYPERPLASIA, REGENERATIVE

PREFERRED TERM	SNOVET CODE	LADB TERMS
HYPERPLASIA, CORTICAL (T-93100)	M-72000	HYPERPLASIA, CORTICAL
HYPERPLASIA, MEDULLARY (T-93200)	M-72000	HYPERPLASIA, CHROMAFFIN CELL
HYPERPLASIA, NODULAR	M-72030	HYPERPLASIA, NODULAR
		NODULE
HYPERPLASIA, SPINDLE CELL	M-72000	HYPERPLASIA, SPINDLE CELL
HYPERTROPHY	M-71000	HYPERTROPHY
INFARCT	M-54700	INFARCT
INFLAMMATION	M-40000	INFLAMMATION
INFLAMMATION, NONSUPPURATIVE	M-47130	INFILTRATION, LYMPHOID
		LYMPHOCYTIC INFLAMMATORY INFILTRATE
		LYMPHOID CELLS
LEIOMYOSARCOMA	M-88903	LEIOMYOSARCOMA
LEUKEMIA	M-98003	LEUKEMIA
		LEUKEMIC INFILTRATION
LEUKEMIA, GRANULOCYTIC	M-98603	GRANULOCYTIC LEUKEMIA
LEUKEMIA, HISTIOCYTIC	M-98903	LEUKEMIC HISTIOCYTIC SARCOMA
LEUKEMIA, LYMPHOBLASTIC	M-98213	LYMPHOBLASTIC LEUKEMIA
LEUKEMIA, LYMPHOCYTIC	M-98203	LEUKEMIC LYMPHOSARCOMA
		LYMPHATIC LEUKEMIA
		LYMPHOCYTIC LEUKEMIA
LEUKEMIA, STEM CELL	M-98013	STEM CELL LEUKEMIA
LIPID DEPOSITION	M-55200	DEGENERATION, LIPOID
		LIPOIDOSIS
LYMPHOMA	M-95903	LYMPHOMA
		MALIGNANT LYMPHOMA
LYMPHOMA, LYMPHOCYTIC	M-96203	MALIGNANT LYMPHOMA, LYMPHOCYTIC TYPE
LYMPHOMA, RETICULUM CELL	M-96403	RETICULUM CELL SARCOMA
MASTOCYTOSIS	M-97413	MASTOCYTOSIS
MESOTHELIOMA	M-90503	MESOTHELIOMA
METAPLASIA, OSSEOUS	M-73400	METAPLASIA, OSSEOUS
MINERALIZATION	M-55400	CALCIFICATION
		CALCIFICATION, DYSTROPHIC
		CALCIUM DEPOSIT
NECROSIS	M-54000	NECROSIS
		NECROSIS, HEMORRHAGIC
NEUROFIBROSARCOMA	M-95403	NEUROFIBROSARCOMA
OSSIFICATION	F-89340	OSSIFICATION, HETEROPLASTIC
PERIADENITIS (T-93300)	M-40000	PERIADENITIS
PHEOCHROMOCYTOMA (T-93...)	M-87000	PHEOCHROMOCYTOMA
PHEOCHROMOCYTOMA, MALIGNANT (T-93...)	M-87003	PHEOCHROMOCYTOMA, MALIGNANT
PIGMENT DEPOSITION	M-57000	PIGMENTATION
PIGMENT DEPOSITION, LIPOFUSCIN	M-57740	LIPOFUSCIN
SYMPATHICOBLASTOMA	M-95003	SYMPATHICOBLASTOMA
THROMBUS	M-35100	THROMBOSIS
		THROMBUS
THROMBUS, ORGANIZED	M-35103	THROMBOSIS, ORGANIZED
THROMBUS, SINUSOIDAL		THROMBOSIS, SINUSOIDAL

PREFERRED TERM	SNOVET CODE	LADB TERMS
ABSCESS	M-41740	MICROABSCESS
ADENOCARCINOMA, MUCINOUS	M-84803	MUCINOUS ADENOCARCINOMA
ADENOMA	M-81400	ADENOMA
ADENOMA, ACINAR CELL	M-85501	ACINAR-CELL ADENOMA
ADENOMA, ISLET CELL	M-81500	ISLET CELL ADENOMA
		ISLET-CELL ADENOMA
AMYLOID DEPOSITION		AMYLOIDOSIS
ANGIOSARCOMA	M-91203	HEMANGIOSARCOMA
ARTERIOSCLEROSIS	M-52000	ARTERIOSCLEROSIS
ATROPHY	M-58000	ATROPHY
		ATROPHY, FATTY
		ATROPHY, LOBULAR
CARCINOMA	M-80103	CARCINOMA
CARCINOMA, ACINAR CELL	M-85503	ACINAR-CELL CARCINOMA
CARCINOMA, ISLET CELL	M-81503	ISLET CELL CARCINOMA
		ISLET-CELL CARCINOMA
CAST, PROTEINACEOUS	M-33030	CAST
CELLULAR CHANGE, VACUOLAR	M-66040	FOCUS OF VACUOLES
		VACUOLIZATION
CYST	M-33400	CYST
		CYSTIC DUCTS
CYST, EMBRYONAL DUCT		EMBRYONAL DUCT CYST
CYST, HEMORRHAGIC	M-33540	HEMORRHAGIC CYST
DEGENERATION	M-50000	DEGENERATION
		DEGENERATION, CYSTIC
		DEGENERATION, GRANULAR
		DEGENERATION, HYDROPIC
DILATATION	M-32100	DILATATION
		DISTENTION
ECTASIA, DUCT (T-59010)	M-32100	ECTASIA
		ECTASIA, DUCTAL
EDEMA	M-36500	EDEMA
FATTY CHANGE	M-50080	DEGENERATION, FATTY
		METAMORPHOSIS, FATTY
FIBROCYSTIC DISEASE	M-74320	DISEASE, FIBROCYSTIC
FIBROMA	M-81000	FIBROMA
FIBROSARCOMA	M-88103	FIBROSARCOMA
FIBROSIS	M-49000	FIBROSIS
HEMORRHAGE	M-37000	HEMORRHAGE
HEMOSIDERIN DEPOSITION	M-57500	HEMOSIDERIN
HISTIOCYTOMA, FIBROUS	M-88300	FIBROUS HISTIOCYTOMA
HYPERPLASIA	M-72000	HYPERPLASIA
HYPERPLASIA, ADENOMATOUS	M-72420	HYPERPLASIA, ADENOMATOUS
HYPERPLASIA, INTRADUCTAL	M-72170	HYPERPLASIA, DUCTULAR
		HYPERPLASIA, INTRADUCTAL
HYPERPLASIA, ISLET CELL (T-9900C)	M-72000	HYPERPLASIA, ISLET CELL
HYPERPLASIA, LYMPHOID	M-72200	HYPERPLASIA, LYMPHOID
HYPERPLASIA, NODULAR	M-72030	HYPERPLASIA, NODULAR
		HYPERPLASIA, NODULAR, ADENOMATOID
HYPERTROPHY	M-71000	HYPERTROPHY

PREFERRED TERM	SNOVET CODE	LADB TERMS
INFARCT	M-54700	INFARCT
INFILTRATION, FATTY		INFILTRATION, FATTY
INFLAMMATION	M-40000	INFLAMMATION
		INFLAMMATION, INTERSTITIAL
		PANCREATITIS
		PANCREATITIS, INTERSTITIAL
		PANCREATITIS, RELAPSING
		PANCREATITIS, RESOLVING
		PANCREATITIS, UREMIC
		PERIPANCREATITIS
INFLAMMATION, EOSINOPHILIC		PANCREATITIS WITH INFILTRATION BY EOSINOPHIL
INFLAMMATION, FAT (T-1X010)	M-40000	STEATITIS
INFLAMMATION, GRANULOMATOUS	M-44000	INFLAMMATION, GRANULOMATOUS
		INFLAMMATION, PYOGRANULOMATOUS
INFLAMMATION, NONSUPPURATIVE	M-47130	INFILTRATION, LYMPHOID
		INFILTRATION, LYMPHOPLASMACYTIC
		INFILTRATION, MONONUCLEAR CELL
		INFILTRATION, PLASMA CELL
		INFLAMMATION, LYMPHOCYTIC
		INFLAMMATION, NONSUPPURATIVE
		LYMPHOCYTES, FOCUS OF
		LYMPHOCYTIC INFLAMMATORY INFILTRATE
		LYMPHOCYTOSIS
		LYMPHOID COLLECTION
		PANCREATITIS, LYMPHOCYTIC
		PANCREATITIS, NONSUPPURATIVE
INFLAMMATION, SUPPURATIVE	M-40400	INFLAMMATION, SUPPURATIVE
ISLET CELL TUMOR (T-99...)	M-81500	ISLET CELL TUMOR
		PANCREATIC ISLET CELL TUMOR
LEIOMYOSARCOMA	M-88903	LEIOMYOSARCOMA
LEUKEMIA	M-98003	LEUKEMIA
		LEUKEMIC INFILTRATION
LEUKEMIA, GRANULOCYTIC	M-98603	GRANULOCYTIC LEUKEMIA
LEUKEMIA, LYMPHOBLASTIC	M-98213	LYMPHOBLASTIC LEUKEMIA
LEUKEMIA, LYMPHOCYTIC	M-98203	LEUKEMIC LYMPHOSARCOMA
		LYMPHATIC LEUKEMIA
LYMPHOMA	M-95903	LYMPHOMA
		LYMPHOSARCOMATOUS INFILTRATION
		MALIGNANT LYMPHOMA
LYMPHOMA, RETICULUM CELL	M-96403	RETICULUM CELL SARCOMA
MESOTHELIOMA	M-90503	MESOTHELIOMA
MINERALIZATION	M-55400	CALCIFICATION
MINERALIZATION, VASCULAR (T-40000)	M-55400	CALCIFICATION, MEDIAL
NECROSIS	M-54000	NECROSIS
NECROSIS, FAT	M-54110	NECROSIS, FAT
		NECROSIS, LIPID
NEUROFIBROSARCOMA	M-95403	NEUROFIBROSARCOMA
PARASITISM	D-06500	PARASITISM
PERIVASCULITIS/VASCULITIS	M-47800	ARTERIOLITIS
		PANARTERITIS

PREFERRED TERM	SNOVET CODE	LADB TERMS
PERIVASCULITIS/VASCULITIS	M-47800	PERIARTERITIS
		PERIARTERITIS NODOSA
		PERIARTERITIS, NONSUPPURATIVE
		PERIVASCULITIS
		POLYARTERITIS
RHABDOMYOSARCOMA	M-89003	RHABDOMYOSARCOMA
SARCOMA, HISTIOCYTIC	M-96403	HISTIOCYTIC SARCOMA
THROMBUS	M-35100	THROMBOSIS
		THROMBUS

PREFERRED TERM	SNOVET CODE	LADB TERMS
ADENOMA	M-81400	ADENOMA
		PARATHYROID ADENOMA
ADENOMA, CHIEF CELL	M-83210	CHIEF CELL ADENOMA
AMYLOID DEPOSITION		AMYLOIDOSIS
CYST	M-33400	CYST
HYPERPLASIA	M-72000	HYPERPLASIA
		HYPERPLASIA, BENIGN
HYPERPLASIA, ADENOMATOUS	M-72420	HYPERPLASIA, ADENOMATOUS
HYPERPLASIA, NODULAR	M-72030	HYPERPLASIA, NODULAR
HYPERPLASIA, SECONDARY	M-72020	HYPERPLASIA, SECONDARY
LEUKEMIA, GRANULOCYTIC	M-98603	GRANULOCYTIC LEUKEMIA
LEUKEMIA, LYMPHOCYTIC	M-98203	LEUKEMIC LYMPHOSARCOMA
MINERALIZATION	M-55400	CALCIFICATION

ORGAN SYSTEM: ENDOCRINE SYSTEM   ORGAN: PINEAL GLAND (T-92000)

PREFERRED TERM	SNOVET CODE	LADB TERMS
MINERALIZATION	M-55400	CALCIFICATION

PREFERRED TERM	SNOVET CODE	LADB TERMS
ADENOCARCINOMA	M-81403	ADENOCARCINOMA
ADENOMA	M-81400	ADENOMA
		SINUSOIDAL ADENOMA
ADENOMA, ACIDOPHIL	M-82800	ACIDOPHIL ADENOMA
ADENOMA, BASOPHIL	M-83000	BASOPHIL ADENOMA
ADENOMA, CHROMOPHOBE	M-82700	CHROMOPHOBE ADENOMA
		CHROMOPHOBE ADENOMA, MIXED TYPE
		CHROMOPHOBE ADENOMA, SINUSOIDAL TYPE
ANGIECTASIS	M-76300	ANGIECTASIS
		TELANGIECTASIS
CARCINOMA	M-80103	CARCINOMA
CARCINOMA, CHROMOPHOBE	M-82703	CHROMOPHOBE ADENOCARCINOMA
		CHROMOPHOBE CARCINOMA
CELLULAR CHANGE, VACUOLAR	M-66040	CYTOPLASMIC VACUOLIZATION
		VACUOLIZATION
CONGESTION	M-36100	CONGESTION
CYST	M-33400	CYST
		MULTIPLE CYSTS
CYST, COLLOID	M-33790	CYST, COLLOID
CYST, HEMORRHAGIC	M-33540	HEMORRHAGIC CYST
EPENDYMOMA (T-X16...)	M-93913	EPENDYMOMA
HEMORRHAGE	M-37000	HEMORRHAGE
HEMOSIDERIN DEPOSITION	M-57500	HEMOSIDEROSIS
HYPERPLASIA	M-72000	HYPERPLASIA
HYPERPLASIA, ADENOMATOUS	M-72420	HYPERPLASIA, ADENOMATOUS
HYPERPLASIA, BASOPHILIC (T-91130)	M-72000	HYPERPLASIA, BASOPHILIC
HYPERPLASIA, CHROMOPHOBE CELL (T-91170)	M-72000	HYPERPLASIA, CHROMOPHOBE
		HYPERPLASIA, CHROMOPHOBE CELL
HYPERPLASIA, EPITHELIAL	M-72000	HYPERPLASIA, GLANDULAR, CYSTIC
HYPERPLASIA, NODULAR	M-72030	HYPERPLASIA, NODULAR
HYPERTROPHY	M-71000	ENLARGED
		HYPERTROPHY
INFARCT, HEMORRHAGIC	M-54730	INFARCT, HEMORRHAGIC
INFLAMMATION	M-40000	INFLAMMATION
LEUKEMIA	M-98003	LEUKEMIA
LEUKEMIA, LYMPHOCYTIC	M-98203	LEUKEMIC LYMPHOSARCOMA
LEUKEMIA, STEM CELL	M-98013	STEM CELL LEUKEMIA
LYMPHOMA	M-95903	LYMPHOMA
		MALIGNANT LYMPHOMA
LYMPHOMA, RETICULUM CELL	M-96403	RETICULUM CELL SARCOMA
MINERALIZATION	M-55400	CALCIFICATION
		MINERALIZATION
NECROSIS	M-54000	NECROSIS
		NECROSIS, HEMORRHAGIC
NODULE	M-03010	NODULE
PERIVASCULITIS/VASCULITIS	M-47800	PERIVASCULITIS
PIGMENT DEPOSITION	M-57000	PIGMENTATION

PREFERRED TERM	SNOVET CODE	LADB TERMS
ADENOCARCINOMA	M-81403	ADENOCARCINOMA
ADENOCARCINOMA, PAPILLARY	M-82603	PAPILLARY ADENOCARCINOMA
ADENOMA	M-81400	ADENOMA
		FETAL ADENOMA
ADENOMA, C CELL (T-96050)	M-81400	C-CELL ADENOMA
		C-CELL TUMOR
ADENOMA, FOLLICULAR CELL	M-83300	FOLLICULAR ADENOMA
		FOLLICULAR-CELL ADENOMA
ADENOMA, HURTHLE CELL (T-96...)	M-82900	HURTHLE CELL ADENOMA
		HURTHLE CELL TUMOR
ADENOMA, PAPILLARY	M-82600	PAPILLARY ADENOMA
AMYLOID DEPOSITION		AMYLOIDOSIS
ATROPHY	M-58000	ATROPHY
		ATROPHY, SENILE
CARCINOMA	M-80103	CARCINOMA
CARCINOMA, C CELL	M-85103	C-CELL CARCINOMA
CARCINOMA, FOLLICULAR CELL	M-83303	FOLLICULAR CARCINOMA
		FOLLICULAR-CELL CARCINOMA
CARCINOMA, HURTHLE CELL (T-96...)	M-82903	HURTHLE CELL CARCINOMA
CARCINOMA, PAPILLARY	M-80503	PAPILLARY CARCINOMA
CARCINOMA, UNDIFFERENTIATED	M-80203	UNDIFFERENTIATED CARCINOMA
CYST	M-33400	CYST
CYST, COLLOID	M-33790	COLLOID CYST
CYST, FOLLICULAR	M-33500	CYSTIC FOLLICLES
		FOLLICULAR CYST
CYST, THYROGLOSSAL DUCT (T-96500)	M-26500	CYST, THYROGLOSSAL DUCT
CYST, ULTIMOBRANCHIAL DUCT (T-89502)	M-33400	CYST, ULTIMOBRANCHIAL DUCT
		ULTIMOBRANCHIAL CYST
CYSTADENOCARCINOMA, PAPILLARY	M-84503	PAPILLARY CYSTADENOCARCINOMA
CYSTADENOMA	M-84400	CYSTADENOMA
		PAPILLARY CYST ADENOMA
DEGENERATION, CYSTIC	M-50030	DEGENERATION, CYSTIC
FIBROSIS	M-49000	FIBROSIS
		SCLEROSIS
HYPERPLASIA	M-72000	HYPERPLASIA
HYPERPLASIA, ADENOMATOUS	M-72420	HYPERPLASIA, ADENOMATOUS
HYPERPLASIA, C CELL (T-96050)	M-72000	HYPERPLASIA, C-CELL
HYPERPLASIA, CYSTIC	M-72060	HYPERPLASIA, CYSTIC
HYPERPLASIA, FOLLICULAR CELL (T-96030)	M-72000	HYPERPLASIA, FOLLICULAR CELL
HYPERPLASIA, NODULAR	M-72030	NODULE
		SCATTERED SOLID EPITHELIAL NODULES
HYPERPLASIA, PAPILLARY	M-72050	HYPERPLASIA, PAPILLARY
HYPERTROPHY	M-71000	HYPERTROPHY
INFLAMMATION	M-40000	INFLAMMATION
INFLAMMATION, NECROTIZING	M-40700	INFLAMMATION, NECROTIZING
INFLAMMATION, NONSUPPURATIVE	M-47130	FOCUS OF LYMPHOCYTES
		INFLAMMATION, LYMPHOCYTIC
		LYMPHOCYTIC INFLAMMATORY INFILTRATE
		THYROIDITIS, LYMPHOCYTIC
LEUKEMIA, GRANULOCYTIC	M-98603	GRANULOCYTIC LEUKEMIA

PREFERRED TERM	SNOVET CODE	LADB TERMS
LEUKEMIA, LYMPHOCYTIC	M-98203	LEUKEMIC LYMPHOSARCOMA
LYMPHOMA	M-95903	LYMPHOMA
		MALIGNANT LYMPHOMA
LYMPHOMA, RETICULUM CELL	M-96403	RETICULUM CELL SARCOMA
METAPLASIA, SQUAMOUS	M-73220	METAPLASIA, SQUAMOUS
NEUROFIBROSARCOMA	M-95403	NEUROFIBROSARCOMA
PERIVASCULITIS/VASCULITIS	M-47800	PERIARTERITIS
PIGMENT DEPOSITION	M-57000	PIGMENT
		PIGMENTATION
RHABDOMYOSARCOMA	M-89003	RHABDOMYOSARCOMA

PREFERRED TERM	SNOVET CODE	LADB TERMS
ACANTHOSIS	M-72710	ACANTHOSIS
CHOLESTEROL DEPOSITION	M-55240	CHOLESTEROL DEPOSIT
CYST	M-33400	CYST
HYPERPLASIA, BASAL CELL	M-72120	HYPERPLASIA, BASAL CELL
INFLAMMATION	M-40000	INFLAMMATION
INFLAMMATION, SUPPURATIVE	M-40400	INFLAMMATION, SUPPURATIVE
PAPILLOMA, SQUAMOUS CELL	M-80520	SQUAMOUS CELL PAPILLOMA
POLYP, INFLAMMATORY	M-76820	POLYP, INFLAMMATORY

PREFERRED TERM	SNOVET CODE	LADB TERMS
ABSCESS	M-41740	ABSCESS
ADENOCARCINOMA	M-81403	ADENOCARCINOMA
ADENOMA	M-81400	ADENOMA
CARCINOMA	M-80103	CARCINOMA
INFLAMMATION, GRANULOMATOUS	M-44000	INFLAMMATION, GRANULOMATOUS
METAPLASIA, SQUAMOUS	M-73220	METAPLASIA, SQUAMOUS
PAPILLOMA, SQUAMOUS CELL	M-80520	SQUAMOUS CELL PAPILLOMA

PREFERRED TERM	SNOVET CODE	LADB TERMS
ABSCESS	M-41740	ABSCESS
ADENOCARCINOMA	M-81403	ADENOCARCINOMA
		CYSTADENOCARCINOMA
ADENOMA	M-81400	ADENOMA
		PAPILLARY ADENOMA
		PAPILLARY CYSTADENOMA
ADENOMA, FOLLICULAR CELL (T-96...)	M-83300	FOLLICULAR-CELL ADENOMA
ADENOMA, TUBULAR	M-82110	TUBULAR ADENOMA
ANGIECTASIS	M-76300	ANGIECTASIS
ANGIOMA	M-91200	ANGIOMA
ARRHENOBLASTOMA	M-86301	ARRHENOBLASTOMA
ATROPHY	M-58000	ATROPHY
CARCINOMA	M-80103	CARCINOMA
CARCINOMA, GRANULOSA CELL (T-87..)	M-86203	GRANULOSA-CELL CARCINOMA
CONGESTION	M-36100	CONGESTION
CORPORA AMYLACEA	M-55030	CORPORA AMYLACEA
CYST	M-33400	CYST
		CYST, SOLITARY
		MULTILOCULAR CYST
		SEROUS CYSTOMA
CYST, DEGENERATIVE	M-50030	DEGENERATION, CYSTIC
CYST, FOLLICULAR	M-33500	CYST, FOLLICULAR
		CYSTIC FOLLICLES
		FOLLICULAR CYST
CYST, HEMORRHAGIC	M-33540	HEMORRHAGIC CYST
CYST, LUTEAL (T-87670)	M-33520	CORPUS LUTEUM CYST
CYST, MESOTHELIAL	M-33680	CYST, MESOTHELIAL
CYST, PARAOVARIAN (T-86510)	M-33400	PARAOVARIAN CYST
CYSTADENOMA	M-84400	CYSTADENOMA
DYSGERMINOMA	M-90603	SEMINOMA/DYSGERMINOMA
FIBROMA	M-81000	FIBROMA
FIBROSIS	M-49000	FIBROSIS
GRANULOSA CELL TUMOR (T-87...)	M-86201	GRANULOSA CELL TUMOR
		GRANULOSA-CELL TUMOR
GRANULOSA THECAL CELL TUMOR (T-87...)	M-86211	LUTEAL GRANULOSA THECAL CELL TUMOR
HAMARTOMA	M-75500	HAMARTOMA
HEMORRHAGE	M-37000	HEMORRHAGE
HYPERPLASIA	M-72000	HYPERPLASIA
HYPERPLASIA, GRANULOSA CELL (T-87660)	M-72000	HYPERPLASIA, GRANULOSA-CELL
HYPERPLASIA, INTERSTITIAL CELL (T-87710)	M-72000	HYPERPLASIA, INTERSTITIAL CELL
INFARCT	M-54700	INFARCT
INFLAMMATION	M-40000	INFLAMMATION
INFLAMMATION, GRANULOMATOUS	M-44000	INFLAMMATION, GRANULOMATOUS
INFLAMMATION, HEMORRHAGIC	M-40600	INFLAMMATION, HEMORRHAGIC
INFLAMMATION, NONSUPPURATIVE	M-47130	LYMPHOCYTIC INFLAMMATORY INFILTRATE
INFLAMMATION, SUPPURATIVE	M-40400	INFLAMMATION, SUPPURATIVE
LEIOMYOSARCOMA	M-88903	LEIOMYOSARCOMA
LEUKEMIA	M-98003	LEUKEMIA
LEUKEMIA, GRANULOCYTIC	M-98603	GRANULOCYTIC LEUKEMIA
LEUKEMIA, LYMPHOBLASTIC	M-98213	LYMPHOBLASTIC LEUKEMIA

PREFERRED TERM	SNOVET CODE	LADB TERMS
LEUKEMIA, LYMPHOCYTIC	M-98203	LYMPHATIC LEUKEMIA
LIPOMA	M-88500	LIPOMA
LUTEOMA (T-87...)	M-86100	LUTEOMA
LYMPHOMA	M-95903	LYMPHOMA
		MALIGNANT LYMPHOMA
		MALIGNANT LYMPHOMA, UNDIFFERENTIATED TYPE
LYMPHOMA, HISTIOCYTIC	M-96403	MALIGNANT LYMPHOMA, HISTIOCYTIC TYPE
LYMPHOMA, RETICULUM CELL	M-96403	RETICULUM CELL SARCOMA
MINERALIZATION	M-55400	MINERALIZATION
NECROSIS	M-54000	NECROSIS
NECROSIS, FAT	M-54110	NECROSIS, FAT
PAPILLOMA	M-80500	PAPILLOMA
PERIVASCULITIS/VASCULITIS	M-47800	PERIARTERITIS
PIGMENT DEPOSITION	M-57000	PIGMENTATION
POLYCISTIC OVARY (T-87000)	M-33800	POLYCYSTIC OVARY
SERTOLI CELL TUMOR	M-86400	SERTOLI-CELL TUMOR
THECOMA (T-87...)	M-86000	THECOMA
THROMBUS	M-35100	THROMBOSIS

PREFERRED TERM	SNOVET CODE	LADB TERMS
ABSCESS	M-41740	ABSCESS
ADENOMA, PAPILLARY	M-82600	PAPILLARY ADENOMA
CYST	M-33400	CYST
DEGENERATION	M-50000	DEGENERATION
FIBROSIS	M-49000	FIBROSIS
GRANULOMA	M-44000	GRANULOMA
INFLAMMATION	M-40000	INFLAMMATION
INFLAMMATION, NONSUPPURATIVE	M-47130	LYMPHOCYTIC INFLAMMATORY INFILTRATE
INFLAMMATION, SUPPURATIVE	M-40400	INFLAMMATION, SUPPURATIVE
		SALPINGITIS, PURULENT
LYMPHOMA	M-95903	LYMPHOMA
LYMPHOMA, RETICULUM CELL	M-96403	RETICULUM CELL SARCOMA
NECROSIS	M-54000	NECROSIS
NEOPLASM	M-80001	NEOPLASM
PAPILLOMA, INTRADUCTAL	M-85030	INTRADUCTAL PAPILLOMA
RETENTION FLUID	M-33300	RETENTION FLUID

PREFERRED TERM	SNOVET CODE	LADB TERMS
ABSCESS	M-41740	ABSCESS
		ABSCESS, SECONDARY
ADENOCARCINOMA	M-81403	ADENOCARCINOMA
ADENOMA	M-81400	ADENOMA
ADENOMA, PAPILLARY	M-82600	PAPILLARY ADENOMA
AMYLOID DEPOSITION		AMYLOID
ANGIECTASIS	M-76300	ANGIECTASIS
ANGIOMA	M-91200	ANGIOMA
		HEMANGIOMA
ANGIOSARCOMA	M-91203	ANGIOSARCOMA
		HEMANGIOENDOTHELIOMA
		HEMANGIOSARCOMA
CARCINOMA	M-80103	CARCINOMA
CARCINOMA, TRANSITIONAL CELL	M-81203	TRANSITIONAL-CELL CARCINOMA
CYST	M-33400	CHANGE, CYSTIC
		CYST
		MULTILOCULAR CYST
		MULTIPLE CYSTS
CYST, HEMORRHAGIC	M-33540	HEMORRHAGIC CYST
DEGENERATION, CYSTIC	M-50030	DEGENERATION, CYSTIC
DILATATION	M-32100	DILATATION
EDEMA	M-36500	EDEMA
ENDOMETRIAL STROMAL POLYP (T-84000)	M-72040	ENDOMETRIAL POLYP
		ENDOMETRIAL STROMAL POLYP
		POLYPOID HYPERPLASIA
EROSION	M-14110	EROSION
FIBROMA	M-81000	FIBROMA
FIBROSARCOMA	M-88103	FIBROSARCOMA
FIBROSIS	M-49000	FIBROSIS
HEMATOMA	M-37100	HEMATOMA
HEMORRHAGE	M-37000	HEMOMETRA
		HEMORRHAGE
HEMOSIDERIN DEPOSITION	M-57500	HEMOSIDERIN
		HEMOSIDEROSIS
HYDROMETRA	M-33300	HYDROMETRA
		HYDROMETRIA
HYPERPLASIA	M-72000	HYPERPLASIA
HYPERPLASIA, ADENOMATOUS	M-72420	ADENOMYOSIS
		ADENOSIS
		HYPERPLASIA, ADENOMATOUS
HYPERPLASIA, CYSTIC	M-72060	HYPERPLASIA, CYSTIC
HYPERPLASIA, ENDOMETRIAL, CYSTIC (T-84000)	M-72060	HYPERPLASIA, CYSTIC ENDOMETRIAL
		HYPERPLASIA, GLANDULAR, CYSTIC
HYPERPLASIA, EPITHELIAL	M-72000	HYPERPLASIA, EPITHELIAL
HYPERPLASIA, NODULAR	M-72030	ENDOMETRIAL NODULE
HYPERPLASIA, STROMAL	M-72430	HYPERPLASIA, STROMAL
HYPERTROPHY	M-71000	HYPERTROPHY
INFLAMMATION	M-40000	ENDOMETRITIS
		ENDOMETRITIS, CYSTIC
		ENDOMETRITIS, CYSTIC, POLYPOID

PREFERRED TERM	SNOVET CODE	LADB TERMS
INFLAMMATION	M-40000	INFLAMMATION
		INFLAMMATION, VESICULAR
		METRITIS, CYSTIC
		MYOMETRITIS
INFLAMMATION, GRANULOMATOUS	M-44000	INFLAMMATION, GRANULOMATOUS
		INFLAMMATION, PYOGRANULOMATOUS
		LIPOGRANULOMA
INFLAMMATION, HEMORRHAGIC	M-40600	ENDOMETRITIS, HEMORRHAGIC
INFLAMMATION, NECROTIZING	M-40700	INFLAMMATION, NECROTIZING
INFLAMMATION, NONSUPPURATIVE	M-47130	INFLAMMATION, LYMPHOCYTIC
		LYMPHOCYTIC INFLAMMATORY INFILTRATE
INFLAMMATION, SUPPURATIVE	M-40400	ENDOMETRITIS, SUPPURATIVE
		ENDOMETRITIS, SUPPURATIVE, CYSTIC
		INFLAMMATION, SUPPURATIVE
		METRITIS, SUPPURATIVE
INFLAMMATION, ULCERATIVE	M-40030	ULCER
LEIOMYOMA	M-88900	LEIOMYOMA
		LEIOMYOMA, CYSTIC
LEIOMYOSARCOMA	M-88903	LEIOMYOSARCOMA
LEUKEMIA	M-98003	LEUKEMIA
LEUKEMIA, LYMPHOCYTIC	M-98203	LYMPHATIC LEUKEMIA
		LYMPHOCYTIC LEUKEMIA
LIPOMA	M-88500	LIPOMA
LYMPHOMA	M-95903	LYMPHOMA
		MALIGNANT LYMPHOMA
		MALIGNANT LYMPHOMA, UNDIFFERENTIATED TYPE
LYMPHOMA, HISTIOCYTIC	M-96403	MALIGNANT LYMPHOMA, HISTIOCYTIC TYPE
LYMPHOMA, RETICULUM CELL	M-96403	RETICULUM CELL SARCOMA
METAPLASIA	M-73000	METAPLASIA
METAPLASIA, OSSEOUS	M-73400	METAPLASIA, OSSEOUS
METAPLASIA, SQUAMOUS	M-73220	METAPLASIA, SQUAMOUS
MINERALIZATION	M-55400	CALCIFICATION
NECROSIS	M-54000	NECROSIS
NECROSIS, FAT	M-54110	NECROSIS, FAT
NEOPLASM, MALIGNANT	M-80003	NEOPLASM, MALIGNANT
NEURILEMOMA		NEURILEMOMA
PIGMENT DEPOSITION	M-57000	PIGMENTATION
POLYP	M-76800	POLYP
		POLYPOSIS
POLYP, INFLAMMATORY	M-76820	POLYP, INFECTED
		POLYP, INFLAMMATORY
PYOMETRA	M-40460	PYOMETRA
SARCOMA	M-88003	SARCOMA
SARCOMA, ENDOMETRIAL STROMAL	M-89303	ENDOMETRIAL STROMAL SARCOMA
THROMBUS	M-35100	THROMBOSIS
THROMBUS, ORGANIZED	M-35103	THROMBUS, ORGANIZED

PREFERRED TERM	SNOVET CODE	LADB TERMS
CARCINOMA, SQUAMOUS CELL	M-80703	SQUAMOUS CELL CARCINOMA
CYST	M-33400	CYST
CYST, EPIDERMAL INCLUSION	M-33410	EPIDERMAL INCLUSION CYST
DYSPLASIA, EPIDERMOID	M-74000	DYSPLASIA, EPIDERMOID
EDEMA	M-36500	EDEMA
FIBROSARCOMA	M-88103	FIBROSARCOMA
FIBROSIS	M-49000	FIBROSIS
HYPERPLASIA, EPITHELIAL	M-72000	HYPERPLASIA, EPITHELIAL
INFARCT	M-54700	INFARCT
INFLAMMATION	M-40000	INFLAMMATION
INFLAMMATION, NONSUPPURATIVE	M-47130	INFLAMMATION, LYMPHOCYTIC
INFLAMMATION, SUPPURATIVE	M-40400	INFLAMMATION, SUPPURATIVE
INFLAMMATION, ULCERATIVE	M-40030	ULCERATION
LEIOMYOSARCOMA	M-88903	LEIOMYOSARCOMA
LYMPHANGIOSARCOMA	M-91703	LYMPHANGIOSARCOMA
LYMPHOMA	M-95903	LYMPHOMA
LYMPHOMA, RETICULUM CELL	M-96403	RETICULUM CELL SARCOMA
MYOSARCOMA	M-88953	MYOSARCOMA
NECROSIS	M-54000	NECROSIS
NECROSIS, FAT	M-54110	NECROSIS, FAT
PAPILLOMA	M-80500	PAPILLOMA
POLYP	M-76800	POLYP
POLYP, INFLAMMATORY	M-76820	POLYP, INFLAMMATORY
PROLAPSE, UTERINE (T-82000)	M-31050	PROLAPSE, UTERINE
PROLAPSE, VAGINAL	M-31050	PROLAPSE, VAGINAL
SARCOMA	M-88003	SARCOMA
SARCOMA, ENDOMETRIAL STROMAL	M-89306	ENDOMETRIAL STROMAL SARCOMA

 ORGAN SYSTEM: GENITAL SYSTEM; MALES ORGAN: DUCTUS DEFERENS (T-79200)

PREFERRED TERM	SNOVET CODE	LADB TERMS
ABSCESS	M-41740	ABSCESS
DYSPLASIA	M-74000	DYSPLASIA
GRANULOMA	M-44000	GRANULOMA
INFLAMMATION	M-40000	INFLAMMATION
INFLAMMATION, NONSUPPURATIVE	M-47130	INFLAMMATION, NONSUPPURATIVE
KARYOMEGALY	M-65110	KARYOMEGALY

PREFERRED TERM	SNOVET CODE	LADB TERMS
ABSCESS	M-41740	ABSCESS
ADENOCARCINOMA	M-81403	ADENOCARCINOMA
ATROPHY	M-58000	ATROPHY
CELLULAR CHANGE, VACUOLAR	M-66040	VACUOLATION
		VACUOLIZATION
CYST	M-33400	CYST
		DISTENTION, CYSTIC
DILATATION, TUBULAR	M-32100	DILATATION
DYSPLASIA	M-74000	DYSPLASIA
FIBROSIS	M-49000	FIBROSIS
GRANULOMA	M-44000	GRANULOMA, SPERMATIC
		SPERMATIC GRANULOMA
HEMORRHAGE	M-37000	HEMORRHAGE
HYPERPLASIA	M-72000	HYPERPLASIA
HYPERTROPHY	M-71000	HYPERTROPHY
INFLAMMATION	M-40000	EPIDIDYMITIS
		INFLAMMATION
INFLAMMATION, FAT (T-1X010)	M-40000	STEATITIS
INFLAMMATION, NONSUPPURATIVE	M-47130	EPIDIDYMITIS, LYMPHOCYTIC
		EPIDIDYMITIS, NONSUPPURATIVE
		INFLAMMATION, LYMPHOCYTIC
		LYMPHOCYTIC INFLAMMATORY INFILTRATE
INFLAMMATION, SUPPURATIVE	M-40400	INFLAMMATION, SUPPURATIVE
INTERSTITIAL CELL TUMOR	M-86506	INTERSTITIAL-CELL TUMOR
KARYOMEGALY	M-65110	KARYOMEGALY
LIPOMA	M-88500	LIPOMA
LYMPHOMA	M-95903	LYMPHOMA
LYMPHOMA, RETICULUM CELL	M-96403	RETICULUM CELL SARCOMA
MESOTHELIOMA	M-90503	MESOTHELIOMA
MINERALIZATION	M-55400	CALCIFICATION
		CALCIUM DEPOSIT
		MINERALIZATION, DYSTROPHIC
MYOBLASTOMA, GRANULAR CELL	M-95800	GRANULAR CELL MYOBLASTOMA
NECROSIS, FAT	M-54110	NECROSIS, FAT
NEUROFIBROSARCOMA	M-95403	NEUROFIBROSARCOMA
PERIVASCULITIS/VASCULITIS	M-47800	PERIARTERITIS
PIGMENT DEPOSITION	M-57000	PIGMENTATION
SPERMATOCELE (T-78300)'	M-33400	SPERMATOCELE

PREFERRED TERM	SNOVET CODE	LADB TERMS
ABSCESS	M-41740	ABSCESS
ADENOCARCINOMA, MUCINOUS	M-84803	MUCINOUS ADENOCARCINOMA
ADENOMA, ACINAR CELL	M-85501	ACINAR-CELL ADENOMA
ADENOMA, PAPILLARY	M-82600	PAPILLARY ADENOMA
ANGIOMA	M-91200	HEMANGIOMA
ANGIOSARCOMA	M-91203	HEMANGIOSARCOMA
ATROPHY	M-58000	ATROPHY
CALCULI	M-30000	CALCULUS
		CRYSTALS
		LITHIASIS
CARCINOMA, ACINAR CELL	M-85503	ACINAR-CELL CARCINOMA
CARCINOMA, PROSTATE (T-77100)	M-80103	PROSTATIC CARCINOMA
CAST	M-33010	CAST
CHOLESTEROL DEPOSITION	M-55240	CHOLESTEROL DEPOSIT
CYST	M-33400	CYST
		DILATATION, CYSTIC
CYTOMEGALY	M-69220	CYTOMEGALY
DEGENERATION	M-50000	DEGENERATION
DILATATION	M-32100	DILATATION
		DISTENTION
FIBROMA	M-81000	FIBROMA
FIBROSIS	M-49000	FIBROSIS
HEMORRHAGE	M-37000	HEMORRHAGE
HERNIATION, SCROTAL (T-79400)	M-31600	HERNIATION, SCROTAL
HYPERPLASIA	M-72000	HYPERPLASIA
HYPERPLASIA, ADENOMATOUS	M-72420	HYPERPLASIA, ADENOMATOUS
HYPERPLASIA, CYSTIC	M-72060	HYPERPLASIA, CYSTIC
HYPERPLASIA, EPITHELIAL	M-72000	HYPERPLASIA, EPITHELIAL
		HYPERPLASIA, PAPILLARY
HYPERPLASIA, LYMPHOID	M-72200	HYPERPLASIA, LYMPHOID
		LYMPHOCYTIC NODULE
HYPERPLASIA, NODULAR	M-72030	HYPERPLASIA, NODULAR
HYPERTROPHY	M-71000	HYPERTROPHY
INFLAM. NONSUP., SEMINAL VESICLE (T-77500)	M-47130	SEMINAL VESICULITIS, NONSUPPURATIVE
INFLAM. WITH FIBROSIS, PROSTATE (T-77100)	M-45000	PROSTATITIS, FIBROSING
INFLAM., NONSUP., PROSTATE (T-77100)	M-47130	PROSTATITIS, LYMPHOCYTIC
		PROSTATITIS, NONSUPPURATIVE
INFLAM., SUPP., SEMINAL VESICLE (T-77500)	M-40400	SEMINAL VESICULITIS, SUPPURATIVE
INFLAM., SUPPURATIVE, PROSTATE (T-77100)	M-40400	PROSTATITIS, PURULENT
INFLAMMATION	M-40000	INFLAMMATION
INFLAMMATION, GRANULOMATOUS	M-44000	INFLAMMATION, GRANULOMATOUS
		INFLAMMATION, PYOGRANULOMATOUS
INFLAMMATION, NONSUPPURATIVE	M-47130	INFILTRATION, LYMPHOID
		INFLAMMATION, NONSUPPURATIVE
		LYMPHOCYTIC INFLAMMATORY INFILTRATE
INFLAMMATION, PROSTATE (T-77100)	M-40000	PROSTATITIS
		PROSTATITIS, SUBACUTE
INFLAMMATION, SEMINAL VESICLE (T-77500)	M-40000	SEMINAL VESICULITIS
INFLAMMATION, SUPPURATIVE	M-40400	INFLAMMATION, SUPPURATIVE
LEIOMYOMA	M-88900	LEIOMYOMA

PREFERRED TERM	SNOVET CODE	LADB TERMS
LEIOMYOSARCOMA	M-88903	LEIOMYOSARCOMA
LEUKEMIA	M-98003	LEUKEMIA
LEUKEMIA, STEM CELL	M-98013	STEM CELL LEUKEMIA
LYMPHOMA	M-95903	LYMPHOMA
		MALIGNANT LYMPHOMA
LYMPHOMA, HISTIOCYTIC	M-96403	MALIGNANT LYMPHOMA, HISTIOCYTIC TYPE
LYMPHOMA, RETICULUM CELL	M-96403	RETICULUM CELL SARCOMA
MESOTHELIOMA	M-90503	MESOTHELIOMA
METAPLASIA, SQUAMOUS	M-73220	METAPLASIA, SQUAMOUS
MINERALIZATION	M-55400	CALCIFICATION
		MINERALIZATION
MYOSARCOMA	M-88953	MYOSARCOMA
NECROSIS	M-54000	NECROSIS
NEUROFIBROSARCOMA	M-95403	NEUROFIBROSARCOMA
PAPILLOMA	M-80500	PAPILLOMA
PARAGANGLIOMA	M-86801	PARAGANGLIOMA
PERIVASCULITIS/VASCULITIS	M-47800	PERIARTERITIS
		POLYARTERITIS

PREFERRED TERM	SNOVET CODE	LADB TERMS
HYPERPLASIA	M-72000	HYPERPLASIA
INFLAMMATION	M-40000	INFLAMMATION
INFLAMMATION, ULCERATIVE	M-40030	ULCERATION
LYMPHOMA	M-95903	LYMPHOMA
PROLAPSE	M-31050	PROLAPSE
THROMBUS	M-35100	THROMBOSIS

PREFERRED TERM	SNOVET CODE	LADB TERMS
ABSCESS	M-41740	ABSCESS
ACANTHOSIS	M-72710	ACANTHOSIS
ADENOCARCINOMA	M-81403	ADENOCARCINOMA
		ADNEXAL CARCINOMA
ADENOMA	M-81400	ADENOMA
		CYSTIC ADENOMA
ADENOMA, SEBACEOUS (T-02...)	M-84100	SEBACEOUS ADENOMA
ATROPHY	M-58000	ATROPHY
CALCULI	M-30000	CALCULUS
CARCINOMA	M-80103	CARCINOMA
CARCINOMA, SQUAMOUS CELL	M-80703	SQUAMOUS CELL CARCINOMA
CYST	M-33400	CYST
CYST, EPIDERMAL INCLUSION	M-33410	CYST, INCLUSION
DILATATION	M-32100	DILATATION
ECTASIA	M-32100	ECTASIA
HYPERKERATOSIS	M-72600	HYPERKERATOSIS
HYPERPLASIA	M-72000	HYPERPLASIA
INFLAMMATION	M-40000	INFLAMMATION
INFLAMMATION, GRANULOMATOUS	M-44000	INFLAMMATION, GRANULOMATOUS
INFLAMMATION, NONSUPPURATIVE	M-47130	INFLAMMATION, LYMPHOCYTIC
		INFLAMMATION, NONSUPPURATIVE
		INFLAMMATION, NONSUPPURATIVE, REACTIVE
INFLAMMATION, SUPPURATIVE	M-40400	DISTENTION, SEPTIC
		INFLAMMATION, SUPPURATIVE
		POSTHITIS, NECROPURULENT
INFLAMMATION, ULCERATIVE	M-40030	ULCER
LYMPHOMA	M-95903	LYMPHOMA
LYMPHOMA, RETICULUM CELL	M-96403	RETICULUM CELL SARCOMA
METAPLASIA	M-73000	METAPLASIA
METAPLASIA, SQUAMOUS	M-73220	METAPLASIA, SQUAMOUS
NECROSIS	M-54000	NECROSIS

PREFERRED TERM	SNOVET CODE	LADB TERMS
FIBROMA	M-81000	FIBROMA

PREFERRED TERM	SNOVET CODE	LADB TERMS
ANGIOMA	M-91200	ANGIOMA
ARTERIOLOSCLEROSIS	M-52200	ARTERIOLOSCLEROSIS
ARTERIOSCLEROSIS	M-52000	ARTERIOSCLEROSIS
ASPERMATOGENESIS	F-30960	ASPERMATOGENESIS
ATROPHY	M-58000	ATROPHY
		ATROPHY, SENILE
ATROPHY, TUBULAR (T-78100)	M-58000	ATROPHY, SELECTIVE TUBULAR
CARCINOMA, EMBRYONAL	M-90703	EMBRYONAL CARCINOMA
CRYPTORCHID	M-26200	CRYPTORCHID
CYST	M-33400	CYST
		CYST, RETENTION
DEGENERATION	M-50000	DEGENERATION
		DEGENERATION, HYALINE
DEGENERATION, VACUOLAR	M-66040	DEGENERATION, VACUOLAR
EDEMA	M-36500	EDEMA
FIBROSIS	M-49000	FIBROSIS
GRANULOMA	M-44000	GRANULOMA
		GRANULOMA, SPERMATIC
HYPERPLASIA	M-72000	HYPERPLASIA
HYPERPLASIA, INTERSTITIAL CELL (T-78210)	M-72000	HYPERPLASIA, INTERSTITIAL CELL
		HYPERPLASIA, INTERSTITIAL CELLS
HYPERPLASIA, NODULAR	M-72030	HYPERPLASIA, NODULAR
		NODULE, HYPERPLASTIC
HYPERTROPHY	M-71000	HYPERTROPHY
HYPOPLASIA	M-75300	ARREST, MATURATION
		HYPOSPERMATOGENESIS
HYPOSPERMIA	F-30960	HYPOSPERMIA
INFARCT	M-54700	INFARCT
INFLAMMATION	M-40000	ORCHITIS
INFLAMMATION, FAT (T-1X010)	M-40000	STEATITIS
INFLAMMATION, GRANULOMATOUS	M-44000	INFLAMMATION, GRANULOMATOUS
INFLAMMATION, NONSUPPURATIVE	M-47130	INFLAMMATION, NONSUPPURATIVE
		ORCHITIS, LYMPHOCYTIC
INFLAMMATION, SUPPURATIVE	M-40400	INFLAMMATION, SUPPURATIVE
INTERSTITIAL CELL TUMOR	M-86501	INTERSTITIAL CELL ADENOMA
		INTERSTITIAL CELL TUMOR
		INTERSTITIAL-CELL TUMOR
INTERSTITIAL CELL TUMOR, MALIGNANT	M-86503	INTERSTITIAL-CELL TUMOR, MALIGNANT
KARYOMEGALY	M-65110	KARYOMEGALY
LEUKEMIA	M-98003	LEUKEMIA
LEUKEMIA, GRANULOCYTIC	M-98603	GRANULOCYTIC LEUKEMIA
LEUKEMIA, LYMPHOCYTIC	M-98203	LEUKEMIC LYMPHOSARCOMA
LYMPHOMA	M-95903	LYMPHOMA
MESOTHELIOMA	M-90503	MESOTHELIOMA
		MESOTHELIOMA, MALIGNANT
MINERALIZATION	M-55400	CALCIFICATION
		CALCIFICATION, DYSTROPHIC
		CALCIUM DEPOSIT
		MINERALIZATION
		MINERALIZATION, DYSTROPHIC

PREFERRED TERM	SNOVET CODE	LADB TERMS
MINERALIZATION	M-55400	MINERALIZATION, DYSTROPHIC, POST INFARCTIVE
MINERALIZATION, VASCULAR (T-40000)	M-55400	CALCIFICATION, MEDIAL
NECROSIS	M-54000	NECROSIS
NECROSIS, COAGULATIVE	M-54060	NECROSIS, COAGULATION
NECROSIS, FAT	M-54110	NECROSIS, FAT
NECROSIS, VASCULAR, FIBRINOID (T-40000)	M-54090	NECROSIS, VASCULAR, FIBRINOID
NEOPLASM	M-80001	TUMOR
PERIVASCULITIS/VASCULITIS	M-47800	ARTERIOLITIS
		PANARTERITIS
		PERIARTERITIS
		PERIARTERITIS, NONSUPPURATIVE
		PERIVASCULITIS
		POLYARTERITIS
PIGMENT DEPOSITION	M-57000	PIGMENTATION
SEMINOMA	M-90613	SEMINOMA
		SEMINOMA/DYSGERMINOMA
SPERMATOCELE (T-78300)	M-33400	SPERMATOCELE
THROMBUS	M-35100	THROMBUS

PREFERRED TERM	SNOVET CODE	LADB TERMS
ANGIOSARCOMA	M-91203	HEMANGIOSARCOMA
ATYPCIAL WHITE BLOOD CELLS (T-OX140)	M-69700	ATYPICAL WHITE BLOOD CELLS
ATYPICAL RED BLOOD CELLS (T-OX120)	M-69700	ATYPICAL RED BLOOD CELLS
CELLULAR CHANGE, VACUOLAR	M-66040	CYTOPLASMIC VACUOLIZATION
CYTOMEGALY	M-69220	CYTOMEGALY
FATTY CHANGE	M-50080	INFILTRATION, FATTY
FIBROSIS	M-49000	FIBROSIS
HEMATOPOIESIS	F-50010	GRANULOPOIESIS
HEMORRHAGE	M-37000	HEMORRHAGE
HISTIOCYTOSIS	M-77800	HISTIOCYTOSIS
HYPERPLASIA	M-72000	HYPERPLASIA
HYPERPLASIA, EOSINOPHILIC (T-06340)	M-72000	EOSINOPHILIA
HYPERPLASIA, ERYTHROID	M-72000	ERYTHROPOIESIS
		HYPERPLASIA, ERYTHROID
		NORMOBLASTOSIS
HYPERPLASIA, MEGAKARYOCYTIC (T-06450)	M-72000	HYPERPLASIA, MEGAKARYOCYTIC
		MEGAKARYOCYTOSIS
HYPERPLASIA, MYELOID (T-06200)	M-72000	HYPERPLASIA, MYELOID
		MYELOPOIESIS
HYPOPLASIA, HEMATOPOIETIC (T-06000)	M-75300	HYPOPLASIA
LEUKEMIA	M-98003	LEUKEMIA
		LEUKEMIC INFILTRATION
LEUKEMIA, GRANULOCYTIC	M-98603	GRANULOCYTIC LEUKEMIA
LEUKEMIA, LYMPHOBLASTIC	M-98213	LYMPHOBLASTIC LEUKEMIA
LEUKEMIA, LYMPHOCYTIC	M-98203	LYMPHOCYTIC LEUKEMIA
LEUKEMIA, MONONUCLEAR	M-98903	MONOCYTOID LEUKEMIA
		MONONUCLEAR LEUKEMIA
LEUKOCYTOSIS (T-OX140)	M-77600	LEUKOCYTOSIS
LEUKOCYTOSIS, LYMPHOCYTIC (T-OX210)	M-77600	LYMPHOCYTOSIS
LEUKOCYTOSIS, NEUTROPHILIC (T-OX160)	M-77600	HYPERPLASIA, NEUTROPHILIC
		LEUKOCYTOSIS, NEUTROPHILIC
		NEUTROPHILIA
LEUKOPENIA (T-OX140)	M-59100	LEUKOPENIA
LYMPHOMA	M-95903	LYMPHOMA
		MALIGNANT LYMPHOMA
LYMPHOMA, HISTIOCYTIC	M-96403	LYMPHOSARCOMATOUS INFILTRATION, HISTIOCYTIC
LYMPHOMA, RETICULUM CELL	M-96403	RETICULUM CELL SARCOMA
MYELOFIBROSIS (T-06000)	M-49000	MYELOFIBROSIS
MYELOHISTIOCYTOSIS (T-06000)	M-49000	MYELOHISTIOCYTOSIS
NECROSIS	M-54000	KARYORRHEXIS
		NECROSIS
NEOPLASM	M-80001	NEOPLASM
NEUTROPENIA (T-OX160)	M-59100	NEUTROPENIA
OSTEODYSTROPHY, FIBROUS	M-65600	OSTEODYSTROPHY, FIBROUS
OSTEOSCLEROSIS	M-49910	OSTEOSCLEROSIS
PERIVASCULITIS/VASCULITIS	M-47800	PERIARTERITIS
PIGMENT DEPOSITION	M-57000	PIGMENTATION

PREFERRED TERM	SNOVET CODE	LADB TERMS
ABSCESS	M-41740	ABSCESS
ADENOCARCINOMA	M-81403	ADENOCARCINOMA
ADENOCARCINOMA, MUCINOUS	M-84803	MUCINOUS ADENOCARCINOMA
ANGIECTASIS	M-76300	ANGIECTASIS
ANGIOMA	M-91200	ANGIOMA
		HEMANGIOMA
ANGIOSARCOMA	M-91203	ANGIOSARCOMA
		HEMANGIOSARCOMA
ATROPHY	M-58000	ATROPHY
		DEPLETION, LYMPHOID
CARCINOMA, ACINAR CELL	M-85506	ACINAR-CELL CARCINOMA
CARCINOMA, ALVEOLAR BRONCHIOLAR	M-82506	ALVEOLAR/BRONCHIOLAR CARCINOMA
CARCINOMA, BASAL CELL	M-80906	BASAL-CELL CARCINOMA
CARCINOMA, C CELL	M-85106	C-CELL CARCINOMA
CARCINOMA, HEPATOCELLULAR	M-81703	HEPATOCELLULAR CARCINOMA
CARCINOMA, SQUAMOUS CELL	M-80703	SQUAMOUS CELL CARCINOMA
CARCINOMA, UNDIFFERENTIATED	M-80203	UNDIFFERENTIATED CARCINOMA
CELLULAR CHANGE	M-66040	CYTOLOGIC ALTERATION
CONGESTION	M-36100	CONGESTION
		CONGESTION, PASSIVE
		HYPEREMIA
CYST	M-33400	CYST
CYST, HEMORRHAGIC	M-33540	HEMORRHAGIC CYST
DEGENERATION, CYSTIC	M-50030	DEGENERATION, CYSTIC
DILATATION	M-32100	DILATATION
EDEMA	M-36500	EDEMA
ERYTHROPHAGOCYTOSIS (T-0X120)	F-42000	ERYTHROPHAGOCYTOSIS
FATTY CHANGE	M-50080	METAMORPHOSIS, FATTY
FIBROSIS	M-49000	FIBROSIS
GRANULOMA	M-44000	REACTION, FOREIGN BODY
HEMATOPOIESIS	F-50010	GRANULOPOIESIS
		HEMATOPOIESIS
		HYPERPLASIA, HEMATOPOIETIC
		METAPLASIA, MYELOID
		MYELOPOIESIS, EXTRAMEDULLARY
HEMORRHAGE	M-37000	HEMORRHAGE
		HEMORRHAGE, PETECHIAL
HEMOSIDERIN DEPOSITION	M-57500	HEMOSIDERIN
		HEMOSIDEROSIS
HISTIOCYTOMA, MALIGNANT	M-88323	MALIGNANT HISTIOCYTOSIS
HYPERPLASIA	M-72000	HYPERPLASIA
		HYPERPLASIA, REACTIVE
		LYMPHADENOPATHY
HYPERPLASIA, GRANULOCYTIC (T-06200)	M-72000	HYPERPLASIA, GRANULOCYTIC
HYPERPLASIA, HISTIOCYTIC (T-05120)	M-72000	HISTIOCYTOSIS
HYPERPLASIA, LYMPHOID	M-72200	HYPERPLASIA, LYMPHOID
HYPERPLASIA, MEGAKARYOCYTIC	M-72000	MEGAKARYOCYTOSIS
HYPERPLASIA, PLASMA CELL	M-72150	HYPERPLASIA, PLASMA CELL
		HYPERPLASIA, PLASMOCYTIC
		INFILTRATION, PLASMA CELL

PREFERRED TERM	SNOVET CODE	LADB TERMS
HYPERPLASIA, PLASMA CELL	M-72150	PLASMACYTOSIS
HYPERPLASIA, RETICULOENDOTHELIAL	M-72180	HYPERPLASIA, RETICULOENDOTHELIAL
		RETICULOCYTOSIS
		RETICULOENDOTHELIOSIS
HYPERPLASIA, RETICULUM CELL	M-72180	HYPERPLASIA, RE CELL
		HYPERPLASIA, RETICULUM CELL
		RETICULOHISTIOCYTOSIS
HYPERTROPHY	M-71000	HYPERTROPHY
INFLAMMATION	M-40000	ADENITIS
		INFLAMMATION
INFLAMMATION, FAT (T-1X010)	M-40000	STEATITIS
INFLAMMATION, GRANULOMATOUS	M-44000	INFLAMMATION, GRANULOMATOUS
INFLAMMATION, HEMORRHAGIC	M-40600	INFLAMMATION, HEMORRHAGIC
INFLAMMATION, NECROTIZING	M-40700	INFLAMMATION, NECROTIZING
INFLAMMATION, NONSUPPURATIVE	M-47130	INFLAMMATION, NONSUPPURATIVE
INFLAMMATION, SUPPURATIVE	M-40400	INFLAMMATION, SUPPURATIVE
		LYMPHADENITIS, PURULENT
INTERSTITIAL CELL TUMOR	M-86506	INTERSTITIAL-CELL TUMOR
LEIOMYOSARCOMA	M-88903	LEIOMYOSARCOMA
LEUKEMIA	M-98003	LEUKEMIA
		LEUKEMIC INFILTRATION
LEUKEMIA, GRANULOCYTIC	M-98603	GRANULOCYTIC LEUKEMIA
LEUKEMIA, LYMPHOBLASTIC	M-98213	LYMPHOBLASTIC LEUKEMIA
LEUKEMIA, LYMPHOCYTIC	M-98203	LEUKEMIC LYMPHOSARCOMA
		LYMPHATIC LEUKEMIA
		LYMPHOCYTIC LEUKEMIA
LEUKEMIA, STEM CELL	M-98013	STEM CELL LEUKEMIA
LIPOCHROMATOSIS	M-57740	LIPOCHROMATOSIS
LYMPHANGIECTASIS	M-76400	LYMPHANGIECTASIS
LYMPHEDEMA	M-36660	LYMPHEDEMA
LYMPHOMA	M-95903	LYMPHOMA
		MALIGNANT LYMPHOMA
		MALIGNANT LYMPHOMA, MIXED TYPE
		MALIGNANT LYMPHOMA, UNDIFFERENTIATED TYPE
LYMPHOMA, HISTIOCYTIC	M-96403	HISTIOCYTIC LYMPHOSARCOMA
		MALIGNANT LYMPHOMA, HISTIOCYTIC TYPE
LYMPHOMA, LYMPHOCYTIC	M-96203	MALIGNANT LYMPHOMA, LYMPHOCYTIC TYPE
LYMPHOMA, RETICULUM CELL	M-96403	RETICULUM CELL SARCOMA
MESOTHELIOMA	M-90503	MESOTHELIOMA
MINERALIZATION	M-55400	CALCIFICATION
		CALCIFICATION, DYSTROPHIC
NECROSIS	M-54000	NECROSIS
		NECROSIS, COAGULATIVE
NEUROFIBROSARCOMA	M-95403	NEUROFIBROSARCOMA
PERIVASCULITIS/VASCULITIS	M-47800	POLYARTERITIS
PIGMENT DEPOSITION	M-57000	PIGMENTATION
SARCOMA	M-88003	SARCOMA
SARCOMA, HISTIOCYTIC	M-96403	HISTIOCYTIC SARCOMA
THROMBUS	M-35100	THROMBOSIS

PREFERRED TERM	SNOVET CODE	LADB TERMS
ABSCESS	M-41740	ABSCESS
ACCESSORY SPLEEN	M-22300	ACCESSORY SPLEEN
ADENOCARCINOMA	M-81403	ADENOCARCINOMA
ADENOCARCINOMA, MUCINOUS	M-84803	MUCINOUS ADENOCARCINOMA
AMYLOID DEPOSITION		AMYLOIDOSIS
ANGIECTASIS	M-76300	ANGIECTASIS
ANGIOMA	M-91200	ANGIOMA
		HEMANGIOMA
ANGIOMA, CAPILLARY	M-91310	CAPILLARY HEMANGIOMA
ANGIOSARCOMA	M-91203	ANGIOSARCOMA
		HEMANGIOSARCOMA
ATROPHY	M-58000	ATROPHY
ATROPHY, FIBROUS		ATROPHY, FIBROUS
ATROPHY, LYMPHOID		ATROPHY, LYMPHOID
		DEPLETION, LYMPHOID
		LYMPHOID DEPLETION
CARCINOMA, ACINAR CELL	M-85506	ACINAR-CELL CARCINOMA
CARCINOMA, SQUAMOUS CELL	M-80703	SQUAMOUS CELL CARCINOMA
CONGESTION	M-36100	CONGESTION
		CONGESTION, PASSIVE
CYST	M-33400	CYST
CYST, HEMORRHAGIC	M-33540	HEMORRHAGIC CYST
DEGENERATION	M-50000	DEGENERATION, HYALINE
FIBROSARCOMA	M-88103	FIBROSARCOMA
FIBROSIS	M-49000	FIBROSIS
		FIBROSIS, SPLENIC
		INFLAMMATION WITH FIBROSIS
		SCARRING
		SCLEROSIS
GRANULOMA	M-44000	REACTION, FOREIGN BODY
HEMATOMA	M-37100	HEMATOMA
HEMATOMA, ORGANIZED	M-37120	HEMATOMA, ORGANIZED
HEMATOPOIESIS	F-50010	ERYTHROPOIESIS
		GRANULOPOIESIS
		HEMATOPOIESIS
		HEMATOPOIESIS, EXTRAMEDULLARY
		HYPERPLASIA, EXTRAMEDULLARY
		HYPERPLASIA, HEMATOPOIETIC
		MYELOPOIESIS
HEMORRHAGE	M-37000	HEMORRHAGE
HEMOSIDERIN DEPOSITION	M-57500	HEMOSIDERIN
		HEMOSIDEROSIS
HISTIOCYTOMA, MALIGNANT	M-88323	MALIGNANT HISTIOCYTOSIS
HYPERPLASIA	M-72000	HYPERPLASIA
HYPERPLASIA, ERYTHROID	M-72000	HYPERPLASIA, ERYTHROID
HYPERPLASIA, LYMPHOID	M-72200	HYPERPLASIA, LYMPHOID
		LYMPHOCYTOSIS
HYPERPLASIA, MEGAKARYOCYTIC	M-72000	HYPERPLASIA, MEGAKARYOCYTIC
HYPERPLASIA, MYELOID (T-06200)	M-72000	LEUKEMOID REACTION
HYPERPLASIA, PLASMA CELL	M-72150	HYPERPLASIA, PLASMA CELL

PREFERRED TERM	SNOVET CODE	LADB TERMS
HYPERPLASIA, RETICULOENDOTHELIAL	M-72180	HYPERPLASIA, RE CELL
		HYPERPLASIA, RETICULOENDOTHELIAL
		HYPERPLASIA, RETICULUM CELL
		RETICULOCYTOSIS
HYPERPLASIA, VASCULAR	M-72570	VASCULARIZATION
HYPOPLASIA	M-75300	HYPOPLASIA
INFARCT	M-54700	INFARCT
INFARCT, HEALED	M-54750	INFARCT, FIBROTIC
		INFARCT, HEALED
INFARCT, HEMORRHAGIC	M-54730	INFARCT, HEMORRHAGIC
INFLAMMATION	M-40000	INFLAMMATION
		SPLENITIS
INFLAMMATION, GRANULOMATOUS	M-44000	INFLAMMATION, GRANULOMATOUS
INFLAMMATION, NONSUPPURATIVE	M-47130	INFILTRATION, MONONUCLEAR CELL
INFLAMMATION, SUPPURATIVE	M-40400	INFLAMMATION, SUPPURATIVE
		SPLENITIS, NEUTROPHILIC
LEIOMYOSARCOMA	M-88903	LEIOMYOSARCOMA
LEUKEMIA	M-98003	LEUKEMIA
		LEUKEMIC INFILTRATION
		UNDIFFERENTIATED LEUKEMIA
LEUKEMIA, GRANULOCYTIC	M-98603	GRANULOCYTIC LEUKEMIA
LEUKEMIA, LYMPHOBLASTIC	M-98213	LYMPHOBLASTIC LEUKEMIA
LEUKEMIA, LYMPHOCYTIC	M-98203	LEUKEMIC LYMPHOSARCOMA
		LYMPHATIC LEUKEMIA
		LYMPHOCYTIC LEUKEMIA
LEUKEMIA, MYELOMONOCYTIC	M-98603	MYELOMONOCYTIC LEUKEMIA
LEUKEMIA, STEM CELL	M-98013	STEM CELL LEUKEMIA
LEUKOCYTOSIS (T-OX140)	M-77600	LEUKOCYTOSIS
LIPID DEPOSITION		LIPOIDOSIS
LYMPHOMA	M-95903	LYMPHOMA
		MALIGNANT LYMPHOMA
		MALIGNANT LYMRHOMA, MIXED TYPE
		MALIGNANT LYMPHOMA, UNDIFFERENTIATED TYPE
LYMPHOMA, HISTIOCYTIC	M-96403	HISTIOCYTIC LYMPHOSARCOMA
		MALIGNANT LYMPHOMA, HISTIOCYTIC TYPE
LYMPHOMA, LYMPHOCYTIC	M-96203	MALIGNANT LYMPHOMA, LYMPHOCYTIC TYPE
LYMPHOMA, RETICULUM CELL	M-96403	RETICULUM CELL SARCOMA
MASTOCYTOSIS	M-97413	MASTOCYTOSIS
MESENCHYMOMA, BENIGN	M-89900	MESENCHYMOMA, BENIGN
MESOTHELIOMA	M-90503	MESOTHELIOMA
METAPLASIA, OSSEOUS	M-73400	METAPLASIA, OSSEOUS
MINERALIZATION	M-55400	CALCIFICATION
		CALCIUM DEPOSIT
MINERALIZATION, VASCULAR (T-40000)	M-55400	CALCIFICATION, MEDIAL
NECROSIS	M-54000	NECROSIS
		NECROSIS, CASEOUS
		NECROSIS, COAGULATIVE
NEUROFIBROSARCOMA	M-95403	NEUROFIBROSARCOMA
OSTEOSARCOMA	M-91806	OSTEOSARCOMA
PERIVASCULITIS/VASCULITIS	M-47800	PERIARTERITIS

PREFERRED TERM	SNOVET CODE	LADB TERMS
PERIVASCULITIS/VASCULITIS	M-47800	POLYARTERITIS
PIGMENT DEPOSITION	M-57000	PIGMENT
		PIGMENTATION
SARCOMA	M-88003	SARCOMA
SARCOMA, HISTIOCYTIC	M-96403	HISTIOCYTIC SARCOMA
THROMBUS	M-35100	THROMBOSIS

PREFERRED TERM	SNOVET CODE	LADB TERMS
ADENOMA	M-81400	ADENOMA
ATROPHY	M-58000	ATROPHY
CARCINOMA	M-80103	CARCINOMA
CONGESTION	M-36100	CONGESTION
CYST	M-33400	CYST
FIBROSARCOMA	M-88103	FIBROSARCOMA
FIBROSIS	M-49000	FIBROSIS
HEMORRHAGE	M-37000	HEMORRHAGE
HEMOSIDERIN DEPOSITION	M-57500	HEMOSIDERIN
		HEMOSIDEROSIS
HYPERPLASIA	M-72000	HYPERPLASIA
HYPERPLASIA, EPITHELIAL	M-72000	HYPERPLASIA, EPITHELIAL
HYPERPLASIA, LYMPHOID	M-72200	HYPERPLASIA, LYMPHOID
HYPERPLASIA, RETICULOENDOTHELIAL	M-72180	HYPERPLASIA, RETICULOENDOTHELIAL
INFLAMMATION	M-40000	INFLAMMATION
INFLAMMATION, SUPPURATIVE	M-40400	INFLAMMATION, SUPPURATIVE
		THYMITIS, PURULENT
LEUKEMIA	M-98003	LEUKEMIA
LEUKEMIA, GRANULOCYTIC	M-98603	GRANULOCYTIC LEUKEMIA
LEUKEMIA, HISTIOCYTIC	M-98903	LEUKEMIC HISTIOCYTIC SARCOMA
LEUKEMIA, LYMPHOBLASTIC	M-98213	LYMPHOBLASTIC LEUKEMIA
LEUKEMIA, LYMPHOCYTIC	M-98203	LEUKEMIC LYMPHOSARCOMA
		LYMPHATIC LEUKEMIA
		LYMPHOCYTIC LEUKEMIA
LEUKEMIA, STEM CELL	M-98013	STEM CELL LEUKEMIA
LIPID DEPOSITION		LIPOIDOSIS
LYMPHOMA	M-95903	LYMPHOMA
		MALIGNANT LYMPHOMA
LYMPHOMA, HISTIOCYTIC	M-96403	MALIGNANT LYMPHOMA, HISTIOCYTIC TYPE
LYMPHOMA, LYMPHOCYTIC	M-96203	MALIGNANT LYMPHOMA, LYMPHOCYTIC TYPE
LYMPHOMA, RETICULUM CELL	M-96403	RETICULUM CELL SARCOMA
MASTOCYTOSIS	M-97413	MASTOCYTOSIS
MINERALIZATION	M-55400	CALCIFICATION
NECROSIS	M-54000	NECROSIS
OSTEOSARCOMA	M-91806	OSTEOSARCOMA
PERIVASCULITIS/VASCULITIS	M-47800	ANGIITIS
PIGMENT DEPOSITION	M-57000	PIGMENTATION
SARCOMA, HISTIOCYTIC	M-96403	HISTIOCYTIC SARCOMA
THROMBUS	M-35100	THROMBOSIS
THYMOMA	M-85800	THYMOMA
		THYMOMA, MALIGNANT
		THYMOMA, SPINDLE CELL

PREFERRED TERM	SNOVET CODE	LADB TERMS
ATROPHY, SEROUS	M-58080	ATROPHY, SEROUS
FIBROSIS	M-49000	FIBROSIS
INFARCT	M-54700	INFARCT
INFLAMMATION	M-40000	ADIPOSE TISSUE WITH INFLAMMATION, ENCAPSULAT
		INFLAMMATION
		STEATITIS
INFLAMMATION, GRANULOMATOUS	M-44000	INFLAMMATION, GRANULOMATOUS
		LIPOGRANULOMA
INFLAMMATION, NONSUPPURATIVE	M-47130	LYMPHOCYTIC INFLAMMATORY INFILTRATE
INFLAMMATION, SUPPURATIVE	M-40400	INFLAMMATION, SUPPURATIVE
LEUKEMIA	M-98003	LEUKEMIA
LEUKEMIA, LYMPHOCYTIC	M-98203	LYMPHATIC LEUKEMIA
LIPOMA	M-88500	LIPOMA
		PEDUNCULATED LIPOMA
LYMPHOMA	M-95903	MALIGNANT LYMPHOMA
MESOTHELIOMA	M-90503	MESOTHELIOMA
NECROSIS, FAT	M-54110	NECROSIS, FAT
NEUROFIBROSARCOMA	M-95403	NEUROFIBROSARCOMA

PREFERRED TERM	SNOVET CODE	LADB TERMS
ADENOCARCINOMA	M-81403	ADENOCARCINOMA
ADENOCARCINOMA, MUCINOUS	M-84803	MUCINOUS ADENOCARCINOMA
AMYLOID DEPOSITION		AMYLOIDOSIS
ANGIOMA	M-91200	HEMANGIOMA
ANGIOSARCOMA	M-91203	HEMANGIOSARCOMA
ARTERIOSCLEROSIS	M-52000	ARTERIOSCLEROSIS
ATROPHY	M-58000	ATROPHY
CARCINOMA	M-80103	CARCINOMA
CARCINOMA, C CELL	M-85106	C-CELL CARCINOMA
CARCINOMA, ISLET CELL	M-81503	ISLET-CELL CARCINOMA
CONGESTION	M-36100	CONGESTION
CYTOMEGALY	M-69220	CYTOMEGALY
ERYTHROLEUKEMIA	M-98403	ERYTHROLEUKEMIA
FIBROSARCOMA	M-88103	FIBROSARCOMA
HEMATOPOIESIS	F-50010	HEMATOPOIESIS
HISTIOCYTOMA, FIBROUS, MALIGNANT		FIBROUS HISTIOCYTOMA, MALIGNANT
HYPERPLASIA	M-72000	HYPERPLASIA
HYPERPLASIA, GRANULOCYTIC (T-06200)	M-72000	HYPERPLASIA, GRANULOCYTIC
HYPERPLASIA, LYMPHOID	M-72200	HYPERPLASIA, LYMPHOID
HYPERPLASIA, PLASMA CELL	M-72150	HYPERPLASIA, PLASMA CELL
HYPERPLASIA, RETICULUM CELL	M-72180	HYPERPLASIA, RETICULUM CELL
INFLAMMATION, GRANULOMATOUS	M-44000	INFLAMMATION, GRANULOMATOUS
INFLAMMATION, NONSUPPURATIVE	M-47130	LYMPHOCYTIC INFLAMMATORY INFILTRATE
LEIOMYOSARCOMA	M-88903	LEIOMYOSARCOMA
LEUKEMIA	M-98003	LEUKEMIA
		UNDIFFERENTIATED LEUKEMIA
LEUKEMIA, GRANULOCYTIC	M-98603	GRANULOCYTIC LEUKEMIA
LEUKEMIA, LYMPHOBLASTIC	M-98213	LEUKEMIC LYMPHOBLASTOMA
LEUKEMIA, LYMPHOCYTIC	M-98203	LYMPHOCYTIC LEUKEMIA
LEUKEMIA, MONONUCLEAR	M-98903	MONOCYTIC LEUKEMIA
LEUKEMIA, MYELOMONOCYTIC	M-98603	MYELOMONOCYTIC LEUKEMIA
LYMPHOMA	M-95903	MALIGNANT LYMPHOMA
		MALIGNANT LYMPHOMA, UNDIFFERENTIATED TYPE
LYMPHOMA, HISTIOCYTIC	M-96403	MALIGNANT LYMPHOMA, HISTIOCYTIC TYPE
		MALIGNANT LYMPHOMA, MIXED TYPE
LYMPHOMA, LYMPHOCYTIC	M-96203	MALIGNANT LYMPHOMA, LYMPHOCYTIC TYPE
MASTOCYTOMA	M-97401	MAST-CELL TUMOR
MESOTHELIOMA	M-90503	MESOTHELIOMA
		MESOTHELIOMA, MALIGNANT
MINERALIZATION	M-55400	CALCIFICATION
		MINERALIZATION
NEUROFIBROSARCOMA	M-95403	NEUROFIBROSARCOMA
OSTEOSARCOMA	M-91806	OSTEOSARCOMA
PERIVASCULITIS/VASCULITIS	M-47800	PERIARTERITIS
		PERIVASCULITIS
SARCOMA	M-88003	SARCOMA
SARCOMA, GRANULOCYTIC	M-99303	GRANULOCYTIC SARCOMA
SARCOMA, OSTEOGENIC	M-91806	OSTEOGENIC SARCOMA

PREFERRED TERM	SNOVET CODE	LADB TERMS
ADENOCARCINOMA, MUCINOUS	M-84803	MUCINOUS ADENOCARCINOMA
ANGIOMA	M-91200	HEMANGIOMA
ANGIOSARCOMA	M-91203	ANGIOSARCOMA
		HEMANGIOSARCOMA
ATROPHY	M-58000	ATROPHY
CARCINOSARCOMA, MAMMARY	M-89806	MAMMARY CARCINOSARCOMA
DUAL STERNUM (T-10310)	M-22360	DUAL STERNUM
DYSPLASIA, EPITHELIAL	M-74000	DYSPLASIA, EPITHELIAL
DYSPLASIA, FIBROUS	M-74910	FIBROUS DYSPLASIA
FATTY CHANGE	M-50080	METAMORPHOSIS, FATTY
FIBROADENOMA	M-90100	FIBROADENOMA
FIBROSARCOMA	M-88103	FIBROSARCOMA
FIBROSIS	M-49000	FIBROSIS
FRACTURE	M-12000	FRACTURE
FRACTURE, HEALED	M-12500	FRACTURE, HEALED
GRANULOMA	M-44000	GRANULOMA
HEMORRHAGE	M-37000	HEMORRHAGE
HYPERPLASIA	M-72000	HYPERPLASIA
		HYPERPLASIA, BONE
HYPERPLASIA, ERYTHROID (T-06100)	M-72000	HYPERPLASIA, ERYTHROID
HYPERPLASIA, GRANULOCYTIC (T-06200)	M-72000	HYPERPLASIA, GRANULOCYTIC
HYPERPLASIA, HEMATOPOIETIC (T-06000)	M-72000	HYPERPLASIA, HEMATOPOIETIC
HYPERPLASIA, MEGAKARYOCYTIC (T-06540)	M-72000	MEGAKARYOCYTOSIS
HYPOPLASIA, HEMATOPOIETIC (T-06000)	M-75300	HYPOPLASIA, HEMATOPOIETIC
INFLAMMATION, FAT (T-1X010)	M-40000	STEATITIS
INFLAMMATION, SUPPURATIVE	M-40400	OSTEOMYELITIS, PURULENT
LEUKEMIA	M-98003	LEUKEMIA
LEUKEMIA, LYMPHOCYTIC	M-98203	LEUKEMIC LYMPHOSARCOMA
LYMPHOMA	M-95903	LYMPHOMA
		MALIGNANT LYMPHOMA
MINERALIZATION	M-55400	MINERALIZATION
MYELOFIBROSIS	M-49000	MYELOFIBROSIS
NECROSIS	M-54000	NECROSIS
OSTEOCHONDROMA	M-92100	OSTEOCHONDROMA
OSTEODYSTROPHY, FIBROUS	M-65600	FIBROUS OSTEODYSTROPHY
		OSTEITIS FIBROSA CYSTICA
		OSTEODYSTROPHY, FIBROUS
OSTEOMA	M-91800	OSTEOMA
OSTEOPETROSIS	M-22540	OSTEOPETROSIS
OSTEOPHYTE	M-71430	OSTEOPHYTE
OSTEOPOROSIS	M-50210	OSTEOPOROSIS
OSTEOSARCOMA	M-91803	OSTEOSARCOMA
OSTEOSCLEROSIS	M-49910	OSTEOSCLEROSIS
RESORPTION	M-51700	RESORPTION
RHABDOMYOSARCOMA	M-89003	RHABDOMYOSARCOMA
SARCOMA, GRANULOCYTIC	M-99303	GRANULOCYTIC SARCOMA
SARCOMA, HISTIOCYTIC	M-96403	HISTIOCYTIC SARCOMA

PREFERRED TERM	SNOVET CODE	LADB TERMS
CYST	M-33400	CYST

PREFERRED TERM	SNOVET CODE	LADB TERMS
ARTHROPATHY, DEGENERATIVE	M-50230	ARTHROPATHY, DEGENERATIVE
EXOSTOSIS	M-71440	EXOSTOSIS
INFLAMMATION, SUPPURATIVE	M-40400	INFLAMMATION, SUPPURATIVE
MINERALIZATION	M-55400	MINERALIZATION
		MINERALIZATION, GENERAL
MYODEGENERATION (T-13...)	M-54400	MYODEGENERATION
OSSIFICATION, NODULAR		OSSIFICATION, NODULAR
OSTEOPHYTE	M-71430	OSTEOPHYTE

PREFERRED TERM	SNOVET CODE	LADB TERMS
ABSCESS	M-41740	ABSCESS
ANGIOSARCOMA	M-91203	ANGIOSARCOMA
		HEMANGIOSARCOMA
ARTERIOLOSCLEROSIS	M-52200	ARTERIOLOSCLEROSIS
ATROPHY	M-58000	ATROPHY
ATROPHY, SENILE	M-58110	ATROPHY. SENILE
CARCINOSARCOMA, MAMMARY (T-04000)	M-89806	MAMMARY CARCINOSARCOMA
DEGENERATION	M-50000	DEGENERATION
		DEGENERATION, MYOFIBER
		DEGENERATION, SENILE
		DEGENERATION, ZENKERS
DYSTROPHY	M-56000	DYSTROPHY
FATTY CHANGE	M-50080	INFILTRATION, FATTY
FIBROSARCOMA	M-88103	FIBROSARCOMA
FIBROSIS	M-49000	FIBROSIS
HEMORRHAGE	M-37000	HEMORRHAGE
HERNIA	M-31600	HERNIA
HISTIOCYTOMA, FIBROUS, MALIGNANT	M-88303	FIBROUS HISTIOCYTOMA, MALIGNANT
HYPERPLASIA, LYMPHOID	M-72200	LYMPHOCYTIC NODULE
INFLAMMATION	M-40000	INFLAMMATION
		MYOSITIS
INFLAMMATION, GRANULOMATOUS	M-44000	INFLAMMATION, GRANULOMATOUS
INFLAMMATION, NONSUPPURATIVE	M-47130	INFILTRATION, LYMPHOID
		MYOSITIS, NONSUPPURATIVE
LEUKEMIA	M-98003	LEUKEMIC INFILTRATION
LEUKEMIA, LYMPHOBLASTIC	M-98213	LYMPHOBLASTIC LEUKEMIA
LIPOMA	M-88500	LIPOMA
LYMPHOMA	M-95903	LYMPHOMA
		MALIGNANT LYMPHOMA
LYMPHOMA, RETICULUM CELL	M-96403	RETICULUM CELL SARCOMA
MESOTHELIOMA	M-90503	MESOTHELIOMA
MINERALIZATION	M-55400	CALCIFICATION
		CALCIFICATION, DYSTROPHIC
		MINERALIZATION
		MINERALIZATION, DYSTROPHIC
MINERALIZATION, VASCULAR (T-40000)	M-55400	CALCIFICATION, MEDIAL
NECROSIS	M-54000	NECROSIS
OSTEOSARCOMA	M-91806	OSTEOSARCOMA
PARASITISM	D-06500	PARASITISM
PERIVASCULITIS/VASCULITIS	M-47800	ANGIITIS
		PERIARTERITIS
REGENERATION	M-79900	REGENERATION
SARCOMA	M-88003	SARCOMA
SARCOMA, HISTIOCYTIC	M-96403	HISTIOCYTIC SARCOMA
SARCOMA, UNDIFFERENTIATED	M-88003	UNDIFFERENTIATED SARCOMA

PREFERRED TERM	SNOVET CODE	LADB TERMS
ABSCESS	M-41740	ABSCESS
ADENOCARCINOMA, PITUITARY	M-81406	PITUITARY ADENOCARCINOMA
ASTROCYTOMA (T-X2...)	M-94003	ASTROCYTOMA
		PROTOPLASMIC ASTROCYTOMA
ASTROCYTOMA, FIBROUS (T-X2...)	M-94230	FIBROUS ASTROCYTOMA
ATROPHY	M-58000	ATROPHY
ATROPHY, PRESSURE	M-58120	ATROPHY, PRESSURE
CARCINOMA	M-80103	CARCINOMA
CARCINOMA, CERUMINOUS	M-84203	CERUMINOUS CARCINOMA
CARCINOMA, CHROMOPHOBE	M-82703	CHROMOPHOBE CARCINOMA
CARCINOMA, SQUAMOUS CELL	M-80703	SQUAMOUS CELL CARCINOMA
CELLULAR CHANGE, VACUOLAR	M-66040	CYTOPLASMIC VACUOLIZATION
		VACUOLATION
		VACUOLATION, NEURONAL
CHOLESTEROL DEPOSITION	M-55240	CHOLESTEROL DEPOSITION
CONGESTION	M-36100	CONGESTION
CORPORA AMYLACEA	M-55030	CORPORA AMYLACEA
CYST	M-33400	CYST
CYST, EPIDERMAL INCLUSION	M-33410	CYST, EPIDERMOID
		EPIDERMAL INCLUSION CYST
DEGENERATION	M-50000	DEGENERATION
		DEGENERATION, WALLERIAN
DEMYELINATION	M-50410	DEGENERATION, MYELIN
		DEMYELINIZATION
EPENDYMOMA (T-X16...)	M-93913	EPENDYMOMA
EPENDYMOMA, NEUROHYPOPHYSEAL (T-91...)	M-93913	NEUROHYPOPHYSEAL EPENDYMOMA
FIBROSIS	M-49000	FIBROSIS
GLIOBLASTOMA MULTIFORMI	M-94403	GLIOBLASTOMA MULTIFORMI
GLIOMA	M-93803	GLIOMA
GLIOSIS	M-49160	GLIOSIS
HEMORRHAGE	M-37000	CEREBRAL HEMORRHAGE
		HEMORRHAGE
		HEMORRHAGE, CEREBRAL
HISTIOCYTOMA, MALIGNANT	M-88323	MALIGNANT HISTIOCYTOSIS
HYDROCEPHALUS (T-X1600)	M-33320	DILATATION
		DILATED
		HYDROCEPHALUS
		HYDROCEPHALUS, INTERNAL
INFARCT, HEMORRHAGIC	M-54730	INFARCT, HEMORRHAGIC
INFLAM., GRANULOMATOUS, MENINGES (T-X1110)	M-44000	LEPTOMENINGITIS, GRANULOMATOUS
INFLAMMATION	M-40000	INFLAMMATION
INFLAMMATION, MENINGES (T-X1110)	M-40000	LEPTOMENINGITIS
		MENINGITIS
INFLAMMATION, NONSUPPURATIVE	M-47130	ENCEPHALITIS, NONSUPPURATIVE
		INFILTRATION, LYMPHOID
		INFLAMMATION, LYMPHOCYTIC
		LYMPHOCYTIC INFLAMMATORY INFILTRATE
INFLAMMATION, SUPPURATIVE	M-40400	ABSCESS, SECONDARY
		INFLAMMATION, SUPPURATIVE
LEUKEMIA	M-98003	LEUKEMIA

PREFERRED TERM	SNOVET CODE	LADB TERMS
LEUKEMIA, GRANULOCYTIC	M-98603	GRANULOCYTIC LEUKEMIA
LEUKEMIA, HISTIOCYTIC	M-98903	LEUKEMIC HISTIOCYTIC SARCOMA
LEUKEMIA, LYMPHOBLASTIC	M-98213	LYMPHOBLASTIC LEUKEMIA
LEUKEMIA, LYMPHOCYTIC	M-98203	LEUKEMIC LYMPHOSARCOMA
		LYMPHATIC LEUKEMIA
LYMPHOMA	M-95903	LYMPHOMA
		MALIGNANT LYMPHOMA
		MALIGNANT LYMPHOMA, MIXED TYPE
LYMPHOMA, HISTIOCYTIC	M-96403	MALIGNANT LYMPHOMA, HISTIOCYTIC TYPE
LYMPHOMA, RETICULUM CELL	M-96403	RETICULUM CELL SARCOMA
MALACIA	M-54400	ENCEPHALOMALACIA
		MALACIA
MEDULLOBLASTOMA (T-X6...)	M-94703	MEDULLOBLASTOMA
MELANOSIS	M-57210	MELANOSIS
MENINGIOMA (T-X1110)	M-95300	MENINGIOMA
		MICROMENINGIOMA
MENINGOVENTRICULAR COMMUNICATION		MENINGOVENTRICULAR COMMUNICATION
MESOTHELIOMA	M-90503	MESOTHELIOMA
MINERALIZATION	M-55400	CALCIFICATION
		CALCINOSIS
		CALCINOSIS, ECTOPIC
		MINERALIZATION
		MINERALIZATION, DYSTROPHIC
MIXED GLIAL TUMOR	M-93823	MIXED GLIAL TUMOR
NECROSIS	M-54000	NECROSIS
NECROSIS, HEMORRHAGIC	M-54040	NECROSIS, HEMORRHAGIC
NEUROFIBROMA, ACOUSTIC (T-X8500)	M-95400	ACOUSTIC NEUROFIBROMA
OLIGODENDROGLIOMA	M-94503	OLIGODENDROGLIOMA
OSTEOSARCOMA	M-91806	OSTEOSARCOMA
PERIVASCULITIS/VASCULITIS	M-47800	PERIARTERITIS
		PERIVASCULITIS
PSAMMOMA BODY	M-55590	PSAMMOMA BODIES
RADICULONEUROPATHY	D-8X350	RADICULONEUROPATHY
SARCOMA	M-88003	SARCOMA
STATUS SPONGIOSIS	M-36620	STATUS SPONGIOSIS
THROMBUS	M-35100	THROMBOSIS
THYMOMA	M-85806	THYMOMA
VASCULAR ANOMALY (T-40000)	M-24600	VASCULAR ANOMALY

PREFERRED TERM	SNOVET CODE	LADB TERMS
CELLULAR CHANGE, VACUOLAR	M-66040	VACUOLATION
DEGENERATION	M-50000	DEGENERATION, CALCIFIC
		DEGENERATION, PSEUDOCYSTIC
		DEGENERATION, SENILE
INFLAMMATION	M-40000	INFLAMMATION
		PERINEURITIS
LYMPHOMA, LYMPHOCYTIC	M-96203	MALIGNANT LYMPHOMA, LYMPHOCYTIC TYPE
MINERALIZATION	M-55400	CALCIFICATION
		CALCIFICATION, DYSTROPHIC
NEUROFIBROSARCOMA	M-95403	NEUROFIBROSARCOMA
SARCOMA, HISTIOCYTIC	M-96403	HISTIOCYTIC SARCOMA
SWELLING	M-02570	SWELLING

PREFERRED TERM	SNOVET CODE	LADB TERMS
ABSCESS	M-41740	ABSCESS
ADENOMA, SEBACEOUS (T-02...)	M-84100	SEBACEOUS ADENOMA
ADENOMA, SEBACEOUS SQUAMOUS	M-84100	SEBACEOUS SQUAMOUS ADENOMA
CARCINOMA	M-80103	CARCINOMA
CARCINOMA, CERUMINOUS	M-84203	CERUMINOUS CARCINOMA
CARCINOMA, SQUAMOUS CELL	M-80703	SQUAMOUS CELL CARCINOMA
CONGESTION	M-36100	CONGESTION
DYSPLASIA, CHONDROECTODERMAL	D-52150	CHONDROECTODERMAL DYSPLASIA
EDEMA	M-36500	EDEMA
FIBROMA	M-81000	FIBROMA
FIBROMA, ULCERATIVE		ULCERATIVE FIBROMA
FIBROSARCOMA	M-88103	FIBROSARCOMA
HEMORRHAGE	M-37000	HEMORRHAGE
HISTIOCYTOMA, FIBROUS	M-88300	FIBROUS HISTIOCYTOMA
HYPERPLASIA	M-72000	HYPERPLASIA
INFLAMMATION	M-40000	INFLAMMATION
		OTITIS INTERNA
INFLAMMATION, GRANULOMATOUS	M-44000	DERMATITIS, PYOGRANULOMATOUS
INFLAMMATION, SUPPURATIVE	M-40400	INFLAMMATION, SUPPURATIVE
		OTITIS, SUPPURATIVE
KERATOACANTHOMA	M-72860	KERATOACANTHOMA
MELANOMA, MALIGNANT	M-87203	MALIGNANT MELANOMA
PAPILLOMA, SQUAMOUS CELL	M-80520	SQUAMOUS CELL PAPILLOMA
		SQUAMOUS PAPILLOMA
RHABDOMYOSARCOMA	M-89003	RHABDOMYOSARCOMA
THROMBUS	M-35100	THROMBOSIS

PREFERRED TERM	SNOVET CODE	LADB TERMS
INFLAMMATION	M-40000	OTITIS MEDIA
INFLAMMATION, SUPPURATIVE	M-40400	OTITIS MEDIA, SUPPURATIVE

PREFERRED TERM	SNOVET CODE	LADB TERMS
ABSCESS	M-41740	ABSCESS
CARCINOMA, SQUAMOUS CELL	M-80703	SQUAMOUS CELL CARCINOMA
CATARACT	M-51100	CATARACT
		CATARACT, CALCIFIED
		CATARACT, CORTICAL
		CATARACT, SENILE
		NUCLEAR CATARACT
HEMORRHAGE	M-37000	HEMORRHAGE
HERNIATION	M-31600	HERNIATION
HYPERKERATOSIS	M-72600	HYPERKERATOSIS
HYPERPLASIA, EPITHELIAL	M-72000	HYPERPLASIA, EPITHELIAL
INFLAMMATION	M-40000	PANOPHTHALMITIS
INFLAMMATION, INTERSTITIAL (T-XX095)	M-40000	INFLAMMATION, INTERSTITIAL
INFLAMMATION, NONSUPPURATIVE	M-47130	INFLAMMATION, NONSUPPURATIVE
INFLAMMATION, SUPPURATIVE	M-40400	HYPOPYON
		INFLAMMATION, SUPPURATIVE
		IRIDOCYCLITIS, SUPPURATIVE
		PANOPHTHALMITIS, PURULENT
		PUS
INFLAMMATION, ULCERATIVE	M-40030	ULCER
		ULCERATION
LEUKEMIA, LYMPHOBLASTIC	M-98213	LYMPHOBLASTIC LEUKEMIA
LEUKEMIA, LYMPHOCYTIC	M-98203	LEUKEMIC LYMPHOSARCOMA
LEUKEMIA, STEM CELL	M-98013	STEM CELL LEUKEMIA
LYMPHOMA	M-95903	LYMPHOMA
		MALIGNANT LYMPHOMA
LYMPHOMA, RETICULUM CELL	M-96403	RETICULUM CELL SARCOMA
MALAKOPLAKIA	M-43180	MALAKOPLAKIA
MELANOMA, MALIGNANT	M-87203	MALIGNANT MELANOMA
MINERALIZATION	M-55400	CALCIFICATION
		CALCIFICATION, DYSTROPHIC
		DEPOSITION, CALCIFEROUS
		MINERALIZATION
		MINERALIZATION, DYSTROPHIC
PHTHISIS BULBI	M-58500	PHTHISIS BULBI
SYNECHIA	M-49460	SYNECHIA
SYNECHIA, ANTERIOR	M-49470	SYNECHIA, ANTERIOR
SYNECHIA, POSTERIOR	M-49480	SYNECHIA, POSTERIOR
VASCULARIZATION	M-73070	VASCULARIZATION

PREFERRED TERM	SNOVET CODE	LADB TERMS
ADENOMA	M-81400	ADENOMA
		PAPILLARY ADENOMA
CYST	M-33400	CYST
CYSTADENOMA	M-84400	CYSTADENOMA
		PAPILLARY CYSTADENOMA
FIBROSIS	M-49000	FIBROSIS
HYPERPLASIA	M-72000	HYPERPLASIA
HYPERPLASIA, EPITHELIAL	M-72000	HYPERPLASIA, EPITHELIAL
INFLAMMATION	M-40000	INFLAMMATION
INFLAMMATION, NONSUPPURATIVE	M-47130	INFLAMMATION, LYMPHOCYTIC
LYMPHOMA	M-95903	LYMPHOMA
LYMPHOMA, RETICULUM CELL	M-96403	RETICULUM CELL SARCOMA
NECROSIS	M-54000	NECROSIS
PIGMENT DEPOSITION	M-57000	PIGMENTATION

PREFERRED TERM	SNOVET CODE	LADB TERMS
INFLAMMATION	M-40000	KERATITIS
		KERATITIS, MINERALIZING
INFLAMMATION, SUPPURATIVE	M-40400	KERATITIS, SUPPURATIVE
INFLAMMATION, ULCERATIVE	M-40030	KERATITIS, ULCERATIVE
		ULCERATION, CORNEAL
KERATOPATHY, BAND	M-55480	KERATOPATHY, BAND
VASCULARIZATION	M-73070	PANNUS

PREFERRED TERM	SNOVET CODE	LADB TERMS
ATROPHY	M-58000	AREA OF LOSS OF ROD AND CONE CELLS
		ATROPHY
		ATROPHY OF OUTER LIMBS
		ATROPHY, RETINAL
		THINNING RETINA
DEGENERATION	M-50000	DEGENERATION
		DEGENERATION, SEGMENTAL
INFLAMMATION	M-40000	INFLAMMATION
		RETINITIS
INFOLDING OF RETINA		IN-FOLDING OF RETINA
RETINAL ROSETTE		RETINAL ROSETTE

PREFERRED TERM	SNOVET CODE	LADB TERMS
INFLAMMATION	M-40000	INFLAMMATION

PREFERRED TERM	SNOVET CODE	LADB TERMS
ABSCESS	M-41740	ABSCESS
		ABSCESS, METASTATIC
ADENOCARCINOMA	M-81403	ADENOCARCINOMA
ADENOCARCINOMA, ALVEOLAR BRONCHIOLAR	M-82503	BRONCHIOLAR-ALVEOLAR ADENOCARCINOMA
		BRONCHO-ALVEOLAR ADENOCARCINOMA
ADENOCARCINOMA, MAMMARY (T-C4000)	M-81406	MAMMARY ADENOCARCINOMA
		MAMMARY GLAND ADENOCARCINOMA
ADENOCARCINOMA, PAPILLARY	M-82603	PAPILLARY ADENOCARCINOMA
ADENOCARCINOMA, SEBACEOUS	M-84103	SEBACEOUS ADENOCARCINOMA
ADENOFIBROSARCOMA, MAMMARY (T-04000)	M-90136	MAMMARY ADENOFIBROSARCOMA
ADENOMA, ALVEOLAR BRONCHIOLAR	M-82510	ALVEOLAR/BRONCHIOLAR ADENOMA
		BRONCHIOLAR-ALVEOLAR ADENOMA
		BRONCHO-ALVEOLAR ADENOMA
ALVEOLAR CELL TUMOR (T-28010)	M-80001	ALVEOLAR CELL TUMOR
ANGIOSARCOMA	M-91203	ANGIOSARCOMA
		HEMANGIOSARCOMA
ARTERIOSCLEROSIS	M-52000	ARTERIOSCLEROSIS
ATELECTASIS	M-34310	ATELECTASIS
CALCULI	M-30000	CRYSTAL
		CRYSTALS
CARCINOMA, ACINAR CELL	M-85506	ACINAR-CELL CARCINOMA
CARCINOMA, ALVEOLAR BRONCHIOLAR	M-82503	ALVEOLAR/BRONCHIOLAR CARCINOMA
CARCINOMA, C CELL	M-85106	C-CELL CARCINOMA
CARCINOMA, CORTICAL	M-82703	CORTICAL CARCINOMA
CARCINOMA, HEPATOCELLULAR	M-81703	HEPATOCELLULAR CARCINOMA
CARCINOMA, SEBACEOUS (T-02...)	M-84103	SEBACEOUS CARCINOMA
CARCINOMA, SQUAMOUS CELL	M-80703	SQUAMOUS CELL CARCINOMA
CARCINOMA, TRANSITIONAL CELL	M-81203	TRANSITIONAL-CELL CARCINOMA
CARCINOSARCOMA, MAMMARY (T-04000)	M-89806	MAMMARY CARCINOSARCOMA
CONGESTION	M-36100	CONGESTION
		CONGESTION, PASSIVE
		CONGESTION, PULMONARY
CYTOMEGALY	M-69220	CYTOMEGALY
EDEMA	M-36500	EDEMA
		EDEMA, PULMONARY
		EFFUSION
		EFFUSION, SEROUS
		PULMONARY EDEMA
EMPHYSEMA	M-32800	EMPHYSEMA
FIBROSARCOMA	M-88103	FIBROSARCOMA
FIBROSIS	M-49000	FIBROSIS
GRANULOMA	M-44000	GRANULOMA
		GRANULOMA, FOREIGN BODY
		GRANULOMA, NONSPECIFIC
HEMANGIOPERICYTOMA	M-91501	HEMANGIOPERICYTOMA
HEMORRHAGE	M-37000	HEMORRHAGE
HEMOSIDERIN DEPOSITION	M-57500	HEMOSIDERIN
		HEMOSIDEROSIS
HEPATIC NEOPLASM	M-81706	HEPATIC NEOPLASM
HISTIOCYTOMA, FIBROUS	M-88300	FIBROUS HISTIOCYTOMA

PREFERRED TERM	SNOVET CODE	LADB TERMS
HISTIOCYTOMA, MALIGNANT	M-88323	MALIGNANT HISTIOCYTOSIS
HYPERPLASIA	M-72000	HYPERPLASIA
HYPERPLASIA, ALVEOLAR (T-28010)	M-72000	HYPERPLASIA, ALVEOLAR EPITHELIUM
HYPERPLASIA, EPITHELIAL	M-72000	ADENOMATOSIS
		ADENOMATOSIS, PULMONARY
		ADENOMATOUS PROLIFERATION
		EPITHELIALIZATION
		HYPERPLASIA, ADENOMATOUS
		HYPERPLASIA, EPITHELIAL
		PULMONARY ADENOMATOSIS
HYPERPLASIA, HISTIOCYTIC	M-77800	HISTIOCYTOSIS
HYPERPLASIA, LYMPHOID	M-72200	HYPERPLASIA, LYMPHOID
		LYMPHOCYTIC NODULE
HYPERPLASIA, MYELOID	M-72000	LEUKEMOID REACTION
HYPERPLASIA, PAPILLARY	M-72050	HYPERPLASIA, PAPILLARY
HYPERPLASIA, PLASMA CELL	M-72150	HYPERPLASIA, PLASMA CELL
HYPERTROPHY	M-71000	HYPERTROPHY
INFLAMMATION	M-40000	BRONCHOPNEUMONIA
		INFLAMMATION
		INFLAMMATION, HYPERPLASTIC
		PNEUMONIA
		PNEUMONIA, LOBAR
		PNEUMONIA, RESOLVING
		PNEUMONIA, SEROUS, PROLIFERATIVE
		PNEUMONITIS
		PNEUMONITIS, UREMIC
INFLAMMATION, ATYPICAL/ADENOMATOID	M-72420	PNEUMONIA, ATYPICAL, ADENOMATOID
INFLAMMATION, GRANULOMATOUS	M-44000	PNEUMONIA, GRANULOMATOUS
		PNEUMONIA, LIPID
INFLAMMATION, HISTIOCYTIC	M-47180	ALVEOLAR MACROPHAGES
		ERYTHROPHAGOCYTOSIS
		FOAM CELL
		FOAM-CELL
		FOCUS OF MACROPHAGES
		HYPERPLASIA, MACROPHAGE
		HYPERPLASIA, RETICULOENDOTHELIAL
		PHAGOCYTOSIS
		PHAGOCYTOSIS, ALVEOLAR
		RETICULOENDOTHELIOSIS
INFLAMMATION, NECROTIZING	M-40700	BRONCHOPNEUMONIA, NECROTIZING
		PNEUMONIA, NECROTIZING
INFLAMMATION, NONSUPPURATIVE	M-47130	INFILTRATION, LYMPHOID
		INFILTRATION, NONSUPPURATIVE
		INFLAMMATION, LYMPHOCYTIC
		LYMPHOCYTE
		LYMPHOCYTIC CUFFING
		LYMPHOCYTIC INFLAMMATORY INFILTRATE
		LYMPHOCYTOSIS
		LYMPHOID COLLECTION
		PNEUMONIA, NONSUPPURATIVE

PREFERRED TERM	SNOVET CODE	LADB TERMS
INFLAMMATION, SUPPURATIVE	M-40400	BRONCHOPNEUMONIA, FIBRINOPURULENT
		BRONCHOPNEUMONIA, PURULENT
		BRONCHOPNEUMONIA, SUPPURATIVE
		PNEUMONIA, SUPPURATIVE
INTERSTITIAL CELL TUMOR	M-86506	INTERSTITIAL-CELL TUMOR
LEIOMYOSARCOMA	M-88903	LEIOMYOSARCOMA
LEUKEMIA	M-98003	LEUKEMIA
		LEUKEMIC INFILTRATION
LEUKEMIA, GRANULOCYTIC	M-98603	GRANULOCYTIC LEUKEMIA
LEUKEMIA, LYMPHOBLASTIC	M-98213	LYMPHOBLASTIC LEUKEMIA
LEUKEMIA, LYMPHOCYTIC	M-98203	LEUKEMIC LYMPHOSARCOMA
		LYMPHATIC LEUKEMIA
		LYMPHOCYTIC LEUKEMIA
LEUKEMIA, STEM CELL	M-98013	STEM CELL LEUKEMIA
LIPOSARCOMA	M-88503	LIPOSARCOMA
LYMPHOMA	M-95903	LYMPHOMA
		LYMPHOSARCOMATOUS INFILTRATION
		MALIGNANT LYMPHOMA
LYMPHOMA, HISTIOCYTIC	M-96403	MALIGNANT LYMPHOMA, HISTIOCYTIC TYPE
LYMPHOMA, LYMPHOCYTIC	M-96203	MALIGNANT LYMPHOMA, LYMPHOCYTIC TYPE
LYMPHOMA, RETICULUM CELL	M-96403	RETICULUM CELL SARCOMA
MESOTHELIOMA	M-90503	MESOTHELIOMA
METAPLASIA, ALVEOLAR	M-73000	METAPLASIA, ALVEOLAR
METAPLASIA, ALVEOLAR (T-28010)	M-73000	METAPLASIA
METAPLASIA, OSSEOUS	M-73400	METAPLASIA, OSSEOUS
METAPLASIA, SQUAMOUS	M-73220	METAPLASIA, SQUAMOUS
MINERALIZATION	M-55400	CALCIFICATION
		CALCIFICATION, DYSTROPHIC
		CALCIUM DEPOSIT
		MINERALIZATION
	M-57500	MINERALIZATION, DYSTROPHIC
MIXED TUMOR, MALIGNANT	M-89403	MIXED TUMOR, MALIGNANT
NECROSIS	M-54000	NECROSIS
NEOPLASM	M-80001	TUMOR
NEUROFIBROSARCOMA	M-95403	NEUROFIBROSARCOMA
OSTEOMA	M-91800	OSTEOMA
OSTEOSARCOMA	M-91806	OSTEOSARCOMA
PERIVASCULITIS/VASCULITIS	M-47800	ARTERITIS
		CUFFING, PERIVASCULAR
		PERIARTERITIS
		PERIARTERITIS, MACROPHAGES
		PERIARTERITIS, NONSUPPURATIVE
		PERIVASCULAR CUFFING
		PERIVASCULITIS
		PERIVASCULITIS, MIXED LEUKOCYTIC
		PERIVASCULITIS, NONSUPPURATIVE
		POLYARTERITIS
		VASCULITIS
		VASCULITIS, NONSUPPURATIVE
PHEOCHROMOCYTOMA, MALIGNANT	M-87003	PHEOCHROMOCYTOMA

PREFERRED TERM	SNOVET CODE	LADB TERMS
PIGMENT DEPOSITION	M-57000	PIGMENTATION
PNEUMONIA, ASPIRATION (F-79000)	M-40000	PNEUMONIA, ASPIRATION
		PNEUMONITIS, ASPIRATION
PNEUMONIA, CHRONIC MURINE	D-75062	PNEUMONIA, CHRONIC MURINE
		PNEUMONIA, MURINE
PROTEINOSIS, ALVEOLAR (T-28010)	M-55640	PROTEINOSIS, ALVEOLAR
RHABDOMYOSARCOMA	M-89003	RHABDOMYOSARCOMA
SARCOMA	M-88003	SARCOMA
SARCOMA, HISTIOCYTIC	M-96403	HISTIOCYTIC SARCOMA
THROMBUS	M-35100	COAGULATION, INTRAVASCULAR
		THROMBUS
THROMBUS, ORGANIZED	M-35103	THROMBUS, ORGANIZED
THYMOMA	M-85806	THYMOMA

PREFERRED TERM	SNOVET CODE	LADB TERMS
ADENOMA	M-81401	BRONCHIAL ADENOMA
BRONCHIECTASIS	M-32100	BRONCHIECTASIS
		BRONCHIOLECTASIS
CARCINOMA	M-80103	BRONCHIAL CARCINOMA
		CARCINOMA
INFLAMMATION	M-40000	BRONCHITIS
		PERIBRONCHIOLITIS
		PERIBRONCHITIS
INFLAMMATION, GRANULOMATOUS	M-44000	INFLAMMATION, GRANULOMATOUS
INFLAMMATION, NECROTIZING	M-40700	INFLAMMATION, NECROTIZING
INFLAMMATION, NONSUPPURATIVE	M-47130	INFILTRATION, LYMPHOID, NODULAR
		LYMPHOCYTIC AND PLASMA CELL BRONCHIAL CUFF
		PERIBRONCHIOLAR LYMPHOCYTOSIS
		PERIBRONCHIOLITIS, LYMPHOCYTIC
		PERIBRONCHITIS, NONSUPPURATIVE
INFLAMMATION, SUPPURATIVE	M-40400	BRONCHIOLITIS, PURULENT
		BRONCHITIS, PURULENT

ORGAN SYSTEM: RESPIRATORY SYSTEM ORGAN: LUNG; INTERSTITIUM (T-28780)

PREFERRED TERM	SNOVET CODE	LADB TERMS
INFLAMMATION	M-40000	INFLAMMATION, INTERSTITIAL
		PNEUMONIA, INTERSTITIAL
		PNEUMONIA, VIRAL
		PNEUMONITIS, INTERSTITIAL
INFLAMMATION, NONSUPPURATIVE	M-47130	INFLAMMATION, NONSUPPURATIVE
		PNEUMONIA, INTERSTITIAL, NONSUPPURATIVE
INFLAMMATION, SUPPURATIVE	M-40400	INFLAMMATION, SUPPURATIVE
		PNEUMONIA, INTERSTITIAL, EXUDATIVE

PREFERRED TERM	SNOVET CODE	LADB TERMS
INFLAMMATION	M-40000	INFLAMMATION
INFLAMMATION, SUPPURATIVE	M-40400	INFLAMMATION, SUPPURATIVE
		NASOPHARYNGITIS, SUPPURATIVE

PREFERRED TERM	SNOVET CODE	LADB TERMS
ADENOCARCINOMA	M-81403	ADENOCARCINOMA
CARCINOMA	M-80103	CARCINOMA
CARCINOMA IN SITU	M-80102	CARCINOMA-IN-SITU
CARCINOMA, SQUAMOUS CELL	M-80703	SQUAMOUS CELL CARCINOMA
CYST	M-33400	CYST
DILATATION	M-32100	DILATATION
FIBROMA	M-81000	FIBROMA
HYPERPLASIA, LYMPHOID	M-72200	HYPERPLASIA, LYMPHOID
HYPERPLASIA, PLASMA CELL	M-72150	HYPERPLASIA, PLASMA CELL
INFLAMMATION	M-40000	INFLAMMATION
		TRACHEITIS
INFLAMMATION, NONSUPPURATIVE	M-47130	INFLAMMATION, NONSUPPURATIVE
		LYMPHOCYTIC INFLAMMATORY INFILTRATE
		TRACHEITIS, LYMPHOID HETEROPHILIC
		TRACHEITIS, NONSUPPURATIVE
INFLAMMATION, SUPPURATIVE	M-40400	INFLAMMATION, SUPPURATIVE
		TRACHEITIS, FIBRINOPURULENT
		TRACHEITIS, MUCOPURULENT
		TRACHEITIS, SUPPURATIVE
LYMPHOMA	M-95903	LYMPHOMA
LYMPHOMA, RETICULUM CELL	M-96403	RETICULUM CELL SARCOMA
METAPLASIA, SQUAMOUS	M-73220	METAPLASIA, SQUAMOUS
MINERALIZATION	M-55400	MINERALIZATION
POLYP, INFLAMMATORY	M-76820	POLYP, INFLAMMATORY
RHABDOMYOSARCOMA	M-89003	RHABDOMYOSARCOMA
SARCOMA	M-88003	SARCOMA

PREFERRED TERM	SNOVET CODE	LADB TERMS
ADENOMA	M-81400	ADENOMA
ADENOMA, PAPILLARY	M-82600	PAPILLARY ADENOMA
ATROPHY	M-58000	ATROPHY
CARCINOMA	M-80103	CARCINOMA
CARCINOMA, SQUAMOUS CELL	M-80703	SQUAMOUS CELL CARCINOMA
CYST	M-33400	CYST
CYSTADENOMA	M-84400	CYSTADENOMA
CYSTADENOMA, PAPILLARY	M-84500	PAPILLARY CYST ADENOMA
		PAPILLARY CYSTADENOMA
FIBROSIS	M-49000	FIBROSIS
HYPERPLASIA	M-72000	HYPERPLASIA
		HYPERPLASIA, DUCTULAR
INFLAMMATION	M-40000	DACRYOADENITIS
		DACRYOADENITIS, ATROPHIC
		INFLAMMATION
INFLAMMATION, NONSUPPURATIVE	M-47130	DACRYOADENITIS, NONSUPPURATIVE
		INFILTRATION, LYMPHOID
		INFLAMMATION, LYMPHOCYTIC
		SIALITIS, NONSUPPURATIVE
LYMPHOMA	M-95903	LYMPHOMA
LYMPHOMA, RETICULUM CELL	M-96403	RETICULUM CELL SARCOMA

PREFERRED TERM	SNOVET CODE	LADB TERMS
ABSCESS	M-41740	ABSCESS
ADENOCARCINOMA	M-81403	ADENOCARCINOMA
		MAMMARY ADENOCARCINOMA
		PAPILLARY ADENOCARCINOMA
		PAPILLARY CYSTADENOCARCINOMA
ADENOMA	M-81400	ADENOMA
		CYST ADENOMA
		CYSTADENOMA
		DUCTAL ADENOMA
		PAPILLARY CYST ADENOMA
		PAPILLARY CYSTADENOMA
		TUBULAR ADENOMA
ADENOMA, PAPILLARY	M-82600	PAPILLARY ADENOMA
ADENOSIS	M-74200	ADENOSIS
ANGIOMA	M-91200	ANGIOMA
ATROPHY	M-58000	ATROPHY
		INVOLUTION
BASAL CELL TUMOR	M-80901	BASAL-CELL TUMOR
CARCINOMA	M-80103	CARCINOMA
		INFILTRATING DUCT CARCINOMA
CARCINOMA, ACINAR CELL	M-85503	ACINAR-CELL CARCINOMA
CARCINOMA, C CELL	M-85106	C-CELL CARCINOMA
CARCINOMA, MEDULLARY	M-85103	MEDULLARY CARCINOMA
CARCINOMA, UNDIFFERENTIATED	M-80203	UNDIFFERENTIATED CARCINOMA
CARCINOSARCOMA	M-89803	CARCINOSARCOMA
CYST	M-33400	CYST
		CYST, EPIDERMAL INCLUSION
		CYST, RUPTURED
		CYSTIC GLAND
		DILATATION, CYSTIC
CYST, HEMORRHAGIC	M-33540	CYST, HEMAL
		HEMORRHAGIC CYST
DYSPLASIA	M-74000	DYSPLASIA
DYSPLASIA, STROMAL (T-04500)	M-74000	DYSPLASIA, STROMAL
ECTASIA, DUCT	M-32100	DILATATION
ECTASIA, DUCT (T-04300)	M-32100	CYSTIC DUCTS
FIBROADENOMA	M-90100	CRIBIFORM ADENOFIBROMA
		FIBROADENOMA
		FIBROADENOMA, SECRETING
		PERICANALICULAR FIBROADENOMA
FIBROCYSTIC DISEASE	M-74320	DISEASE, FIBROCYSTIC
FIBROMA	M-81000	FIBROMA
FIBROMA, STROMAL (T-04500)	M-88100	STROMAL FIBROMA
FIBROSIS	M-49000	FIBROSIS
GALACTOCELE	M-33220	GALACTOCELE
		GALACTOCELE, RUPTURE
GRANULOMA	M-44000	GRANULOMA, LIPOID
GYNECOMASTIA (T-04040)	M-71000	GYNECOMASTIA
		GYNECOMASTIA, ASYMMETRICAL
HEMORRHAGE	M-37000	HEMORRHAGE

PREFERRED TERM	SNOVET CODE	LADB TERMS
HEMOSIDERIN DEPOSITION	M-57500	HEMOSIDERIN
HISTIOCYTOMA, FIBROUS, MALIGNANT	M-88303	FIBROUS HISTIOCYTOMA, MALIGNANT
HYPERPLASIA	M-72000	HYPERPLASIA
HYPERPLASIA, CYSTIC	M-72060	HYPERPLASIA, CYSTIC
HYPERPLASIA, EPITHELIAL	M-72000	EPITHELIALIZATION
		HYPERPLASIA, EPITHELIAL
		HYPERPLASIA, PAPILLARY
		PAPILLOMATOSIS
HYPERPLASIA, HISTIOCYTIC	M-72000	PIGMENTED MACROPHAGES
HYPERPLASIA, NODULAR	M-72030	HYPERPLASIA, NODULAR
HYPERTROPHY	M-71000	HYPERTROPHY
		HYPERTROPHY, FUNCTIONAL
INFLAMMATION	M-40000	INFLAMMATION
		MASTITIS
INFLAMMATION, GRANULOMATOUS	M-44000	INFLAMMATION, FOREIGN BODY
		MASTITIS, PYOGRANULOMATOUS
INFLAMMATION, NONSUPPURATIVE	M-47130	INFLAMMATION, NONSUPPURATIVE
LEIOMYOSARCOMA	M-88903	LEIOMYOSARCOMA
LEUKEMIA	M-98003	LEUKEMIC INFILTRATION
LEUKEMIA, LYMPHOCYTIC	M-98203	LYMPHATIC LEUKEMIA
LYMPHOMA	M-95903	LYMPHOMA
		MALIGNANT LYMPHOMA
LYMPHOMA, RETICULUM CELL	M-96403	RETICULUM CELL SARCOMA
METAPLASIA, SQUAMOUS	M-73220	METAPLASIA, SQUAMOUS
MINERALIZATION	M-55400	CALCIFICATION
		MINERALIZATION
MYOEPITHELIOMA	M-89820	MYOEPITHELIOMA
NECROSIS	M-54000	NECROSIS
NECROSIS, FAT	M-54110	NECROSIS, FAT
PAPILLOMA, INTRADUCTAL	M-85030	INTRADUCTAL PAPILLOMA

PREFERRED TERM	SNOVET CODE	LADB TERMS
ABSCESS	M-41740	ABSCESS
		ABSCESS, DERMAL
		SMALL ABSCESSES
ACANTHOSIS	M-72710	ACANTHOSIS
ADENOCARCINOMA	M-81403	ADENOCARCINOMA
ADENOCARCINOMA, MAMMARY (T-04000)		MAMMARY ADENOCARCINOMA
ADENOCARCINOMA, SEBACEOUS	M-84103	SEBACEOUS ADENOCARCINOMA
ADENOMA	M-81400	ADENOMA
		HYDROCYSTADENOMA
ADENOMA, APOCRINE GLAND	M-84010	APOCRINE GLAND ADENOMA
		APOCRINE GLAND CYSTADENOMA
ADENOMA, PAPILLARY SEBACEOUS (T-01310)	M-82600	PAPILLARY SEBACEOUS ADENOMA
ADENOMA, SEBACEOUS	M-84100	SEBACEOUS TUMOR
ADENOMA, SEBACEOUS (T-02...)	M-84100	SEBACEOUS ADENOMA
ANGIOMA	M-91200	ANGIOMA
		HEMANGIOMA
ANGIOSARCOMA	M-91203	ANGIOSARCOMA
		HEMANGIOENDOTHELIOMA
		HEMANGIOSARCOMA
ATROPHY	M-58000	ATROPHY
CARCINOMA	M-80103	CARCINOMA
CARCINOMA, BASAL CELL	M-80903	BASAL-CELL CARCINOMA
		BASAL-CELL TUMOR
		KERATOTIC BASAL CELL CARCINOMA
CARCINOMA, BASOSQUAMOUS	M-80943	BASOSQUAMOUS CARCINOMA
CARCINOMA, SEBACEOUS	M-84103	SEBACEOUS CARCINOMA
CARCINOMA, SQUAMOUS CELL	M-80703	SQUAMOUS CELL CARCINOMA
CARCINOMA, UNDIFFERENTIATED	M-80203	UNDIFFERENTIATED CARCINOMA
CARCINOSARCOMA	M-89803	CARCINOSARCOMA
CELLULITIS	M-40200	CELLULITIS
CHONDROMA	M-92200	CHONDROMA
CYST	M-33400	CYST
		CYST, RUPTURED
		CYSTIC FOLLICLE
		MULTIPLE CYSTS
CYST, EPIDERMAL INCLUSION	M-33410	CYST, EPIDERMAL INCLUSION
		CYST, EPIDERMAL INCLUSION, RUPTURED
		DERMAL INCLUSION CYST
		EPIDERMAL INCLUSION CYST
CYST, HEMORRHAGIC	M-33540	HEMORRHAGIC CYST
CYST, SEBACEOUS	M-33430	CYST, SEBACEOUS
		SEBACEOUS CYST
DYSPLASIA, FIBROUS	M-74910	FIBROUS DYSPLASIA
EDEMA	M-36500	EDEMA
EPITHELIOMA, BASAL CELL	M-80903	KERATOTIC BASAL CELL EPITHELIOMA
FIBROADENOMA (T-04...)	M-90100	FIBROADENOMA
FIBROMA	M-81000	DERMAL FIBROMA
		FIBROMA
		FIBROMA DURUM
		SUBCUTANEOUS FIBROMA

PREFERRED TERM	SNOVET CODE	LADB TERMS
FIBROSARCOMA	M-88103	FIBROSARCOMA
FIBROSIS	M-49000	FIBROSIS
		GRANULATION TISSUE
		INFLAMMATION WITH FIBROSIS
		SCAR
GRANULOMA	M-44000	GRANULOMA
		GRANULOMA, FOREIGN BODY
		GRANULOMA, PYOGENIC
		REACTION, FOREIGN BODY
HAMARTOMA	M-75500	HAMARTOMA
HEMANGIOPERICYTOMA	M-91501	HEMANGIOPERICYTOMA
HEMANGIOPERICYTOMA, MALIGNANT	M-91503	HEMANGIOPERICYTOMA, MALIGNANT
HEMATOMA	M-37100	HEMATOMA
HEMORRHAGE	M-37000	HEMORRHAGE
HEMOSIDERIN DEPOSITION	M-57500	HEMOSIDERIN
HISTIOCYTOMA, FIBROUS	M-88300	FIBROUS HISTIOCYTOMA
HISTIOCYTOMA, FIBROUS, MALIGNANT	M-88303	FIBROUS HISTIOCYTOMA, MALIGNANT
HYPERKERATOSIS	M-72600	HYPERKERATOSIS
HYPERPLASIA, BASAL CELL	M-72120	HYPERPLASIA, BASAL CELL
HYPERPLASIA, CYSTIC	M-72060	HYPERPLASIA, CYSTIC
HYPERPLASIA, SEBACEOUS		HYPERPLASIA, SEBACEOUS
INFLAMMATION	M-40000	DERMATITIS
		DERMATITIS, NONSPECIFIC
		FASCIITIS
		HIDRADENITIS
		INFLAMMATION
		PANNICULITIS, HEALED
INFLAMMATION, GRANULOMATOUS	M-44000	INFLAMMATION, GRANULOMATOUS
		INFLAMMATION, NECROTIZING GRANULOMATOUS
		INFLAMMATION, VESICULAR GRANULOMATOUS
		LIPOGRANULOMA
INFLAMMATION, NECROTIZING	M-40700	INFLAMMATION, NECROTIZING
INFLAMMATION, NONSUPPURATIVE	M-47130	INFILTRATION, LYMPHOID
		INFILTRATION, MAST CELL
		LYMPHOCYTIC INFLAMMATORY INFILTRATE
INFLAMMATION, SUPPURATIVE	M-40400	INFLAMMATION, SUPPURATIVE
INFLAMMATION, ULCERATIVE	M-40030	DERMATITIS, ULCERATIVE
		ULCER
		ULCERATION
KELOID	M-49720	DERMAL KELOID
		KELOID
KERATOACANTHOMA	M-72860	KERATOACANTHOMA
LEIOMYOMA	M-88900	LEIOMYOMA
LEIOMYOSARCOMA	M-88903	LEIOMYOSARCOMA
LEUKEMIA	M-98003	LEUKEMIC INFILTRATION
LEUKEMIA, HISTIOCYTIC	M-98903	LEUKEMIC HISTIOCYTIC SARCOMA
LEUKEMIA, LYMPHOBLASTIC	M-98213	LYMPHOBLASTIC LEUKEMIA
LEUKEMIA, STEM CELL	M-98013	STEM CELL LEUKEMIA
LIPOMA	M-88500	LIPOMA
LIPOSARCOMA	M-88503	LIPOSARCOMA

PREFERRED TERM	SNOVET CODE	LADB TERMS
LYMPHOMA	M-95903	LYMPHOMA
		MALIGNANT LYMPHOMA
LYMPHOMA, HISTIOCYTIC	M-96403	MALIGNANT LYMPHOMA, HISTIOCYTIC TYPE
LYMPHOMA, RETICULUM CELL	M-96403	RETICULUM CELL SARCOMA
MAST CELL TUMOR	M-97401	MAST-CELL TUMOR
MELANOMA, MALIGNANT	M-87203	MALIGNANT MELANOMA
METAPLASIA, OSSEOUS	M-73400	METAPLASIA, OSSEOUS
METAPLASIA, SQUAMOUS	M-73220	METAPLASIA, SQUAMOUS
MINERALIZATION	M-55400	CALCIFICATION
		CALCINOSIS
		MINERALIZATION
MYOEPITHELIOMA	M-89820	MYOEPITHELIOMA
MYXOFIBROMA	M-88110	MYXOFIBROMA
MYXOMA	M-88400	MYXOMA
NECROSIS	M-54000	NECROSIS
		NECROSIS, PRESSURE
NECROSIS, FAT	M-54110	NECROSIS, FAT
NEOPLASM	M-80001	TUMOR
NEURILEMMOMA	M-95600	NEURILEMOMA
NEURILEMMOMA, MALIGNANT	M-95603	NEURILEMOMA, MALIGNANT
NEUROFIBROMA	M-95400	NEUROFIBROMA
NEUROFIBROSARCOMA	M-95403	NEUROFIBROSARCOMA
NEUROMA, MALIGNANT	M-95603	MALIGNANT NEURINOMA
OSTEOSARCOMA	M-91806	OSTEOSARCOMA
PAPILLOMA	M-80500	PAPILLOMA
PAPILLOMA, BASOSQUAMOUS	M-72750	BASOSQUAMOUS PAPILLOMA
PAPILLOMA, SEBACEOUS		SEBACEOUS PAPILLOMA
PAPILLOMA, SQUAMOUS CELL	M-80520	SQUAMOUS CELL PAPILLOMA
PARAKERATOSIS	M-74030	PARAKERATOSIS
POLYP, INFLAMMATORY	M-76820	POLYP, INFLAMMATORY
RHABDOMYOSARCOMA	M-89003	RHABDOMYOSARCOMA
SARCOMA	M-88003	SARCOMA
SARCOMA, HISTIOCYTIC	M-96403	HISTIOCYTIC SARCOMA
THROMBUS	M-35100	THROMBOSIS
		THROMBUS
TRICHOEPITHELIOMA	M-81000	TRICHOEPITHELIOMA

PREFERRED TERM	SNOVET CODE	LADB TERMS
ABSCESS	M-41740	ABSCESS
		ABSCESS, METASTATIC
ADENOCARCINOMA	M-81403	ADENOCARCINOMA
		TUBULAR-CELL ADENOCARCINOMA
ADENOMA	M-81400	ADENOMA
ADENOMA, PAPILLARY TUBULAR (T-71300)	M-82600	PAPILLARY TUBULAR ADENOMA
ADENOMA, RENAL TUBULAR	M-82110	RENAL TUBULAR ADENOMA
		TUBULAR-CELL ADENOMA
AMYLOID DEPOSITION		AMYLOID
		AMYLOIDOSIS
ANGIOLIPOMA	M-88610	ANGIOLIPOMA
ANGIOSARCOMA	M-91203	HEMANGIOSARCOMA
ARTERIOLONEPHROSCLEROSIS (T-46670)	M-52200	ARTERIOLONEPHROSCLEROSIS
ARTERIOSCLEROSIS	M-52000	ARTERIOSCLEROSIS
ATROPHY	M-58000	ATROPHY
CALCULI	M-30000	CALCIFIED CASTS
		CALCULUS
		CONCRETION
		LITHIASIS
		MICROCONCRETIONS
		MINERALIZED CASTS
		NEPHROLITHIASIS
		UROLITHIASIS
		UROLITHIASIS, RENAL TUBULAR
CARCINOMA	M-80103	CARCINOMA
	M-83123	RENAL CARCINOMA
		RENAL CELL CARCINOMA
CARCINOMA, ACINAR CELL	M-85503	ACINAR-CELL CARCINOMA
CARCINOMA, HEPATOCELLULAR	M-81703	HEPATOCELLULAR CARCINOMA
CARCINOMA, TRANSITIONAL CELL	M-81203	TRANSITIONAL CELL CARCINOMA
		TRANSITIONAL-CELL CARCINOMA
CAST, PROTEINACEOUS	M-33030	CAST, PROTEINACEOUS
CAST, TUBULE (T-71300)	M-33010	CAST
		CAST, HYALINE
		CAST, TUBULE
		TUBULAR CALCIFIED CASTS
CELLULAR CHANGE	M-66040	BASOPHILIC CYTOPLASMIC CHANGE
CELLULAR CHANGE, VACUOLAR	M-66040	CYTOPLASMIC VACUOLIZATION
CONGESTION	M-36100	CONGESTION
CYST	M-33400	CYST
		CYST, RETENTION
		MULTILOCULAR CYST
		MULTIPLE CYSTS
DEGENERATION	M-50000	DEGENERATION
		DEGENERATION, HYALINE
		DEGENERATIVE CHANGE
		NEPHROSIS
		NEPHROSIS, CHOLEMIC
		PYONEPHROSIS
DILATATION, TUBULAR	M-32100	DILATATION, TUBULAR

PREFERRED TERM	SNOVET CODE	LADB TERMS
DOUBLE PELVIS (T-72000)	M-22360	DOUBLE PELVIS
DYSPLASIA, NODULAR	M-74000	DYSPLASIA, NODULAR
FATTY CHANGE	M-50080	METAMORPHOSIS, FATTY
FIBROSARCOMA	M-88103	FIBROSARCOMA
FIBROSIS	M-49000	FIBROSIS
		SCAR
		SCARRING
		SCLEROSIS
GLOMERULONEPHRITIS (T-71200)	M-40000	GLOMERULONEPHRITIS
GLOMERULONEPHRITIS, MEMBRANOUS (T-71200)	M-46840	GLOMERULONEPHRITIS, MEMBRANOUS
GLOMERULOSCLEROSIS	M-53300	GLOMERULONEPHROSCLEROSIS
		GLOMERULOSCLEROSIS
HAMARTOMA	M-75500	HAMARTOMA
HEMOSIDERIN DEPOSITION	M-57500	HEMOSIDERIN
		HEMOSIDEROSIS
HISTIOCYTOMA, FIBROUS	M-88300	FIBROUS HISTIOCYTOMA
HISTIOCYTOMA, MALIGNANT	M-88323	MALIGNANT HISTIOCYTOSIS
HYALINOSIS	M-50060	HYALINE MEMBRANE
		HYALINOSIS
HYDRONEPHROSIS (T-72000)	M-33300	DILATATION
		HYDRONEPHROSIS
HYPERPLASIA, EPITHELIAL	M-72000	HYPERPLASIA
		HYPERPLASIA, EPITHELIAL
HYPERPLASIA, LYMPHOID	M-72200	HYPERPLASIA, LYMPHOID
HYPERPLASIA, PROLIFERATIVE	M-72020	HYPERPLASIA, PROLIFERATIVE
HYPERPLASIA, RETICULUM CELL	M-72180	HYPERPLASIA, RETICULUM CELL
HYPERPLASIA, TUBULAR CELL (T-71300)	M-72000	HYPERPLASIA, TUBULAR CELL
INFARCT	M-54700	INFARCT
		INFARCT, HEALED
INFLAMMATION	M-40000	NEPHRITIS
		PYELITIS
		PYELONEPHRITIS
		PYELONEPHRITIS, HEALED
INFLAMMATION, HISTIOCYTIC	M-47180	FOAMY TUBULAR CELLS
INFLAMMATION, INTERSTITIAL (T-71040)	M-40000	INFLAMMATION, INTERSTITIAL
		NEPHRITIS, INTERSTITIAL
INFLAMMATION, MEMBRANOUS	M-40540	INFLAMMATION, MEMBRANOUS
INFLAMMATION, NONSUPPURATIVE	M-47130	INFILTRATION, LYMPHOID
		INFILTRATION, LYMPHOPLASMACYTIC
		INFILTRATION, MONONUCLEAR CELL
		INFLAMMATION, LYMPHOCYTIC
		INFLAMMATION, NONSUPPURATIVE
		LYMPHOCYTE
		LYMPHOCYTIC INFLAMMATORY INFILTRATE
		LYMPHOCYTOSIS
		LYMPHOID CELLS
		NEPHRITIS, NONSUPPURATIVE
		PYELITIS, LYMPHOCYTIC
INFLAMMATION, SUPPURATIVE	M-40400	INFLAMMATION, SUPPURATIVE
		PYELITIS, PURULENT

PREFERRED TERM	SNOVET CODE	LADB TERMS
INFLAMMATION, SUPPURATIVE	M-40400	PYELONEPHRITIS, SUPPURATIVE
INFLAMMATION, ULCERATIVE	M-40030	ULCERATION
LEIOMYOSARCOMA	M-88903	LEIOMYOSARCOMA
LEUKEMIA	M-98003	LEUKEMIA
		LEUKEMIC INFILTRATION
LEUKEMIA, GRANULOCYTIC	M-98603	GRANULOCYTIC LEUKEMIA
LEUKEMIA, LYMPHOBLASTIC	M-98213	LYMPHOBLASTIC LEUKEMIA
LEUKEMIA, LYMPHOCYTIC	M-98203	LEUKEMIC LYMPHOSARCOMA
		LYMPHATIC LEUKEMIA
		LYMPHOCYTIC LEUKEMIA
LEUKEMIA, STEM CELL	M-98013	STEM CELL LEUKEMIA
LIPID DEPOSITION		LIPOIDOSIS
		NEPHROSIS, LIPOID
LIPOMA	M-88500	LIPOMA
LIPOSARCOMA	M-88503	LIPOSARCOMA
LYMPHOMA	M-95903	LYMPHOMA
		LYMPHOSARCOMATOUS INFILTRATION
		MALIGNANT LYMPHOMA
LYMPHOMA, HISTIOCYTIC	M-96403	HISTIOCYTIC LYMPHOSARCOMA
		MALIGNANT LYMPHOMA, HISTIOCYTIC TYPE
LYMPHOMA, LYMPHOCYTIC	M-96203	MALIGNANT LYMPHOMA, LYMPHOCYTIC TYPE
LYMPHOMA, RETICULUM CELL	M-96403	RETICULUM CELL SARCOMA
MESOTHELIOMA	M-90503	MESOTHELIOMA
METAPLASIA, OSSEOUS	M-73400	METAPLASIA, OSSEOUS
MINERALIZATION	M-55400	CALCIFICATION
		CALCIFICATION, DYSTROPHIC
		CALCINOSIS
		CALCIUM DEPOSIT
		MINERALIZATION
		MINERALIZATION, DYSTROPHIC
		NEPHROCALCINOSIS
MINERALIZATION, VASCULAR (T-40000)	M-55400	CALCIFICATION, MEDIAL
MIXED TUMOR, MALIGNANT	M-89403	MIXED TUMOR, MALIGNANT
NECROSIS	M-54000	NECROSIS
		NECROSIS, COAGULATION
NECROSIS, FAT	M-54110	NECROSIS, FAT
NECROSIS, ISCHEMIC	M-54200	NECROSIS, ISCHEMIC
NECROSIS, MEDULLARY (T-71070)	M-54000	NECROSIS, MEDULLARY
NEPHROPATHY	M-53150	NEPHROPATHY
NEPHROPATHY, SENILE		NEPHROPATHY, SENILE
NEPHROPATHY, TOXIC	M-53100	NEPHROPATHY, TOXIC
NEPHROSCLEROSIS (T-46600)	M-52000	NEPHROSCLEROSIS
NEPHROSCLEROSIS, SENILE		NEPHROSCLEROSIS, SENILE
NEUROFIBROSARCOMA	M-95403	NEUROFIBROSARCOMA
OSTEOSARCOMA	M-91803	OSTEOSARCOMA
PAPILLOMA	M-80500	PAPILLOMA
PAPILLOMA, TRANSITIONAL CELL	M-81200	TRANSITIONAL-CELL PAPILLOMA
PERIVASCULITIS/VASCULITIS	M-47800	CUFFING, PERIVASCULAR
		PERIARTERITIS
		PERIARTERITIS, NONSUPPURATIVE

PREFERRED TERM	SNOVET CODE	LADB TERMS
PERIVASCULITIS/VASCULITIS	M-47800	PERIVASCULAR CUFFING
		PERIVASCULITIS
		PERIVASCULITIS, NONSUPPURATIVE
PIGMENT DEPOSITION	M-57000	PIGMENT
		PIGMENTATION
PIGMENT DEPOSITION, LIPOCHROME	M-57740	LIPOCHROME PIGMENTATION
PIGMENT DEPOSITION, LIPOFUSCIN	M-57740	LIPOFUSCIN
		PIGMENT, LIPOFUSCIN
PYELONEPHRITIS (T-71900)	M-40000	CELLULAR FOCI
		INFLAMMATION
		PYELITIS, PROLIFERATIVE
REGENERATION	M-79900	REGENERATION
RHABDOMYOSARCOMA	M-89003	RHABDOMYOSARCOMA
SARCOMA	M-88003	SARCOMA
SARCOMA, HISTIOCYTIC	M-96403	HISTIOCYTIC SARCOMA
SARCOMA, UNDIFFERENTIATED	M-88003	UNDIFFERENTIATED SARCOMA
THROMBUS	M-35100	THROMBOSIS
THYMOMA	M-85806	THYMOMA

PREFERRED TERM	SNOVET CODE	LADB TERMS
AMYLOID DEPOSITION		AMYLOIDOSIS
CARCINOMA, TRANSITIONAL CELL	M-81203	TRANSITIONAL CELL CARCINOMA
CELLULAR CHANGE, VACUOLAR	M-66040	VACUOLIZATION
DILATATION	M-32100	DISTENSION
HYPERPLASIA	M-72000	HYPERPLASIA
HYPERPLASIA, EPITHELIAL	M-72000	HYPERPLASIA, EPITHELIAL
INFLAMMATION	M-40000	INFLAMMATION
		URETERITIS
INFLAMMATION, NONSUPPURATIVE	M-47130	INFLAMMATION, LYMPHOCYTIC
LYMPHOMA	M-95903	LYMPHOMA
		MALIGNANT LYMPHOMA
LYMPHOMA, RETICULUM CELL	M-96403	RETICULUM CELL SARCOMA

PREFERRED TERM	SNOVET CODE	LADB TERMS
CALCULI	M-30000	CALCULUS
		CONCRETION
HYPERPLASIA	M-72000	HYPERPLASIA
HYPERPLASIA, EPITHELIAL	M-72000	HYPERPLASIA, EPITHELIAL
INFLAMMATION	M-40000	INFLAMMATION
INFLAMMATION, NONSUPPURATIVE	M-47130	INFLAMMATION, LYMPHOCYTIC
INFLAMMATION, SUPPURATIVE	M-40400	URETHRITIS, SUPPURATIVE
LYMPHOMA	M-95903	LYMPHOMA
LYMPHOMA, RETICULUM CELL	M-96403	RETICULUM CELL SARCOMA
OBSTRUCTION	M-34000	OBSTRUCTION

PREFERRED TERM	SNOVET CODE	LADB TERMS
CALCULI	M-30000	CALCULUS
		CALCULUS, CYSTIC
		CONCRETION
		FOREIGN MATERIAL, REFRACTILE SPHERULES
		SOFT STONE
		UROLITHIASIS
CARCINOMA	M-80103	CARCINOMA
CARCINOMA, PAPILLARY	M-80503	PAPILLARY CARCINOMA
CARCINOMA, TRANSITIONAL CELL	M-81203	TRANSITIONAL-CELL CARCINOMA
CARCINOMA, UNDIFFERENTIATED	M-80203	UNDIFFERENTIATED CARCINOMA
CAST, PROTEINACEOUS	M-33030	CAST
		CAST, EOSINOPHILIC
CELLULAR CHANGE, VACUOLAR	M-66040	VACUOLIZATION
CONGESTION	M-36100	CONGESTION
CYST	M-33400	CYST
		DISTENTION, CYSTIC
		MUCOCELE
DILATATION	M-32100	DISTENTION
DYSPLASIA	M-74000	DYSPLASIA
EROSION	M-14110	EROSION
FIBROSIS	M-49000	FIBROSIS
HEMORRHAGE	M-37000	HEMORRHAGE
		HEMORRHAGE, PETECHIAL
HYPERPLASIA	M-72000	HYPERPLASIA
HYPERPLASIA, EPITHELIAL	M-72000	EPITHELIAL CELLS
		HYPERPLASIA, EPITHELIAL
		PAPILLOMATOSIS
HYPERPLASIA, LYMPHOID	M-72200	HYPERPLASIA, LYMPHOID
		LYMPHOCYTIC NODULE
HYPERPLASIA, POLYPOID	M-72040	HYPERPLASIA, POLYPOID
HYPERPLASIA, PSEUDOPAPILLARY	M-72050	HYPERPLASIA, PSEUDOPAPILLARY
HYPERTROPHY	M-71000	HYPERTROPHY
INFLAMMATION	M-40000	CYSTITIS
		INFLAMMATION
		PERICYSTITIS
INFLAMMATION, HEMORRHAGIC	M-40600	INFLAMMATION, HEMORRHAGIC
INFLAMMATION, NONSUPPURATIVE	M-47130	INFILTRATION, LYMPHOID
		INFLAMMATION, LYMPHOCYTIC
		INFLAMMATION, NONSUPPURATIVE
		LYMPHOCYTIC INFLAMMATORY INFILTRATE
		MYOSITIS, LYMPHOCYTIC
INFLAMMATION, SUPPURATIVE	M-40400	CYSTITIS, PROLIFERATIVE
		CYSTITIS, SUPPURATIVE
		CYSTITIS, SUPPURATIVE, HEMORRHAGIC
		INFLAMMATION, SUPPURATIVE
INFLAMMATION, ULCERATIVE	M-40030	ULCER
		ULCERATION
INTERSTITIAL CELL TUMOR	M-86506	INTERSTITIAL CELL TUMOR
LEIOMYOSARCOMA	M-88903	LEIOMYOSARCOMA
LEUKEMIA	M-98003	LEUKEMIA

PREFERRED TERM	SNOVET CODE	LADB TERMS
LEUKEMIA, LYMPHOBLASTIC	M-98213	LYMPHOBLASTIC LEUKEMIA
LYMPHOMA	M-95903	LYMPHOMA
		LYMPHOSARCOMATOUS INFILTRATION
		MALIGNANT LYMPHOMA
LYMPHOMA, RETICULUM CELL	M-96403	RETICULUM CELL SARCOMA
MESOTHELIOMA	M-90503	MESOTHELIOMA
METAPLASIA, SQUAMOUS	M-73220	METAPLASIA, SQUAMOUS
MINERALIZATION	M-55400	CALCIFICATION
		CALCIUM DEPOSIT
		MINERALIZATION
		MINERALIZATION, DYSTROPHIC
NEOPLASM	M-80001	NEOPLASM
NEUROFIBROSARCOMA	M-95403	NEUROFIBROSARCOMA
NODULE	M-03010	NODULE
PAPILLOMA	M-80500	PAPILLOMA
PAPILLOMA, TRANSITIONAL CELL	M-81200	TRANSITIONAL-CELL PAPILLOMA
PERIVASCULITIS/VASCULITIS	M-47800	ANGIITIS
		PERIARTERITIS
		PERIARTERITIS, NONSUPPURATIVE
		PERIVASCULITIS
POLYP	M-76800	POLYP
POLYP, INFLAMMATORY	M-76820	POLYP, INFLAMMATORY
SARCOMA, HISTIOCYTIC	M-96403	HISTIOCYTIC SARCOMA
SARCOMA, UNDIFFERENTIATED	M-88003	UNDIFFERENTIATED SARCOMA

INDEX

INDEX